ORCCA

Open Resources for Community College Algebra

ORCCA
Open Resources for Community College Algebra

Portland Community College Faculty

July 24, 2018

Project Leads: Ann Cary and Alex Jordan
Technology Engineer: Alex Jordan
Contributing Authors: Ann Cary, Alex Jordan, Ross Kouzes, Scot Leavitt, Cara Lee, Carl Yao, and Ralf Youtz
WeBWorK Problem Coding: Chris Hughes, Alex Jordan, Carl Yao
Other Contributors: Kara Colley
Cover Image: Ralf Youtz

Edition: 1.0

Website: pcc.edu/ORCCA

Acknowledgements

This book has been made possible through Portland Community College's Strategic Investment Funding, approved by PCC's Budget Planning Advisory Council and the Board of Directors. Without significant funding to provide the authors with the adequate time, an ambitious project such as this one would not be possible.

The technology that makes it possible to create synced print, eBook, and WeBWorK content is PreTeXt, created by Rob Beezer. David Farmer and the American Institute of Mathematics have worked to make the PreTeXt eBook layout functional, yet simple. A grant from OpenOregon funded the original bridge between WeBWorK and PreTeXt.

This book uses WeBWorK to provide most of its exercises, which may be used for online homework. WeB-WorK was created by Mike Gage and Arnie Pizer, and has benefited from over 25 years of contributions from open source developers. In 2013, Chris Hughes, Alex Jordan, and Carl Yao programmed most of the WeBWorK questions in this book with a PCC curriculum development grant.

The javascript library MathJax, created and maintained by David Cervone, Volker Sorge, Christian Lawson-Perfect, and Peter Krautzberger allows math content to render nicely on screen in the eBook. Additionally, MathJax makes web accessible mathematics possible.

The print edition (PDF) is built using the typesetting software LaTeX, created by Donald Knuth and enhanced by Leslie Lamport.

Each of these open technologies, along with many that we use but have not listed here, has been enhanced by many additional contributors spanning the past 40 years. We are grateful for all of these contributions.

To All

HTML, PDF, and print This book is available as an eBook, a free PDF, or printed and bound. All versions offer the same content and are synchronized such that cross-references match across versions. They can each be found at pcc.edu/orcca.

There are some differences between the eBook, PDF, and printed versions.

- The eBook is recommended, as it offers interactive elements and easier navigation than print. It requires no more than internet access and a modern web browser.

- A PDF version can be downloaded and then accessed without the internet. Some content is in color, but most of the colorized content from the eBook has been converted to black and white to ensure adequate contrast when printing in black and white. The exceptions are the graphs generated by WeBWorK.

- Printed and bound copies are available online. Up-to-date information about purchasing a copy should be available at pcc.edu/orcca. Contact the authors if you have trouble finding the latest version online. For each online sale, all royalties go to a PCC Foundation account, where roughly half will fund student scholarships, and half will fund continued maintenance of this book and other OER.

Copying Content The graphs and other images that appear in this manual may be copied in various file formats using the eBook version. Below each image are links to `.png`, `.eps`, `.svg`, `.pdf`, and `.tex` files that contain the image.

Mathematical content can be copied from the eBook. To copy math content into *MS Word*, right-click or control-click over the math content, and click to `Show Math As MathML Code`. Copy the resulting code, and `Paste Special` into *Word*. In the `Paste Special` menu, paste it as `Unformatted Text`. To copy math content into LaTeX source, right-click or control-click over the math content, and click to `Show Math As TeX Commands`.

Tables can be copied from the eBook version and pasted into applications like *MS Word*. However, mathematical content within tables will not always paste correctly without a little extra effort as described above.

Accessibility The HTML version is intended to meet or exceed web accessibility standards. If you encounter an accessibility issue, please report it.

- All graphs and images should have meaningful alt text that communicates what a sighted person would see, without necessarily giving away anything that is intended to be deduced from the image.

- All math content is rendered using MathJax. MathJax has a contextual menu that can be accessed in several ways, depending on what operating system and browser you are using. The most common way is to right-click or control-click on some piece of math content.

- In the MathJax contextual menu, you may set options for triggering a zoom effect on math content, and also by what factor the zoom will be. Also in the MathJax contextual menu, you can enable the Explorer, which allows for sophisticated navigation of the math content.

- A screen reader will generally have success verbalizing the math content from MathJax. With certain screen reader and browser combinations, you may need to set some configuration settings in the MathJax contextual menu.

Tablets and Smartphones PreTeXt documents like this book are "mobile-friendly." When you view the HTML version, the display adapts to whatever screen size or window size you are using. A math teacher will always recommend that you do not study from the small screen on a phone, but if it's necessary, the eBook gives you that option.

WeBWorK for Online Homework Most exercises are available in a ready-to-use collection of WeBWorK problem sets. Visit webwork.pcc.edu/webwork2/orcca-demonstration to see a demonstration WeBWorK course where guest login is enabled. Anyone interested in using these problem sets should contact the project leads.

Odd Answers The answers to the odd homework exercises at the end of each section are not contained in the PDF or print versions. As the eBook evolves, they may or may not be contained in an appendix there. In any case, odd answers are available *somewhere*. Check pcc.edu/orcca to see where.

Interactive and Static Examples Traditionally, a math textbook has examples throughout each section. This textbook uses two types of "example":

Static These are labeled "Example." Static examples may or may not be subdivided into a "statement" followed by a walk-through solution. This is basically what traditional examples from math textbooks do.

Active These are labeled " Checkpoint," not to be confused with the exercises that come at the end of a section that might be assigned for homework, etc. In the HTML output, active examples have WeBWorK answer blanks where a reader could try submitting an answer. In the PDF output, active examples are almost indistinguishable from static examples, but there is a WeBWorK icon indicating that a reader could interact more actively using the eBook. Generally, a walk-through solution is provided immediately following the answer blank.

Some HTML readers will skip the opportunity to try an active example and go straight to its solution. Some readers will try an active example once and then move on to the solution. Some readers will tough it out for a period of time and resist reading the solution.

For readers of the PDF, it is expected that they would read the example and its solution just as they would read a static example.

A reader is *not* required to try submitting an answer to an active example before moving on. A reader *is* expected to read the solution to an active example, even if they succeed on their own at finding an answer.

Interspersed through a section there are usually several exercises that are intended as active reading exercises. A reader can work these examples and submit answers to WeBWorK to see if they are correct. The important thing is to keep the reader actively engaged instead of providing another static written example. In most cases, it is expected that a reader will read the solutions to these exercises just as they would be expected to read a more traditional static example.

Pedagogical Decisions

The authors and the greater PCC faculty have taken various stances on certain pedagogical and notational questions that arise in basic algebra instruction. We attempt to catalog these decisions here, although this list is certainly incomplete. If you find something in the book that runs contrary to these decisions, please let us know.

- Interleaving is our preferred approach, compared to a proficiency-based approach. To us, this means that once the book covers a topic, that topic will be appear in subsequent sections and chapters in indirect ways.

- Chapter 1 is mostly written as a *review*, and is not intended to teach all of these topics from first principles.

- We round decimal results to four significant digits, or possibly fewer leaving out trailing zeros. We do this to maintain consistency with the most common level of precision that WeBWorK uses to assess decimal answers. We *round*, not *truncate*. And we use the $\approx$ symbol. For example $\pi \approx 3.142$ and Portland's population is ≈ 609500.

- We offer *alternative* video lessons associated with each section, found in most sections in the eBook. We hope these videos provide readers with an alternative to whatever is in the reading, but there may be discrpancies here and there between the video content and reading content.

- We believe in always trying to open a topic with some level of application rather than abstract examples. From applications and practical questions, we move to motivate more abstract definitions and notation. This approach is perhaps absent in the first chapter, which is intended to be a review only. At first this may feel backwards to some instructors, with some "easier" examples (with no context) appearing after "more difficult" contextual examples.

- Linear inequalities are not strictly separated from linear equations. The section that teaches how to solve $2x + 3 = 8$ is immediately followed by the section teaching how to solve $2x + 3 < 8$.

  Our aim is to not treat inequalities as an add-on optional topic, but rather to show how intimately related they are to corresponding equations.

- When issues of "proper formatting" of student work arise, we value that the reader understand *why* such things help the reader to communicate outwardly. We believe that mathematics is about more than understanding a topic, but also about understanding it well enough to communicate results to others.

  For example we promote progression of equations like

  $$1 + 1 + 1 = 2 + 1$$
  $$= 3$$

  instead of

  $$1 + 1 + 1 = 2 + 1 = 3.$$

  And we want students to *understand* that the former method makes their work easier for a reader to read. It is not simply a matter of "this is the standard and this is how it's done."

- When solving equations (or systems of linear equations), most examples should come with a check, intended to communicate to students that checking is part of the process. In Chapters 1–4, these checks will be complete simplifications using order of operations one step at a time. The later sections will often have more summary checks where either order of operations steps are skipped in groups, or we promote entering expressions into a calculator. Occasionally in later sections the checks will still have finer details, especially when there are issues like with negative numbers squared.

- Within a section, any first example of solving some equation (or system) should summarize with some variant of both "the solution is…" and "the solution set is…." Later examples can mix it up, but always offer at least one of these.

- There is a section on very basic arithmetic (five operations on natural numbers) in an appendix, not in the first chapter. This appendix is only available in the eBook.

- With applications of linear equations (as opposed to linear systems), we limit applications to situations where the setup will be in the form $x + f(x) = C$ and also certain rate problems where the setup will be in the form $5t + 4t = C$. There are other classes of application problem (mixing problems, interest problems, …) which can be handled with a system of two equations, and we reserve these until linear systems are covered.

- With simplifications of rational expressions in one variable, we always include domain restrictions that are lost in the simplification. For example, we would write $\frac{x(x+1)}{x+1} = x$, for $x \neq -1$. With multivariable rational expressions, we are content to ignore domain restrictions lost during simplification.

Entering WeBWorK Answers

This preface offers some guidance with syntax for WeBWorK answers. WeBWorK answer blanks appear in the active reading examples (called "checkpoints") in the HTML version of the book. If you are using WeBWorK for online homework, then you will also enter answers into WeBWorK answer blanks there.

Basic Arithemtic   The five basic arithmetic operations are: addition, subtraction, multiplication, and raising to a power. The symbols for addition and subtraction are $+$ and $-$, and both of these are directly avialable on most keyboards as + and -.

On paper, multiplication is sometimes written using $\times$ and sometimes written using $\cdot$ (a centered dot). Since these symbols are not available on most keyboards, WeBWorK uses $*$ instead, which is often `shift-8` on a full keyboard.

On paper, division is sometimes written using $\div$, sometimes written using a fraction layout like $\frac{4}{2}$, and sometimes written just using a slash, $/$. The slash is available on most full keyboards, near the question mark. WeBWorK uses / to indicate division.

On paper, raising to a power is written using a two-dimensional layout like 4^2. Since we don't have a way to directly type that with a simple keyboard, calculators and computers use the caret character, $\wedge$, as in `4^2`. The character is usually `shift-6`.

Roots and Radicals   On paper, a square root is represented with a radical symbol like $\sqrt{}$. Since a keyboard does not usually have this symbol, WeBWorK and many computer applications use `sqrt( )` instead. For example, to enter $\sqrt{17}$, type `sqrt(17)`.

Higher-index radicals are written on paper like $\sqrt[4]{12}$. Again we have no direct way to write this using most keyboards. In *some* WeBWorK problems it is possible to type something like `root(4, 12)` for the fourth root of twelve. However this is not enabled for all WeBWorK problems.

As an alternative that you may learn about in a later chapter, $\sqrt[4]{12}$ is mathematically equal to $12^{1/4}$, so it can be typed as `12^(1/4)`. Take note of the parentheses, which very much matter.

Common Hiccups with Grouping Symbols   Suppose you wanted to enter $\frac{x+1}{2}$. You might type `x+1/2`, but this is not right. The computer will use the order of operations (see Section 1.4) and do your division first, dividing 1 by 2. So the computer will see $x + \frac{1}{2}$. To address this, you would need to use grouping symbols like parentheses, and type something like `(x+1)/2`.

Suppose you wanted to enter $6^{1/4}$, and you typed `6^1/4`. This is not right. The order of operations places a higher priority on exponentiation than division, so it calculates 6^1 first and then divides the result by 4. That is simply not the same as raising 6 to the $\frac{1}{4}$ power. Again the way to address this is to use grouping symbols, like `6^(1/4)`.

Entering Decimal Answers   Often you will find a decimal answer with decimal places that go on and on. You are allowed to round, but not by too much. WeBWorK generally looks at how many *significant digits* you use, and generally expects you to use *four or more* correct significant digits.

"Significant digits" and "places past the decimal" are not the same thing. To count significant digits, read the number left to right and look for the first nonzero digit. Then count all the digits to the right including that first one.

The number 102.3 has four significant digits, but only one place past the decimal. This number could be a correct answer to a WeBWorK question. The number 0.0003 has one significant digit and four places past the decimal. This number might cause you trouble if you enter it, because maybe the "real" answer was 0.0003091, and rounding to 0.0003 was too much rounding.

Special Symbols   There are a handful of special symbols that are easy to write on paper, but it's not clear how to type them. Here are WeBWorK's expectations.

Symbol	Name	How to Type
∞	infinity	infinity or inf
π	pi	pi
$\cup$	union	U
$\mathbb{R}$	the real numbers	R
$\vert$	such that	| (shift-\, where \ is above the enter key)
$\leq$	less than or equal to	<=
$\geq$	greater than or equal to	>=
$\neq$	not equal to	!=

Contents

Contents

Basic Math Review

This chapter is *mostly* intended to *review* topics from a basic math course, especially Sections 1.1–1.4. These topics are covered differently than they would be covered for a student seeing them for the first time.

1.1 Arithmetic with Negative Numbers

Adding, subtracting, multiplying, dividing, and raising to powers each have peculiarities when using negative numbers. This section reviews arithmetic with signed (both positive and negative) numbers.

1.1.1 Signed Numbers

Is it valid to subtract a large number from a smaller one? It may be hard to imagine what it would mean physically to subtract 8 cars from your garage if you only have 1 car in there in the first place. Nevertheless, mathematics has found a way to give meaning to expressions like $1 - 8$ using **signed numbers**.

In daily life, the signed numbers we might see most often are temperatures. Most people on Earth use the Celsius scale; if you're not familiar with the Celsius temperature scale, think about these examples:

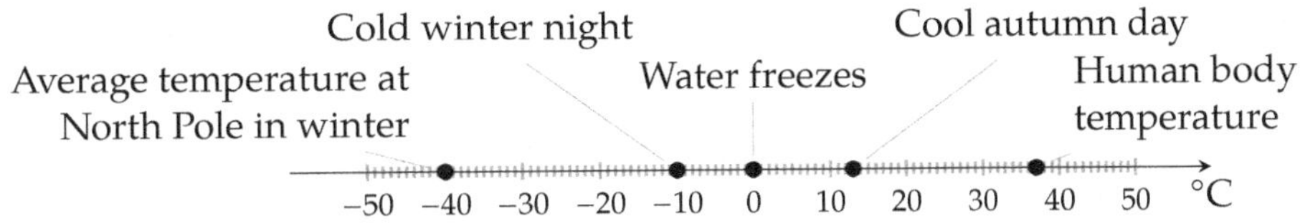

Figure 1.1.2: Number line with interesting Celsius temperatures

Figure 1.1.2 uses a **number line** to illustrate these positive and negative numbers. A number line is a useful device for visualizing how numbers relate to each other and combine with each other. Values to the right of 0 are called **positive** numbers and values to the left of 0 are called **negative numbers**.

Warning 1.1.3 Subtraction Sign versus Negative Sign. Unfortunately, the symbol we use for subtraction looks just like the symbol we use for marking a negative number. It will help to identify when a "minus" sign means "subtract" or means "negative." The key is to see if there is a number to its left, not counting anything farther left than an open parenthesis. Here are some examples.

- -13 has one negative sign and no subtraction sign.

- $20 - 13$ has no negative signs and one subtraction sign.

- $-20 - 13$ has a negative sign and then a subtraction sign.

- $(-20)(-13)$ has two negative signs and no subtraction sign.

Checkpoint 1.1.4. Identify "minus" signs.

In each expression, how many negative signs and subtraction signs are there?

a. $1 - 9$

has ▢ negative signs and ▢ subtraction signs.

b. $-12 + (-50)$

has ▢ negative signs and ▢ subtraction signs.

c. $\dfrac{-13 - (-15) - 17}{23 - 4}$

has ▢ negative signs and ▢ subtraction signs.

Explanation.

a. $1 - 9$ has zero negative signs and one subtraction sign.

b. $-12 + (-50)$ has two negative signs and zero subtraction signs.

c. $\dfrac{-13 - (-15) - 17}{23 - 4}$ has two negative signs and three subtraction signs.

1.1.2 Adding

An easy way to think about adding two numbers with the *same sign* is to simply (at first) ignore the signs, and add the numbers as if they were both positive. Then make sure your result is either positive or negative, depending on what the sign was of the two numbers you started with.

Example 1.1.5 Add Two Negative Numbers. If you needed to add -18 and -7, note that both are negative. Maybe you have this expression in front of you:

$$-18 + -7$$

but that "plus minus" is awkward, and in this book you are more likely to have this expression:

$$-18 + (-7)$$

with extra parentheses. (How many subtraction signs do you see? How many negative signs?)

Since *both* our terms are *negative*, we can add 18 and 7 to get 25 and immediately realize that our final result should be negative. So our result is -25:

$$-18 + (-7) = -25$$

This approach works because adding numbers is like having two people tugging on a rope in one direction or the other, with strength indicated by each number. In Example 1.1.5 we have two people pulling to the left, one with strength 18, the other with strength 7. Their forces combine to pull *left* with strength 25, giving us our total of -25, as illustrated in Figure 1.1.6.

If we are adding two numbers that have *opposite* signs, then the two people tugging the rope are opposing each other. If either of them is using more strength, then the overall effect will be a net pull in that person's direction. And the overall pull on the rope will be the *difference* of the two strengths. This is illustrated in Figure 1.1.7.

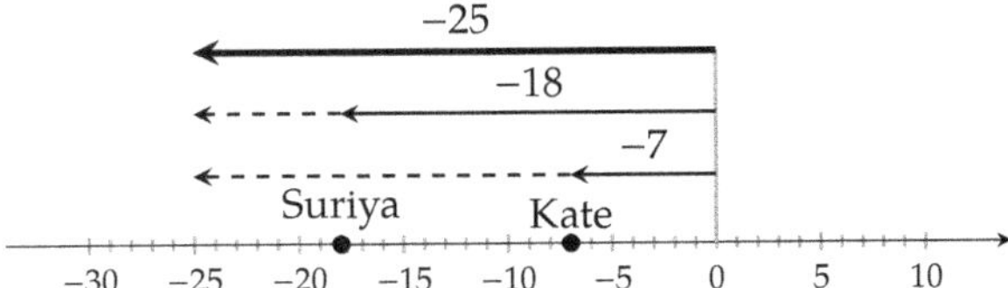

Figure 1.1.6: Working together

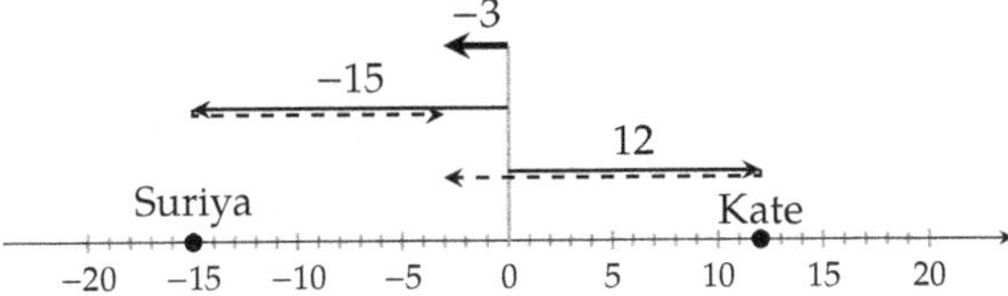

Figure 1.1.7: Working in opposition

Example 1.1.8 Adding One Number of Each Sign. Here are four examples of addition where one number is positive and the other is negative.

a. $-15 + 12$

   We have one number of each sign, with sizes 15 and 12. Their difference is 3. But of the two numbers, the negative number dominated. So the result from adding these is -3.

b. $200 + (-100)$

   We have one number of each sign, with sizes 200 and 100. Their difference is 100. But of the two numbers, the positive number dominated. So the result from adding these is 100.

c. $12.8 + (-20)$

   We have one number of each sign, with sizes 12.8 and 20. Their difference is 7.2. But of the two numbers, the negative number dominated. So the result from adding these is -7.2.

d. $-87.3 + 87.3$

   We have one number of each sign, both with size 87.3. The opposing forces cancel each other, leaving a result of 0.

Checkpoint 1.1.9. Take a moment to practice adding when at least one negative number is involved. The expectation is that readers can make these calculations here without a calculator.

a. Add $-1 + 9$.

b. Add $-12 + (-98)$.

c. Add $100 + (-123)$.

d. Find the sum $-2.1 + (-2.1)$.

e. Find the sum $-34.67 + 81.53$.

Explanation.

a. The two numbers have opposite sign, so we can think to subtract $9 - 1 = 8$. Of the two numbers we added, the positive is larger, so we stick with postive 8 as the answer.

b. The two numbers are both negative, so we can add $12 + 98 = 110$, and take the negative of that as the answer: -110.

c. The two numbers have opposite sign, so we can think to subtract $123 - 100 = 23$. Of the two numbers we added, the negative is larger, so we take the negative of 23 as the answer. That is, the answer is -23.

d. The two numbers are both negative, so we can add $2.1 + 2.1 = 4.2$, and take the negative of that as the answer: -4.2.

e. The two numbers have opposite sign, so we can think to subtract $81.53 - 34.67 = 46.86$. Of the two numbers we added, the positive is larger, so we stick with postive 46.86 as the answer.

1.1.3 Subtracting

Perhaps you can handle a subtraction such as $18 - 5$, where a small positive number is subtracted from a larger number. There are other instances of subtraction that might leave you scratching your head. In such situations, we recommend that you view each subtraction as *adding* the opposite number.

	Original	Adding the Opposite
Subtracting a larger positive number:	$12 - 30$	$12 + (-30)$
Subtracting from a negative number:	$-8.1 - 17$	$-8.1 + (-17)$
Subtracting a negative number:	$42 - (-23)$	$42 + 23$

The benefit is that perhaps you already mastered addition with positive and negative numbers, and this strategy that you convert subtraction to addition means you don't have all that much more to learn. These examples might be computed as follows:

$$12 - 30 = 12 + (-30) \qquad -8.1 - 17 = -8.1 + (-17) \qquad 42 - (-23) = 42 + 23$$
$$= -18 \qquad\qquad = -25.1 \qquad\qquad = 65$$

Checkpoint 1.1.10. Take a moment to practice subtracting when at least one negative number is involved. The expectation is that readers can make these calculations here without a calculator.

a. Subtract -1 from 9.

b. Subtract $32 - 50$.

c. Subtract $108 - (-108)$.

d. Find the difference $-5.9 - (-3.1)$.

e. Find the difference $-12.04 - 17.2$.

Explanation.

a. After writing this as $9 - (-1)$, we can rewrite it as $9 + 1$ and get 10.

b. Subtrcting in the oppsite order with the larger number first, $50 - 32 = 18$. But since we were asked to subtract the larger number *from* the smaller number, the answer is -18.

c. After writing this as $108 - (-108)$, we can rewrite it as $108 + 108$ and get 216.

d. After writing this as $-5.9 - (-3.1)$, we can rewrite it as $-5.9 + 3.1$. Now it is the *sum* of two numbers of opposite sign, so we can subtract $5.9 - 3.1$ to get 2.8. But we were adding numbers where the negative number was larger, so the final answer should be -2.8.

e. Since we are subtracting a positive number from a negative number, the result should be an even more negative number. We can add $12.04 + 17.2$ to get 29.24, but our final answer should be the opposite, -29.24.

1.1.4 Multiplying

Making sense of multiplication of negative numbers isn't quite so straightforward, but it's possible. Should the product of 3 and -7 be a positive number or a negative number? Remembering that we can view multiplication as repeated addition, we can see this result on a number line:

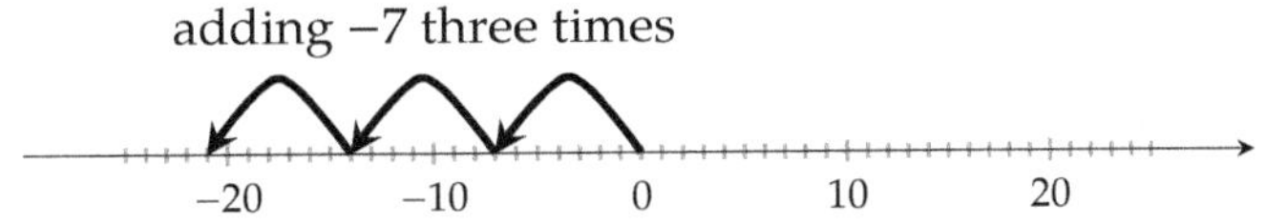

Figure 1.1.11: Viewing $3 \cdot (-7)$ as repeated addition

Figure 1.1.11 illustrates that $3 \cdot (-7) = -21$, and so it would seem that a positive number times a negative number will always give a negative result. (Note that it would not change things if the negative number came first in the product, since the order of multiplication doesn't affect the result.)

What about the product $-3 \cdot (-7)$, where both factors are negative? Should the product be positive or negative? If $3 \cdot (-7)$ can be seen as adding -7 three times as in Figure 1.1.12, then it isn't too crazy to interpret $-3 \cdot (-7)$ as *subtracting* -7 three times, as in Figure 1.1.12.

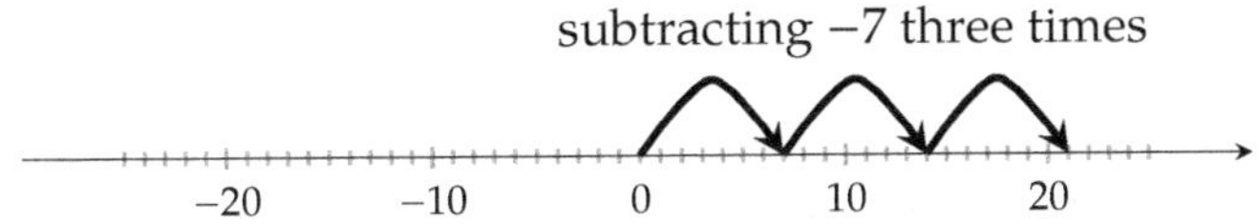

Figure 1.1.12: Viewing $-3 \cdot (-7)$ as repeated subtraction

This illustrates that $-3 \cdot (-7) = 21$, and it would seem that a negative number times a negative number always gives a positive result.

Positive and negative numbers are not the whole story. The number 0 is neither positive nor negative. What happens with multiplication by 0? You can choose to view $7 \cdot 0$ as adding the number 0 seven times. And you can choose to view $0 \cdot 7$ as adding the number 7 zero times. Either way, you really added nothing at all, which is the same as adding 0.

Fact 1.1.13 Multiplication by 0. *Multiplying any number by 0 results in 0.*

Checkpoint 1.1.14. Here are some practice exercises with multiplication and signed numbers. The expectation is that readers can make these calculations here without a calculator.

a. Multiply $-13 \cdot 2$.

b. Find the product of 30 and -50.

c. Compute $-12(-7)$.

d. Find the product $-285(0)$.

Explanation.

a. Since $13 \cdot 2 = 26$, and we are multiplying numbers of opposite signs, the answer is negative: -26.

b. Since $30 \cdot 50 = 1500$, and we are multiplying numbers of opposite signs, the answer is negative: -1500.

c. Since $12 \cdot 7 = 84$, and we are multiplying numbers of the same sign, the answer is positive: 84.

d. Any number multiplied by 0 is 0.

1.1.5 Powers

For early sections of this book the only exponents you will see will be the **natural numbers**: $\{1, 2, 3, \ldots\}$. But negative numbers can and will arise as the *base* of a power.

An exponent is a shorthand for how many times to multiply by the base. For example,

$$\overbrace{(-2)^5 \text{ means } (-2) \cdot (-2) \cdot (-2) \cdot (-2) \cdot (-2)}^{5 \text{ instances}}$$

Will the result here be positive or negative? Since we can view $(-2)^5$ as repeated multiplication, and we now understand that multiplying two negatives gives a positive result, this expression can be thought of this way:

$$\underbrace{\underbrace{(-2) \cdot (-2)}_{\text{positive}} \cdot \underbrace{(-2) \cdot (-2)}_{\text{positive}} \cdot (-2)}_{\text{positive}}$$

and that lone last negative number will be responsible for making the final product negative.

More generally, if the base of a power is negative, then whether or not the result is positive or negative depends on if the exponent is even or odd. It depends on whether or not the factors can all be paired up to "cancel" negative signs, or if there will be a lone factor left by itself.

Once you understand whether the result is positive or negative, for a moment you may forget about signs. Continuing the example, you may calculate that $2^5 = 32$, and then since we know $(-2)^5$ is negative, you can report

$$(-2)^5 = -32$$

Warning 1.1.15 Negative Signs and Exponents. Expressions like -3^4 may not mean what you think they mean. What base do you see here? The correct answer is 3. The exponent 4 *only* applies to the 3, not to -3. So this expression, -3^4, is actually the same as $-(3^4)$, which is -81. Be careful not to treat -3^4 as having base -3. That would make it equivalent to $(-3)^4$, which is *positive* 81.

Checkpoint 1.1.16. Here is some practice with natural exponents on negative bases. The expectation is that readers can make these calculations here without a calculator.

a. Compute $(-8)^2$.

b. Calculate the power $(-1)^{203}$.

c. Find $(-3)^3$.

d. Calculate -5^2.

Explanation.

a. Since 8^2 is 64 and we are raising a negative number to an *even* power, the answer is positive: 64.

b. Since 1^{203} is 1 and we are raising a negative number to an *odd* power, the answer is negative: -1.

c. Since 3^3 is 27 and we are raising a negative number to an *odd* power, the answer is negative: -27.

d. Careful: here we are raising *positive* 5 to the second power to get 25 and *then* negating the result: -25. Since we don't see "$(-5)^2$," the answer is not positive 25.

1.1.6 Summary

Addition Add two negative numbers: add their positive counterparts and make the result negative.

Add a positive with a negative: find their difference using subtraction, and keep the sign of the dominant number.

Subtraction Any subtraction can be converted to addition of the opposite number. For all but the most basic subtractions, this is a useful strategy.

Multiplication Multiply two negative numbers: multiply their positive counterparts and make the result positive.

Multiply a positive with a negative: multiply their positive counterparts and make the result negative.

Multiply any number by 0: the result will be 0.

Division (not discussed in this section) Division by some number is the same as multiplication by its reciprocal. So the multiplication rules can be adopted.

Division of 0 by any nonzero number always results in 0.

Division of any number by 0 is always undefined.

Powers Raise a negative number to an even power: raise the positive counterpart to that power.

Raise a negative number to an odd power: raise the positive counterpart to that power, then make the result negative.

Expressions like -2^4 mean $-(2^4)$, not $(-2)^4$.

Exercises

Add the following.

1. a. $-8 + (-1) =$
    b. $-6 + (-7) =$
    c. $-1 + (-7) =$

2. a. $-10 + (-2) =$
    b. $-5 + (-3) =$
    c. $-2 + (-9) =$

3. a. $2 + (-10) =$
    b. $5 + (-2) =$
    c. $7 + (-7) =$

4. a. $2 + (-6) =$
    b. $8 + (-3) =$
    c. $7 + (-7) =$

5. a. $-8 + 3 =$
    b. $-4 + 10 =$
    c. $-4 + 4 =$

6. a. $-10 + 3 =$
    b. $-1 + 7 =$
    c. $-4 + 4 =$

7. a. $-41 + (-31) =$
    b. $-31 + 58 =$
    c. $64 + (-27) =$

8. a. $-31 + (-63) =$
    b. $-84 + 33 =$
    c. $62 + (-72) =$

Subtract the following.

9. a. $5 - 6 =$ ____

   b. $8 - 4 =$ ____

   c. $4 - 19 =$ ____

10. a. $5 - 9 =$ ____

    b. $5 - 3 =$ ____

    c. $4 - 14 =$ ____

11. a. $-5 - 2 =$ ____

    b. $-8 - 3 =$ ____

    c. $-6 - 6 =$ ____

12. a. $-4 - 1 =$ ____

    b. $-6 - 1 =$ ____

    c. $-6 - 6 =$ ____

13. a. $-2 - (-8) =$ ____

    b. $-6 - (-2) =$ ____

    c. $-4 - (-4) =$ ____

14. a. $-3 - (-6) =$ ____

    b. $-9 - (-2) =$ ____

    c. $-4 - (-4) =$ ____

Perform the given addition and subtraction.

15. a. $-15 - 10 + (-9) =$ ____

    b. $4 - (-15) + (-12) =$ ____

16. a. $-13 - 7 + (-5) =$ ____

    b. $1 - (-15) + (-17) =$ ____

17. a. $-12 - 3 + (-1) =$ ____

    b. $8 - (-15) + (-12) =$ ____

18. a. $-11 - 10 + (-7) =$ ____

    b. $5 - (-16) + (-17) =$ ____

Multiply the following.

19. a. $(-8) \cdot (-2) =$ ____

    b. $(-4) \cdot 3 =$ ____

    c. $7 \cdot (-2) =$ ____

    d. $(-5) \cdot 0 =$ ____

20. a. $(-10) \cdot (-3) =$ ____

    b. $(-7) \cdot 7 =$ ____

    c. $7 \cdot (-6) =$ ____

    d. $(-4) \cdot 0 =$ ____

21. a. $(-3) \cdot (-4) \cdot (-3) =$ ____

    b. $5 \cdot (-8) \cdot (-2) =$ ____

    c. $(-84) \cdot (-52) \cdot 0 =$ ____

22. a. $(-3) \cdot (-5) \cdot (-5) =$ ____

    b. $4 \cdot (-8) \cdot (-4) =$ ____

    c. $(-83) \cdot (-70) \cdot 0 =$ ____

23. a. $(-2)(-1)(-2)(-1) =$ ____

    b. $(-3)(-3)(1)(-1) =$ ____

24. a. $(-2)(-3)(-3)(-3) =$ ____

    b. $(-1)(-1)(-1)(-3) =$ ____

Evaluate the following.

25. a. $\dfrac{-30}{-5} =$

b. $\dfrac{36}{-6} =$

c. $\dfrac{-48}{6} =$

26. a. $\dfrac{-12}{-4} =$

b. $\dfrac{20}{-5} =$

c. $\dfrac{-24}{6} =$

27. a. $\dfrac{-3}{-1} =$

b. $\dfrac{9}{-1} =$

c. $\dfrac{110}{-110} =$

d. $\dfrac{-10}{-10} =$

e. $\dfrac{8}{0} =$

f. $\dfrac{0}{-4} =$

28. a. $\dfrac{-2}{-1} =$

b. $\dfrac{7}{-1} =$

c. $\dfrac{150}{-150} =$

d. $\dfrac{-13}{-13} =$

e. $\dfrac{8}{0} =$

f. $\dfrac{0}{-8} =$

29. a. $(-9)^2 =$

b. $-4^2 =$

30. a. $(-7)^2 =$

b. $-6^2 =$

31. a. $(-3)^3 =$

b. $-1^3 =$

32. a. $(-2)^3 =$

b. $-4^3 =$

33. a. $4^2 =$

b. $2^3 =$

c. $(-4)^2 =$

d. $(-3)^3 =$

34. a. $5^2 =$

b. $4^3 =$

c. $(-3)^2 =$

d. $(-5)^3 =$

35. a. $1^{10} =$

b. $(-1)^{11} =$

c. $(-1)^{12} =$

d. $0^{20} =$

36. a. $1^5 =$

b. $(-1)^{13} =$

c. $(-1)^{18} =$

d. $0^{18} =$

Simplify without using a calculator.

37. $-9.33 + (-21.7) = \boxed{}$ **38.** $-1.97 + (-81.3) = \boxed{}$ **39.** $6.6 - 1.62 = \boxed{}$

40. $6.3 - 3.32 = \boxed{}$ **41.** $-6.12 + 7.9 = \boxed{}$ **42.** $-3.82 + 7.6 = \boxed{}$

43. $-6.52 - (-8.3) = \boxed{}$ **44.** $-2.21 - (-8.9) = \boxed{}$ **45.** $75 - 6.91 = \boxed{}$

46. $82 - 1.71 = \boxed{}$ **47.** $-18 + 5.41 = \boxed{}$ **48.** $-25 + 8.11 = \boxed{}$

49. It's given that $32 \cdot 39 = 1248$. Use this fact to calculate the following without using a calculator:

$3.2(-0.039) = \boxed{}$

50. It's given that $48 \cdot 76 = 3648$. Use this fact to calculate the following without using a calculator:

$4.8(-7.6) = \boxed{}$

51. It's given that $55 \cdot 24 = 1320$. Use this fact to calculate the following without using a calculator:

$(-5.5)(-0.024) = \boxed{}$

52. It's given that $62 \cdot 51 = 3162$. Use this fact to calculate the following without using a calculator:

$(-6.2)(-5.1) = \boxed{}$

Applications

53. Consider the following situation in which you borrow money from your cousin:

- On June 1st, you borrowed 1400 dollars from your cousin.
- On July 1st, you borrowed 460 more dollars from your cousin.
- On August 1st, you paid back 690 dollars to your cousin.
- On September 1st, you borrowed another 960 dollars from your cousin.

How much money do you owe your cousin now?

54. Consider the following scenario in which you study your bank account.

- On Jan. 1, you had a balance of −450 dollars in your bank account.
- On Jan. 2, your bank charged 40 dollar overdraft fee.
- On Jan. 3, you deposited 870 dollars.
- On Jan. 10, you withdrew 650 dollars.

What is your balance on Jan. 11?

55. A mountain is 1100 feet *above* sea level. A trench is 360 feet *below* sea level. What is the difference in elevation between the mountain top and the bottom of the trench?

56. A mountain is 1200 feet *above* sea level. A trench is 420 feet *below* sea level. What is the difference in elevation between the mountain top and the bottom of the trench?

Challenge

57. Select the correct word to make each statement true.

    a. A positive number minus a positive number is   (□ sometimes □ always □ never) negative.

    b. A negative number plus a negative number is   (□ sometimes □ always □ never) negative.

    c. A positive number minus a negative number is   (□ sometimes □ always □ never) positive.

    d. A negative number multiplied by a negative number is   (□ sometimes □ always □ never) negative.

1.2 Fractions and Fraction Arithmetic

The word "fraction" comes from the Latin word *fractio*, which means "break into pieces." For thousands of years, cultures from all over the world have used fractions to understand parts of a whole.

1.2.1 Visualizing Fractions

Parts of a Whole   One approach to understanding fractions is to think of them as parts of a whole.

In Figure 1.2.2, we see 1 whole divided into 7 parts. Since 3 parts are shaded, we have an illustration of the fraction $\frac{3}{7}$. The **denominator** 7 tells us how many parts to cut up the whole; since we have 7 parts, they're called "sevenths." The **numerator** 3 tells us how many sevenths to consider.

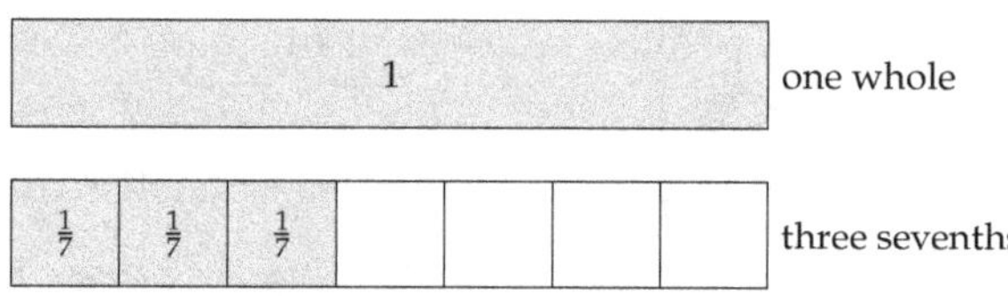

Figure 1.2.2: Representing $\frac{3}{7}$ as parts of a whole.

Checkpoint 1.2.3 A Fraction as Parts of a Whole. To visualize the fraction $\frac{14}{35}$, you might cut a rectangle into ☐ equal parts, and then count up ☐ of them.

Explanation.   You could cut a rectangle into 35 equal pieces, and then 14 of them would represent $\frac{14}{35}$.

We can also locate fractions on number lines. When ticks are equally spread apart, as in Figure 1.2.4, each tick represents a fraction.

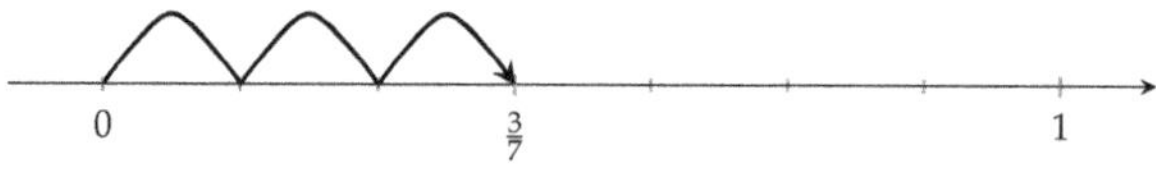

Figure 1.2.4: Representing $\frac{3}{7}$ on a number line.

Checkpoint 1.2.5 A Fraction on a Number Line. In the given number line, what fraction is marked?

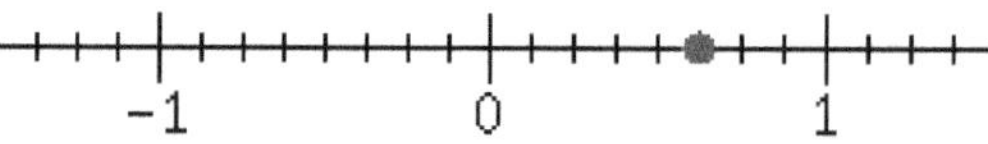

Explanation.   There are 8 subdivisions between 0 and 1, and the mark is at the fifth subdivision. So the mark is $\frac{5}{8}$ of the way from 0 to 1 and therefore represents the fraction $\frac{5}{8}$.

Division   Fractions can also be understood through division.

For example, we can view the fraction $\frac{3}{7}$ as 3 divided into 7 equal parts, as in Figure 1.2.6. Just one of those parts represents $\frac{3}{7}$.

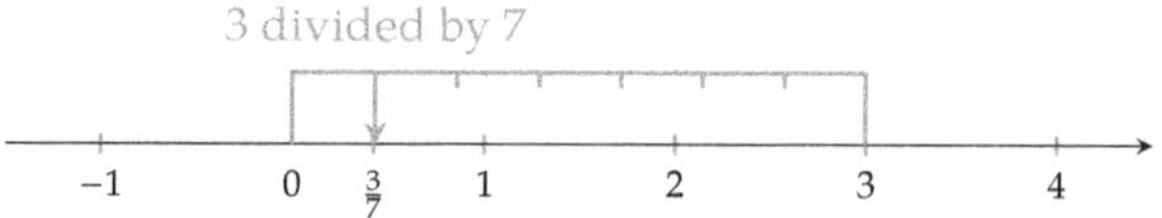

Figure 1.2.6: Representing $\frac{3}{7}$ on a number line.

Checkpoint 1.2.7 Seeing a Fraction as Division Arithmetic. The fraction $\frac{21}{40}$ can be thought of as dividing the whole number [____] into [____] equal-sized parts.

Explanation. Since $\frac{21}{40}$ means the same as $21 \div 40$, it can be thought of as dividing 21 into 40 equal parts.

1.2.2 Equivalent Fractions

It's common to have two fractions that represent the same amount. Consider $\frac{2}{5}$ and $\frac{6}{15}$ represented in various ways in Figures 1.2.8–1.2.10.

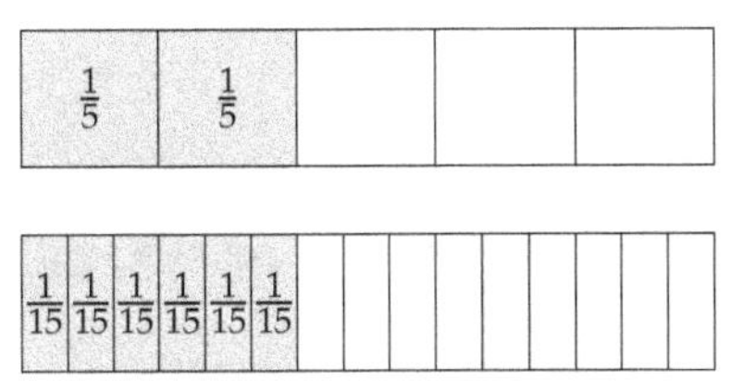

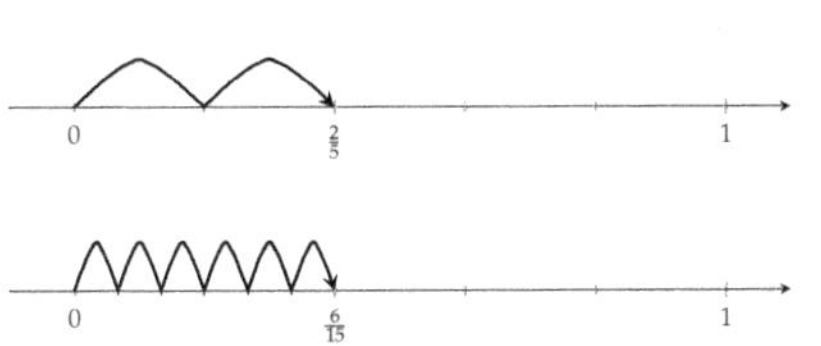

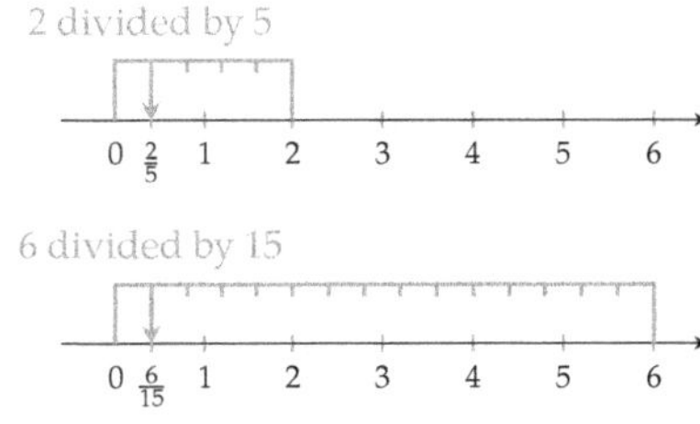

Figure 1.2.8: $\frac{2}{5}$ and $\frac{6}{15}$ as equal parts of a whole

Figure 1.2.9: $\frac{2}{5}$ and $\frac{6}{15}$ as equal on a number line

Figure 1.2.10: $\frac{2}{5}$ and $\frac{6}{15}$ as equal results from division

Those two fractions, $\frac{2}{5}$ and $\frac{6}{15}$ are equal, as those figures demonstrate. Also, because they each equal 0.4 as a decimal. If we must work with this number, the fraction that uses smaller numbers, $\frac{2}{5}$, is preferable. Working with smaller numbers decreases the likelihood of making a human arithmetic error. And it also increases the chances that you might make useful observations about the nature of that number.

So if you are handed a fraction like $\frac{6}{15}$, it is important to try to **reduce** it to "lowest terms." The most important skill you can have to help you do this is to know the multiplication table well. If you know it well, you know that $6 = 2 \cdot 3$ and $15 = 3 \cdot 5$, so you can break down the numerator and denominator that way. Both the numerator and denominator are divisible by 3, so they can be "factored out" and then as factors, cancel out.

$$\begin{aligned} \frac{6}{15} &= \frac{2 \cdot 3}{3 \cdot 5} \\ &= \frac{2 \cdot \cancel{3}}{\cancel{3} \cdot 5} \\ &= \frac{2 \cdot 1}{1 \cdot 5} \\ &= \frac{2}{5} \end{aligned}$$

Checkpoint 1.2.11. Reduce these fractions into lowest terms.

a. $\dfrac{14}{42} =$ [____]

b. $\dfrac{8}{30} =$ [____]

c. $\dfrac{70}{90} =$ [____]

Explanation.

a. With $\frac{14}{42}$, we have $\frac{2 \cdot 7}{2 \cdot 3 \cdot 7}$, which reduces to $\frac{1}{3}$.

c. With $\frac{70}{90}$, we have $\frac{7 \cdot 10}{9 \cdot 10}$, which reduces to $\frac{7}{9}$.

b. With $\frac{8}{30}$, we have $\frac{2 \cdot 2 \cdot 2}{2 \cdot 3 \cdot 5}$, which reduces to $\frac{4}{15}$.

Sometimes it is useful to do the opposite of reducing a fraction, and **build up** the fraction to use larger numbers.

Checkpoint 1.2.12. Sayid scored $\frac{21}{25}$ on a recent exam. Build up this fraction so that the denominator is 100, so that Sayid can understand what percent score he earned.

Explanation. To change the denominator from 25 to 100, it needs to be multiplied by 4. So we calculate

$$\frac{21}{25} = \frac{21 \cdot 4}{25 \cdot 4}$$
$$= \frac{84}{100}$$

So the fraction $\frac{21}{25}$ is equivalent to $\frac{84}{100}$. (This means Sayid scored an 84%.)

1.2.3 Multiplying with Fractions

Example 1.2.13 Suppose a recipe calls for $\frac{2}{3}$ cup of milk, but we'd like to quadruple the recipe (make it four times as big). We'll need four times as much milk, and one way to measure this out is to fill a measuring cup to $\frac{2}{3}$ full, four times:

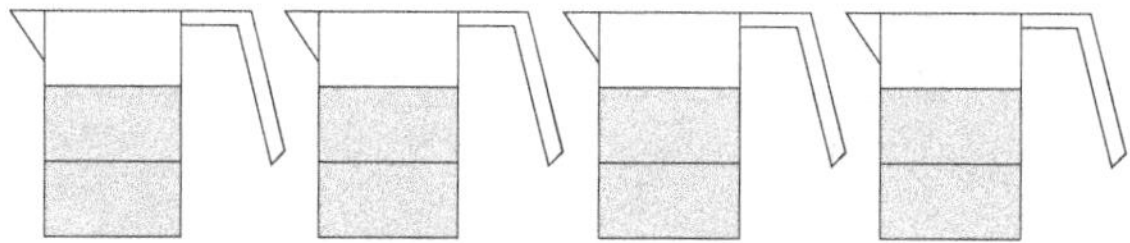

When you count up the shaded thirds, there are eight of them. So multiplying $\frac{2}{3}$ by the whole number 4, the result is $\frac{8}{3}$. Mathematically:

$$4 \cdot \frac{2}{3} = \frac{4 \cdot 2}{3}$$
$$= \frac{8}{3}$$

Fact 1.2.14 Multiplying a Fraction and a Whole Number. *When you multiply a whole number by a fraction, you may just multiply the whole number by the numerator and leave the denominator alone. In other words, as long as d is not 0, then a whole number and a fraction multiply this way:*

$$a \cdot \frac{c}{d} = \frac{a \cdot c}{d}$$

Example 1.2.15 We could also use multiplication to decrease amounts. Suppose we needed to cut the recipe down to just one fifth. Instead of *four* of the $\frac{2}{3}$ cup milk, we need *one fifth* of the $\frac{2}{3}$ cup milk. So instead of multiplying by 4, we multiply by $\frac{1}{5}$. But how much is $\frac{1}{5}$ of $\frac{2}{3}$ cup?

If we cut the measuring cup into five equal vertical strips along with the three equal horizontal strips, then in total there are $3 \cdot 5 = 15$ subdivisions of the cup. Two of those sections represent $\frac{1}{5}$ of the $\frac{2}{3}$ cup. 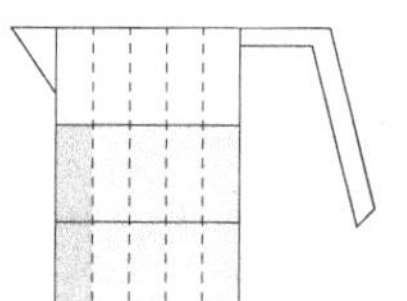

In the end, we have $\frac{2}{15}$ of a cup. The denominator 15 came from multiplying 5 and 3, the denominators of the fractions we had to multiply. The numerator 2 came from multiplying 1 and 2, the numerators of the fractions we had to multiply.

$$\frac{1}{5} \cdot \frac{2}{3} = \frac{1 \cdot 2}{5 \cdot 3}$$
$$= \frac{2}{15}$$

Fact 1.2.16 Multiplication with Fractions. *As long as b and d are not 0, then fractions multiply this way:*

$$\frac{a}{b} \cdot \frac{c}{d} = \frac{a \cdot c}{b \cdot d}$$

Checkpoint 1.2.17. Simplify these fraction products.

a. $\dfrac{1}{3} \cdot \dfrac{10}{7} = \boxed{}$

b. $\dfrac{12}{3} \cdot \dfrac{15}{3} = \boxed{}$

c. $-\dfrac{14}{5} \cdot \dfrac{2}{3} = \boxed{}$

d. $\dfrac{70}{27} \cdot \dfrac{12}{-20} = \boxed{}$

Explanation.

a. Multiplying numerators gives 10, and multiplying denominators gives 21. The answer is $\frac{10}{21}$.

b. Before we multiply fractions, note that $\frac{12}{3}$ reduces to 4, and $\frac{15}{3}$ reduces to 5. So we just have $4 \cdot 5 = 20$.

c. Multiplying numerators gives 28, and multiplying denominators gives 15. The answer is $\frac{28}{15}$.

d. Before we multiply fractions, note that $\frac{12}{-20}$ reduces to $\frac{-3}{5}$. So we have $\frac{70}{27} \cdot \frac{-3}{5}$. Both the numerator of the first fraction and denominator of the second fraction are divisible by 5, so it helps to reduce both fractions accordingly and get $\frac{14}{27} \cdot \frac{-3}{1}$. Both the denominator of the first fraction and numerator of the second fraction are divisible by 3, so it helps to reduce both fractions accordingly and get $\frac{14}{9} \cdot \frac{-1}{1}$. Now we are just multiplying $\frac{14}{9}$ by -1, so the result is $\frac{-14}{9}$.

1.2.4 Division with Fractions

How does division with fractions work? Are we able to compute/simplify each of these examples?

a. $3 \div \frac{2}{7}$
b. $\frac{18}{19} \div 5$
c. $\frac{14}{3} \div \frac{8}{9}$
d. $\dfrac{\frac{2}{4}}{\frac{5}{2}}$

We know that when we divide something by 2, this is the same as multiplying it by $\frac{1}{2}$. Conversely, dividing a number or expression by $\frac{1}{2}$ is the same as multiplying by $\frac{2}{1}$, or just 2. The more general property is that when we divide a number or expression by $\frac{a}{b}$, this is equivalent to multiplying by the reciprocal $\frac{b}{a}$.

Fact 1.2.18 Division with Fractions. *As long as b, c and d are not 0, then division with fractions works this way:*

$$\frac{a}{b} \div \frac{c}{d} = \frac{a}{b} \cdot \frac{d}{c}$$

Example 1.2.19 With our examples from the beginning of this subsection:

a. $\begin{aligned}[t] 3 \div \frac{2}{7} &= 3 \cdot \frac{7}{2} \\ &= \frac{3}{1} \cdot \frac{7}{2} \\ &= \frac{21}{2} \end{aligned}$

c. $\begin{aligned}[t] \frac{14}{3} \div \frac{8}{9} &= \frac{14}{3} \cdot \frac{9}{8} \\ &= \frac{14}{1} \cdot \frac{3}{8} \\ &= \frac{7}{1} \cdot \frac{3}{4} \\ &= \frac{21}{4} \end{aligned}$

b. $\begin{aligned}[t] \frac{18}{19} \div 5 &= \frac{18}{19} \div \frac{5}{1} \\ &= \frac{18}{19} \cdot \frac{1}{5} \\ &= \frac{18}{95} \end{aligned}$

d. $\begin{aligned}[t] \frac{\frac{2}{5}}{\frac{5}{2}} &= \frac{2}{5} \div \frac{5}{2} \\ &= \frac{2}{5} \cdot \frac{2}{5} \\ &= \frac{4}{25} \end{aligned}$

Checkpoint 1.2.20. Simplify these fraction division expressions.

a. $\dfrac{1}{3} \div \dfrac{10}{7} = \boxed{}$ b. $\dfrac{12}{5} \div 5 = \boxed{}$ c. $-14 \div \dfrac{3}{2} = \boxed{}$ d. $\dfrac{70}{9} \div \dfrac{11}{-20} = \boxed{}$

Explanation.

a. $\begin{aligned}[t] \frac{1}{3} \div \frac{10}{7} &= \frac{1}{3} \cdot \frac{7}{10} \\ &= \frac{7}{30} \end{aligned}$

c. $\begin{aligned}[t] -14 \div \frac{3}{2} &= -\frac{1}{14} \cdot \frac{2}{3} \\ &= -\frac{1}{7} \cdot \frac{1}{3} \\ &= -\frac{1}{21} \end{aligned}$

b. $\begin{aligned}[t] \frac{12}{5} \div 5 &= \frac{12}{5} \cdot \frac{1}{5} \\ &= \frac{12}{25} \end{aligned}$

d. $\begin{aligned}[t] \frac{70}{9} \div \frac{11}{-20} &= -\frac{70}{9} \cdot \frac{20}{11} \\ &= -\frac{1400}{99} \end{aligned}$

1.2.5 Adding and Subtracting Fractions

With whole numbers and integers, operations of addition and subtraction are relatively straightforward. The situation is almost as straightforward with fractions *if the two fractions have the same denominator.* Consider

$$\frac{7}{2} + \frac{3}{2} = 7 \text{ halves} + 3 \text{ halves}$$

In the same way that 7 tacos and 3 tacos make 10 tacos, we have:

$$\begin{array}{ccccc} 7 \text{ halves} & + & 3 \text{ halves} & = & 10 \text{ halves} \\ \frac{7}{2} & + & \frac{3}{2} & = & \frac{10}{2} \\ & & & = & 5 \end{array}$$

Fact 1.2.21 Adding/Subtracting with Fractions Having the Same Denominator. *To add or subtract two fractions having the same denominator,* keep *that denominator, and add or subtract the numerators.*

$$\frac{a}{b} + \frac{c}{b} = \frac{a+c}{b} \qquad\qquad \frac{a}{b} - \frac{c}{b} = \frac{a-c}{b}$$

If it's possible, useful, or required of you, simplify the result by reducing to lowest terms.

Checkpoint 1.2.22. Add or subtract these fractions.

a. $\frac{1}{3} + \frac{10}{3} =$ [____________]

b. $\frac{13}{6} - \frac{5}{6} =$ [____________]

Explanation.

a. Since the denominators are both 3, we can add the numerators: $1 + 10 = 11$. The answer is $\frac{11}{3}$.

b. Since the denominators are both 6, we can subtract the numerators: $13 - 5 = 8$. The answer is $\frac{8}{6}$, but that reduces to $\frac{4}{3}$.

Whenever we'd like to combine fractional amounts that don't represent the same number of parts of a whole (that is, when the denominators are different), finding sums and differences is more complicated.

> **Example 1.2.23 Quarters and Dimes.** Find the sum $\frac{3}{4} + \frac{2}{10}$. Does this seem intimidating? Consider this:
>
> - $\frac{1}{4}$ of a dollar is a quarter, and so $\frac{3}{4}$ of a dollar is 75 cents.
>
> - $\frac{1}{10}$ of a dollar is a dime, and so $\frac{2}{10}$ of a dollar is 20 cents.
>
> So if you know what to look for, the expression $\frac{3}{4} + \frac{2}{10}$ is like adding 75 cents and 20 cents, which gives you 95 cents. As a fraction of one dollar, that is $\frac{95}{100}$. So we can report
>
> $$\frac{3}{4} + \frac{2}{10} = \frac{95}{100}.$$
>
> (Although we should probably reduce that last fraction to $\frac{19}{20}$.)

This example was not something you can apply to other fraction addition situations, because the denominators here worked especially well with money amounts. But there is something we can learn here. The fraction $\frac{3}{4}$ was equivalent to $\frac{75}{100}$, and the other fraction $\frac{2}{10}$ was equivalent to $\frac{20}{100}$. These *equivalent* fractions have the same denominator and are therefore "easy" to add. What we saw happen was:

$$\begin{aligned} \frac{3}{4} + \frac{2}{10} &= \frac{75}{100} + \frac{20}{100} \\ &= \frac{95}{100} \end{aligned}$$

This realization gives us a strategy for adding (or subtracting) fractions.

Fact 1.2.24 Adding/Subtracting Fractions with Different Denominators. *To add (or subtract) generic fractions together, use their denominators to find a* **common denominator.** *This means some whole number that is a whole multiple of both of the original denominators. Then rewrite the two fractions as equivalent fractions that use this common denominator. Write the result* keeping *that denominator and adding (or subtracting) the numerators. Reduce the fraction if that is useful or required.*

Example 1.2.25 Let's add $\frac{2}{3} + \frac{2}{5}$. The denominators are 3 and 5, so the number 15 would be a good common denominator.

$$\begin{aligned}\frac{2}{3} + \frac{2}{5} &= \frac{2 \cdot 5}{3 \cdot 5} + \frac{2 \cdot 3}{5 \cdot 3} \\ &= \frac{10}{15} + \frac{6}{15} \\ &= \frac{16}{15}\end{aligned}$$

Checkpoint 1.2.26. A chef had $\frac{2}{3}$ cups of flour and needed to use $\frac{1}{8}$ cup to thicken a sauce. How much flour is left?

Explanation. We need to compute $\frac{2}{3} - \frac{1}{8}$. The denominators are 3 and 8. One common denominator is 24, so we move to rewrite each fraction using 24 as the denominator:

$$\begin{aligned}\frac{2}{3} - \frac{1}{8} &= \frac{2 \cdot 8}{3 \cdot 8} - \frac{1 \cdot 3}{8 \cdot 3} \\ &= \frac{16}{24} - \frac{3}{24} \\ &= \frac{13}{24}\end{aligned}$$

The numerical result is $\frac{13}{24}$, but a pure number does not answer this question. The amount of flour remaining is $\frac{13}{24}$ *cups.*

1.2.6 Mixed Numbers and Improper Fractions

A simple recipe for bread contains only a few ingredients:

$1\frac{1}{2}$	tablespoons yeast
$1\frac{1}{2}$	tablespoons kosher salt
$6\frac{1}{2}$	cups unbleached, all-purpose flour (more for dusting)

Each ingredient is listed as a **mixed number** that quickly communicates how many whole amounts and how many parts are needed. It's useful for quickly communicating a practical amount of something you are cooking with, measuring on a ruler, purchasing at the grocery store, etc. But it causes trouble in an algebra class. The number $1\frac{1}{2}$ means "one *and* one half." So really,

$$1\frac{1}{2} = 1 + \frac{1}{2}$$

The trouble is that with $1\frac{1}{2}$, you have two numbers written right next to each other. Normally with two math expressions written right next to each other, they should be *multiplied*, not *added*. But with a mixed number, they *should* be added.

Fortunately we just reviewed how to add fractions. If we need to do any arithmetic with a mixed number like $1\,{}^{1}\!/_{2}$, we can treat it as $1 + \frac{1}{2}$ and simplify to get a "nice" fraction instead: $\frac{3}{2}$. A fraction like $\frac{3}{2}$ is called an **improper fraction** because it's actually larger than 1. And a "proper" fraction would be something small that is only *part* of a whole instead of *more* than a whole.

$$1\frac{1}{2} = 1 + \frac{1}{2}$$
$$= \frac{1}{1} + \frac{1}{2}$$
$$= \frac{2}{2} + \frac{1}{2}$$
$$= \frac{3}{2}$$

Exercises

Review and Warmup

1. Which letter is $-\frac{29}{4}$ on the number line?

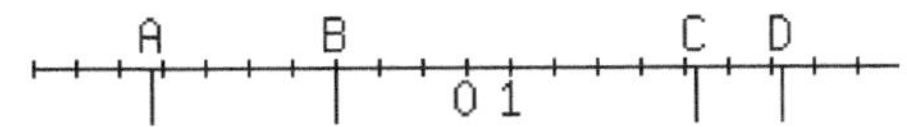

($\square$ A $\square$ B $\square$ C $\square$ D)

2. Which letter is $\frac{23}{4}$ on the number line?

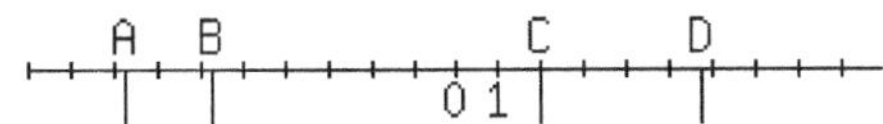

($\square$ A $\square$ B $\square$ C $\square$ D)

3. The dot in the graph can be represented by what fraction?

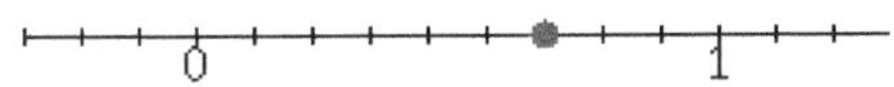

4. The dot in the graph can be represented by what fraction?

5. The dot in the graph can be represented by what fraction?

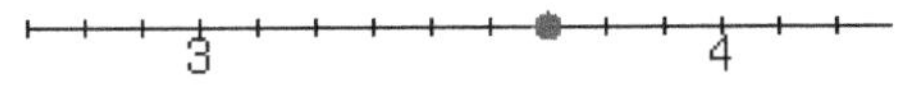

6. The dot in the graph can be represented by what fraction?

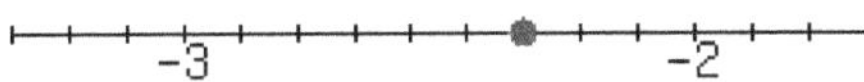

Reducing Fractions

7. Reduce the fraction $\dfrac{7}{70}$.

8. Reduce the fraction $\dfrac{14}{63}$.

9. Reduce the fraction $\dfrac{10}{50}$.

10. Reduce the fraction $\dfrac{20}{33}$.

11. Reduce the fraction $\dfrac{100}{70}$.

12. Reduce the fraction $\dfrac{42}{6}$.

Building Fractions

13. Find an equivalent fraction to $\frac{5}{7}$ with denominator 35.

14. Find an equivalent fraction to $\frac{3}{7}$ with denominator 14.

15. Find an equivalent fraction to $\frac{1}{17}$ with denominator 68.

16. Find an equivalent fraction to $\frac{13}{19}$ with denominator 38.

Multiplying/Dividing Fractions

17. Multiply: $\frac{1}{7} \cdot \frac{3}{8}$

18. Multiply: $\frac{3}{7} \cdot \frac{3}{8}$

19. Multiply: $\frac{6}{11} \cdot \frac{13}{6}$

20. Multiply: $\frac{7}{11} \cdot \frac{3}{14}$

21. Multiply: $3 \cdot \frac{3}{10}$

22. Multiply: $6 \cdot \frac{4}{5}$

23. Multiply: $-\frac{20}{7} \cdot \frac{7}{25}$

24. Multiply: $-\frac{18}{5} \cdot \frac{13}{28}$

25. Multiply: $28 \cdot \left(-\frac{9}{7}\right)$

26. Multiply: $6 \cdot \left(-\frac{1}{3}\right)$

27. Multiply: $\frac{7}{4} \cdot \frac{5}{49} \cdot \frac{6}{25}$

28. Multiply: $\frac{5}{9} \cdot \frac{14}{25} \cdot \frac{3}{4}$

29. Multiply: $\frac{14}{3} \cdot \frac{1}{4} \cdot 15$

30. Multiply: $\frac{7}{5} \cdot \frac{2}{49} \cdot 15$

31. Divide: $\frac{4}{7} \div \frac{7}{3}$

32. Divide: $\frac{4}{7} \div \frac{7}{4}$

33. Divide: $\frac{7}{10} \div \left(-\frac{5}{4}\right)$

34. Divide: $\frac{1}{20} \div \left(-\frac{5}{12}\right)$

35. Divide: $-\frac{3}{2} \div (-15)$

36. Divide: $-\frac{3}{8} \div (-9)$

37. Divide: $20 \div \frac{5}{3}$

38. Divide: $4 \div \frac{2}{3}$

39. Multiply: $3\frac{8}{9} \cdot 2\frac{1}{10}$

40. Multiply: $1\frac{5}{9} \cdot 6\frac{3}{4}$

Adding/Subtracting Fractions

41. Add: $\frac{2}{27} + \frac{1}{27}$

42. Add: $\frac{11}{40} + \frac{1}{40}$

43. Add: $\frac{5}{7} + \frac{11}{21}$

44. Add: $\dfrac{2}{9} + \dfrac{13}{18}$

45. Add: $\dfrac{1}{9} + \dfrac{13}{18}$

46. Add: $\dfrac{5}{6} + \dfrac{11}{30}$

47. Add: $\dfrac{3}{7} + \dfrac{1}{10}$

48. Add: $\dfrac{3}{8} + \dfrac{3}{7}$

49. Add: $\dfrac{1}{6} + \dfrac{1}{10}$

50. Add: $\dfrac{3}{10} + \dfrac{1}{6}$

51. Add: $\dfrac{5}{6} + \dfrac{9}{10}$

52. Add: $\dfrac{4}{5} + \dfrac{7}{10}$

53. Add: $-\dfrac{2}{5} + \dfrac{3}{5}$

54. Add: $-\dfrac{1}{5} + \dfrac{4}{5}$

55. Add: $-\dfrac{3}{7} + \dfrac{11}{14}$

56. Add: $-\dfrac{5}{9} + \dfrac{5}{54}$

57. Add: $-\dfrac{1}{8} + \dfrac{1}{7}$

58. Add: $-\dfrac{1}{8} + \dfrac{1}{5}$

59. Add: $2 + \dfrac{7}{8}$

60. Add: $4 + \dfrac{2}{5}$

61. Add: $\dfrac{3}{10} + \dfrac{1}{6} + \dfrac{1}{5}$

62. Add: $\dfrac{1}{3} + \dfrac{1}{8} + \dfrac{1}{6}$

63. Add: $\dfrac{2}{3} + \dfrac{1}{6} + \dfrac{3}{10}$

64. Add: $\dfrac{1}{3} + \dfrac{1}{6} + \dfrac{3}{5}$

65. Subtract: $\dfrac{16}{21} - \dfrac{4}{21}$

66. Subtract: $\dfrac{33}{32} - \dfrac{13}{32}$

67. Subtract: $\dfrac{4}{7} - \dfrac{33}{35}$

68. Subtract: $\dfrac{4}{9} - \dfrac{43}{45}$

69. Subtract: $\dfrac{1}{18} - \dfrac{5}{6}$

70. Subtract: $\dfrac{39}{40} - \dfrac{5}{8}$

71. Subtract: $-\dfrac{3}{10} - \dfrac{1}{6}$

72. Subtract: $-\dfrac{5}{6} - \dfrac{7}{10}$

73. Subtract: $-\dfrac{3}{10} - \left(-\dfrac{5}{6}\right)$

74. Subtract: $-\dfrac{5}{6} - \left(-\dfrac{3}{10}\right)$

75. Subtract: $-4 - \dfrac{15}{7}$

76. Subtract: $1 - \dfrac{10}{9}$

Applications

77. Michele walked $\frac{3}{10}$ of a mile in the morning, and then walked $\frac{3}{8}$ of a mile in the afternoon. How far did Michele walk altogether?

Michele walked a total of [ ] of a mile.

78. Kurt walked $\frac{1}{11}$ of a mile in the morning, and then walked $\frac{1}{9}$ of a mile in the afternoon. How far did Kurt walk altogether?

Kurt walked a total of [ ] of a mile.

79. Penelope and Kenji are sharing a pizza. Penelope ate $\frac{1}{10}$ of the pizza, and Kenji ate $\frac{3}{8}$ of the pizza. How much of the pizza was eaten in total?

They ate [ ] of the pizza.

80. The pie chart represents a school's student population.

School Population Breakdown by Race

Together, white and black students make up [ ] of the school's population.

81. A trail's total length is $\frac{19}{45}$ of a mile. It has two legs. The first leg is $\frac{2}{9}$ of a mile long. How long is the second leg?

The second leg is [ ] of a mile in length.

82. A trail's total length is $\frac{5}{18}$ of a mile. It has two legs. The first leg is $\frac{1}{9}$ of a mile long. How long is the second leg?

The second leg is [ ] of a mile in length.

83. Jessica is participating in a running event. In the first hour, she completed $\frac{1}{10}$ of the total distance. After another hour, in total she had completed $\frac{17}{70}$ of the total distance.

What fraction of the total distance did Jessica complete during the second hour?

Jessica completed [   ] of the distance during the second hour.

84. Each page of a book is $6\frac{1}{2}$ inches in height, and consists of a header (a top margin), a footer (a bottom margin), and the middle part (the body). The header is $\frac{2}{9}$ of an inch thick and the middle part is $5\frac{8}{9}$ inches from top to bottom.

What is the thickness of the footer?

The footer is [   ] of an inch thick.

85. The pie chart represents a school's student population.

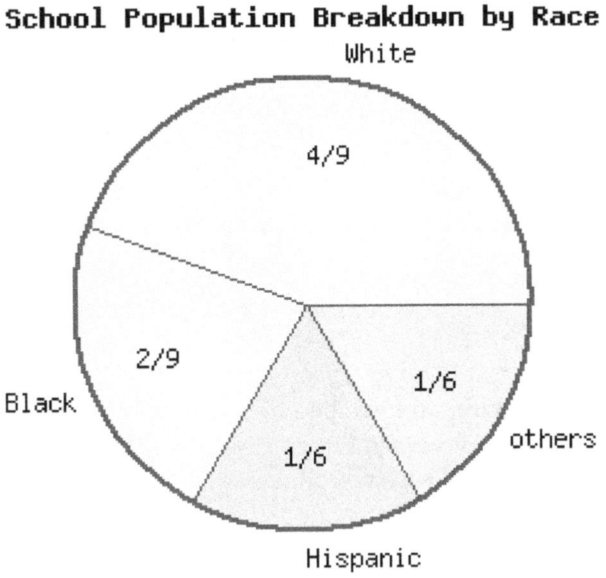

School Population Breakdown by Race

[   ] more of the school is white students than black students.

86. Jessica and Carly are sharing a pizza. Jessica ate $\frac{2}{9}$ of the pizza, and Carly ate $\frac{1}{6}$ of the pizza. How much more pizza did Jessica eat than Carly?

Jessica ate [   ] more of the pizza than Carly ate.

87. Kayla and Blake are sharing a pizza. Kayla ate $\frac{2}{9}$ of the pizza, and Blake ate $\frac{1}{8}$ of the pizza. How much more pizza did Kayla eat than Blake?

Kayla ate [______] more of the pizza than Blake ate.

88. A school had a fund-raising event. The revenue came from three resources: ticket sales, auction sales, and donations. Ticket sales account for $\frac{1}{10}$ of the total revenue; auction sales account for $\frac{5}{7}$ of the total revenue. What fraction of the revenue came from donations?

[______] of the revenue came from donations.

89. A few years back, a car was purchased for \$10,800. Today it is worth $\frac{1}{3}$ of its original value. What is the car's current value?

The car's current value is [______].

90. A few years back, a car was purchased for \$15,000. Today it is worth $\frac{1}{3}$ of its original value. What is the car's current value?

The car's current value is [______].

91. A town has 200 residents in total, of which $\frac{3}{4}$ are white Americans. How many white Americans reside in this town?

There are [______] white Americans residing in this town.

92. A company received a grant, and decided to spend $\frac{20}{21}$ of this grant in research and development next year. Out of the money set aside for research and development, $\frac{3}{5}$ will be used to buy new equipment. What fraction of the grant will be used to buy new equipment?

[______] of the grant will be used to buy new equipment.

93. A food bank just received 28 kilograms of emergency food. Each family in need is to receive $\frac{2}{5}$ kilograms of food. How many families can be served with the 28 kilograms of food?

[______] families can be served with the 28 kilograms of food.

94. A construction team maintains a 52-mile-long sewage pipe. Each day, the team can cover $\frac{4}{7}$ of a mile. How many days will it take the team to complete the maintenance of the entire sewage pipe?

It will take the team [______] days to complete maintaining the entire sewage pipe.

95. A child is stacking up tiles. Each tile's height is $\frac{3}{4}$ of a centimeter. How many layers of tiles are needed to reach 15 centimeters in total height?

To reach the total height of 15 centimeters, ⬚ layers of tiles are needed.

96. A restaurant made 450 cups of pudding for a festival.

Customers at the festival will be served $\frac{1}{10}$ of a cup of pudding per serving. How many customers can the restaurant serve at the festival with the 450 cups of pudding?

The restaurant can serve ⬚ customers at the festival with the 450 cups of pudding.

97. A 2×4 piece of lumber in your garage is $62\frac{17}{32}$ inches long. A second 2×4 is $42\frac{7}{16}$ inches long. If you lay them end to end, what will the total length be?

The total length will be ⬚ inches.

98. A 2×4 piece of lumber in your garage is $39\frac{3}{4}$ inches long. A second 2×4 is $31\frac{3}{4}$ inches long. If you lay them end to end, what will the total length be?

The total length will be ⬚ inches.

99. Each page of a book consists of a header, a footer and the middle part. The header is $\frac{6}{7}$ inches in height; the footer is $\frac{3}{14}$ inches in height; and the middle part is $3\frac{3}{7}$ inches in height.

What is the total height of each page in this book? Use mixed number in your answer if needed.

Each page in this book is ⬚ inches in height.

100. To pave the road on Ellis Street, the crew used $4\frac{1}{4}$ tons of cement on the first day, and used $5\frac{5}{6}$ tons on the second day. How many tons of cement were used in all?

⬚ tons of cement were used in all.

101. When driving on a high way, noticed a sign saying exit to Johnstown is $1\frac{3}{4}$ miles away, while exit to Jerrystown is $3\frac{1}{2}$ miles away. How far is Johnstown from Jerrystown?

Johnstown and Jerrystown are ⬚ miles apart.

102. A cake recipe needs $2\frac{1}{4}$ cups of flour. Using this recipe, to bake 9 cakes, how many cups of flour are needed?

To bake 9 cakes, ⬚ cups of flour are needed.

Sketching Fractions

103. Sketch a number line showing each fraction. (Be sure to carefully indicate the correct number of equal parts of the whole.)

(a) $\dfrac{2}{3}$ (b) $\dfrac{6}{8}$ (c) $\dfrac{5}{4}$ (d) $-\dfrac{4}{5}$

104. Sketch a number line showing each fraction. (Be sure to carefully indicate the correct number of equal parts of the whole.)

(a) $\dfrac{1}{6}$ (b) $\dfrac{3}{9}$ (c) $\dfrac{7}{6}$ (d) $-\dfrac{8}{5}$

105. Sketch a picture of the product $\frac{3}{5} \cdot \frac{1}{2}$, using a number line or rectangles.

106. Sketch a picture of the sum $\frac{2}{3} + \frac{1}{8}$, using a number line or rectangles.

Challenge

107. Given that $a \neq 0$, simplify $\dfrac{6}{a} + \dfrac{5}{a}$.

108. Given that $a \neq 0$, simplify $\dfrac{7}{a} + \dfrac{1}{2a}$.

109. Given that $a \neq 0$, simplify $\dfrac{8}{a} - \dfrac{8}{5a}$.

1.3 Absolute Value and Square Root

In this section, we will learn the basics of **absolute value** and **square root**. These are actions you can *do* to a given number, often changing the number into something else.

1.3.1 Introduction to Absolute Value

Definition 1.3.2. The **absolute value** of a number is the distance between that number and 0 on a number line. For the absolute value of x, we write $|x|$.

Let's look at $|2|$ and $|-2|$, the absolute value of 2 and the absolute value of -2.

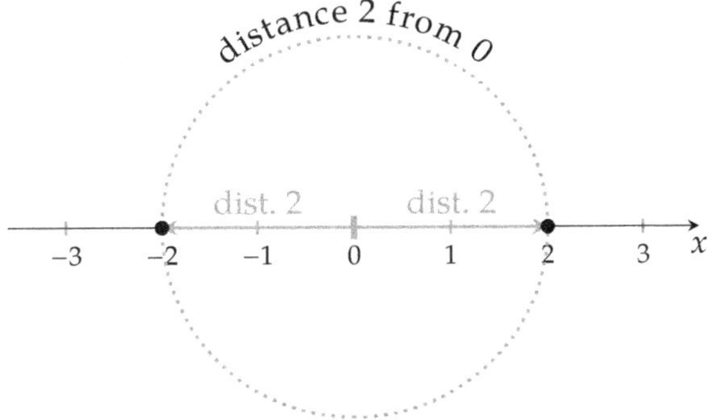

Figure 1.3.3: $|2|$ and $|-2|$

Since the distance between 2 and 0 on the number line is 2 units, the absolute value of 2 is 2. We write $|2| = 2$.

Since the distance between -2 and 0 on the number line is also 2 units, the absolute value of -2 is also 2. We write $|-2| = 2$.

Fact 1.3.4 Absolute Value. *Taking the absolute value of a number results in whatever the "positive version" of that number is. This is because the real meaning of absolute value is its* distance *from zero.*

Checkpoint 1.3.5 Calculating Absolute Value. Try calculating some absolute values.

a. $|57| = \boxed{}$
b. $|-43| = \boxed{}$
c. $\left|\frac{2}{-5}\right| = \boxed{}$

Explanation.

a. 57 is 57 units away from 0 on a number line, so $|57| = 57$. Another way to think about this is that the "positive version" of 57 is 57.

b. -43 is 43 units away from 0 on a number line, so $|-43| = 43$. Another way to think about this is that the "positive version" of -43 is 43.

c. $\frac{2}{-5}$ is $\frac{2}{5}$ units away from 0 on a number line, so $\left|\frac{2}{-5}\right| = \frac{2}{5}$. Another way to think about this is that the "positive version" of $\frac{2}{-5}$ is $\frac{2}{5}$.

Warning 1.3.6 Absolute Value Does Not Exactly "Make Everything Positive". Students may see an expression like $|2 - 5|$ and incorrectly think it is OK to "make everything positive" and write $2 + 5$. This is incorrect since $|2 - 5|$ works out to be 3, not 7, as we are actually taking the absolute value of -3 (the equivalent number inside the absolute value).

1.3.2 Square Root Facts

If you have learned your basic multiplication table, you know:

×	1	2	3	4	5	6	7	8	9
1	1	2	3	4	5	6	7	8	9
2	2	4	6	8	10	12	14	16	18
3	3	6	9	12	15	18	21	24	27
4	4	8	12	16	20	24	28	32	36
5	5	10	15	20	25	30	35	40	45
6	6	12	18	24	30	36	42	48	54
7	7	14	21	28	35	42	49	56	63
8	8	16	24	32	40	48	56	64	72
9	9	18	27	36	45	54	63	72	81

Table 1.3.7: Multiplication table with squares

The numbers along the diagonal are special; they are known as **perfect squares**. And for working with square roots, it will be helpful if you can memorize these first few perfect square numbers.

"Taking a square root" is the opposite action of squaring a number. For example, when you square 3, the result is 9. So when you take the square root of 9, the result is 3. Just knowing that 9 comes about as 3^2 lets us realize that 3 is the square root of 9. This is why memorizing the perfect squares from the multiplication table can be so helpful.

The notation we use for taking a square root is the **radical**, $\sqrt{\ }$. For example, "the square root of 9" is denoted $\sqrt{9}$. And now we know enough to be able to write $\sqrt{9} = 3$.

Tossing in a few extra special square roots, it's advisable to memorize the following:

$$\sqrt{0} = 0 \qquad \sqrt{1} = 1 \qquad \sqrt{4} = 2 \qquad \sqrt{9} = 3$$
$$\sqrt{16} = 4 \qquad \sqrt{25} = 5 \qquad \sqrt{36} = 6 \qquad \sqrt{49} = 7$$
$$\sqrt{64} = 8 \qquad \sqrt{81} = 9 \qquad \sqrt{100} = 10 \qquad \sqrt{121} = 11$$
$$\sqrt{144} = 12 \qquad \sqrt{169} = 13 \qquad \sqrt{196} = 14 \qquad \sqrt{225} = 15$$

1.3.3 Calculating Square Roots with a Calculator

Most square roots are actually numbers with decimal places that go on forever. Take $\sqrt{5}$ as an example:

$$\sqrt{4} = 2 \qquad\qquad \sqrt{5} = ? \qquad\qquad \sqrt{9} = 3$$

Since 5 is between 4 and 9, then $\sqrt{5}$ must be somewhere between 2 and 3. There are no whole numbers between 2 and 3, so $\sqrt{5}$ must be some number with decimal places. If the decimal places eventually stopped, then squaring it would give you another number with decimal places that stop further out. But squaring it gives you 5 with no decimal places. So the only possibility is that $\sqrt{5}$ is a decimal between 2 and 3 that goes on forever. With a calculator, we can see:

$$\sqrt{5} \approx 2.236$$

Actually the decimal will not terminate, and that is why we used the $\approx$ symbol instead of an equals sign. To get 2.236 we rounded down slightly from the true value of $\sqrt{5}$. With a calculator, we can check that $2.236^2 = 4.999696$, a little shy of 5.

1.3.4 Square Roots of Fractions

We can calculate the square root of some fractions by hand, such as $\sqrt{\frac{1}{4}}$. The idea is the same: can you think of a number that you would square to get $\frac{1}{4}$? Being familiar with fraction multiplication, we know that $\frac{1}{2} \cdot \frac{1}{2} = \frac{1}{4}$ and so $\sqrt{\frac{1}{4}} = \frac{1}{2}$.

Checkpoint 1.3.8 Square Roots of Fractions. Try calculating some absolute values.

a. $\sqrt{\dfrac{1}{25}} = \boxed{}$

b. $\sqrt{\dfrac{4}{9}} = \boxed{}$

c. $\sqrt{\dfrac{81}{121}} = \boxed{}$

Explanation.

a. Since $\sqrt{1} = 1$ and $\sqrt{25} = 5$, then $\sqrt{\dfrac{1}{25}} = \dfrac{1}{5}$.

b. Since $\sqrt{4} = 2$ and $\sqrt{9} = 3$, then $\sqrt{\dfrac{4}{9}} = \dfrac{2}{3}$.

c. Since $\sqrt{81} = 9$ and $\sqrt{121} = 11$, then $\sqrt{\dfrac{81}{121}} = \dfrac{9}{11}$.

1.3.5 Square Root of Negative Numbers

Can we find the square root of a negative number, such as $\sqrt{-25}$? That would mean that there is some number out there that multiplies by itself to make -25. Would $\sqrt{-25}$ be positive or negative? Either way, once you square it (multiply it by itself) the result would be positive. So it couldn't possibly square to -25. So there is no square root of -25 or of any negative number for that matter.

If you are confronted with an expression like $\sqrt{-25}$, or any other square root of a negative number, you can state that "there is no real square root" or that the result "does not exist" (as a real number).

Imaginary Numbers. Mathematicians imagined a new type of number, neither positive nor negative, that would square to a negative result. But that is beyond the scope of this chapter.

Exercises

Review and Warmup

1. Evaluate the expressions.

   a. 1^2 c. 5^2 e. 9^2

   b. 3^2 d. 7^2 f. 11^2

2. Evaluate the expressions.

   a. 2^2 c. 6^2 e. 10^2

   b. 4^2 d. 8^2 f. 12^2

Absolute Value   Evaluate the following.

3. $|-3| =$

4. $|0| =$

5. $|31.69| =$

6. $|-15.76| =$

7. $\left|\dfrac{21}{8}\right| =$

8. $\left|-\dfrac{17}{9}\right| =$

9. a. $|2| =$
    b. $|-2| =$
    c. $-|2| =$
    d. $-|-2| =$

10. a. $|3| =$
    b. $|-3| =$
    c. $-|3| =$
    d. $-|-3| =$

11. a. $|3| =$
    b. $|-8| =$
    c. $|0| =$
    d. $|16 + (-6)| =$
    e. $|-9 - (-1)| =$

12. a. $|4| =$
    b. $|-2| =$
    c. $|0| =$
    d. $|20 + (-8)| =$
    e. $|-9 - (-3)| =$

13. a. $-|3 - 7| =$
    b. $|-3 - 7| =$
    c. $-3|7 - 3| =$

14. a. $-|1 - 10| =$
    b. $|-1 - 10| =$
    c. $-3|10 - 1| =$

15.   Which of the following are square numbers? There may be more than one correct answer.

☐ 81   ☐ 34   ☐ 9   ☐ 116   ☐ 55   ☐ 121

16.   Which of the following are square numbers? There may be more than one correct answer.

☐ 15   ☐ 100   ☐ 64   ☐ 45   ☐ 32   ☐ 36

Square Roots   Evaluate the following.

17. a. $\sqrt{144} =$
    b. $\sqrt{4} =$
    c. $\sqrt{1} =$

18. a. $\sqrt{1} =$
    b. $\sqrt{100} =$
    c. $\sqrt{36} =$

19. a. $\sqrt{\dfrac{4}{121}} =$
    b. $\sqrt{-\dfrac{16}{9}} =$

20.

   a. $\sqrt{\dfrac{16}{25}} = \boxed{}$

   b. $\sqrt{-\dfrac{144}{49}} = \boxed{}$

21. Do not use a calculator.

   a. $\sqrt{25} = \boxed{}$

   b. $\sqrt{0.25} = \boxed{}$

   c. $\sqrt{2500} = \boxed{}$

22. Do not use a calculator.

   a. $\sqrt{49} = \boxed{}$

   b. $\sqrt{0.49} = \boxed{}$

   c. $\sqrt{4900} = \boxed{}$

23. Do not use a calculator.

   a. $\sqrt{64} = \boxed{}$

   b. $\sqrt{6400} = \boxed{}$

   c. $\sqrt{640000} = \boxed{}$

24. Do not use a calculator.

   a. $\sqrt{81} = \boxed{}$

   b. $\sqrt{8100} = \boxed{}$

   c. $\sqrt{810000} = \boxed{}$

25. Do not use a calculator.

   a. $\sqrt{100} = \boxed{}$

   b. $\sqrt{1} = \boxed{}$

   c. $\sqrt{0.01} = \boxed{}$

26. Do not use a calculator.

   a. $\sqrt{121} = \boxed{}$

   b. $\sqrt{1.21} = \boxed{}$

   c. $\sqrt{0.0121} = \boxed{}$

27. Use a calculator to approximate with a decimal.

$\sqrt{13} \approx \boxed{}$

28. Use a calculator to approximate with a decimal.

$\sqrt{17} \approx \boxed{}$

29. $\sqrt{\dfrac{9}{100}} = \boxed{}$.

30. $\sqrt{\dfrac{16}{121}} = \boxed{}$.

31. $-\sqrt{49} = \boxed{}$.

32. $-\sqrt{64} = \boxed{}$.

33. $\sqrt{-81} = \boxed{}$.

34. $\sqrt{-100} = \boxed{}$.

35. $\sqrt{-\dfrac{121}{144}} = \boxed{}$.

36. $\sqrt{-\dfrac{1}{49}} = \boxed{}$.

37. $-\sqrt{\dfrac{4}{25}} = \boxed{}$.

38. $-\sqrt{\dfrac{9}{121}} = \boxed{}$.

39.

   a. $\sqrt{100} - \sqrt{36} = \boxed{}$

   b. $\sqrt{100 - 36} = \boxed{}$

40.

   a. $\sqrt{25} - \sqrt{9} = \boxed{}$

   b. $\sqrt{25 - 9} = \boxed{}$

41. $\dfrac{5}{\sqrt{81}} = \boxed{}$.

42. $\dfrac{7}{\sqrt{36}} = \boxed{}$.

1.4 Order of Operations

Mathematical symbols are a means of communication, and it's important that when you write something, everyone else knows exactly what you intended. For example, if we say in English, "two times three squared," do we mean that:

- 2 is multiplied by 3, and then the result is squared?

- or that 2 is multiplied by the result of squaring 3?

English is allowed to have ambiguities like this. But mathematical language needs to be precise and mean the same thing to everyone reading it. For this reason, a standard **order of operations** has been adopted, which we review here.

1.4.1 Grouping Symbols

Consider the math expression $2 \cdot 3^2$. There are two mathematical operations here: a multiplication and an exponentiation. The result of this expression will change depending on which operation you decide to execute first: the multiplication or the exponentiation. If you multiply $2 \cdot 3$, and then square the result, you have 36. If you square 3, and then multiply 2 by the result, you have 18. If we want all people everywhere to interpret $2 \cdot 3^2$ in the same way, then only *one* of these can be correct.

The first tools that we have to tell readers what operations to execute first are grouping symbols, like parentheses and brackets. If you *intend* to execute the multiplication first, then writing

$$(2 \cdot 3)^2$$

clearly tells your reader to do that. And if you *intend* to execute the power first, then writing

$$2 \cdot \left(3^2\right)$$

clearly tells your reader to do that.

To visualize the difference between $2 \cdot \left(3^2\right)$ or $(2 \cdot 3)^2$, consider these garden plots:

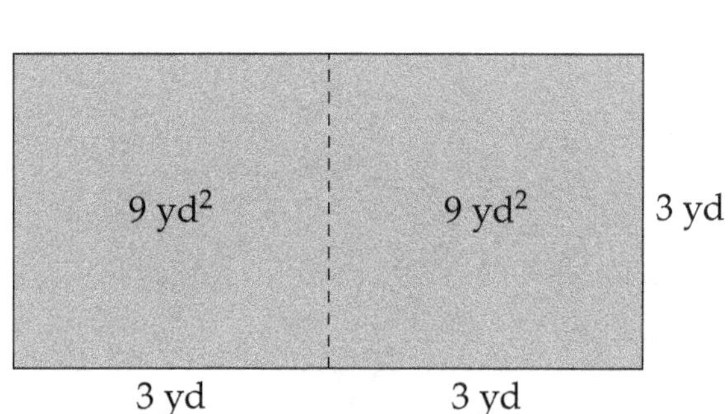

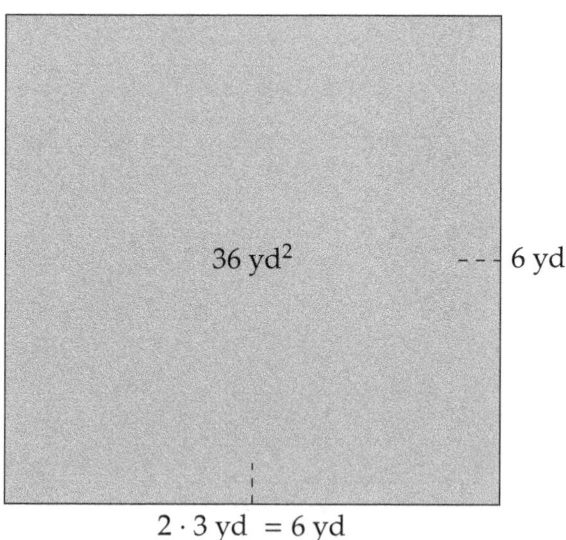

Figure 1.4.2: 3 yd is squared, then doubled: $2 \cdot \left(3^2\right)$

Figure 1.4.3: 3 yd is doubled, then squared: $(2 \cdot 3)^2$

If we calculate 3^2, we have the area of one of the small square garden plots on the left. If we then double that, we have $2 \cdot (3^2)$, the area of the left garden plot.

But if we calculate $(2 \cdot 3)^2$, then first we are doubling 3. So we are calculating the area of a square garden plot whose sides are twice as long. We end up with the area of the garden plot on the right.

The point is that these amounts are different.

Checkpoint 1.4.4. Calculate the value of $30 - ((2+3) \cdot 2)$, respecting the order that the grouping symbols are telling you to execute the arithmetic operations.

Explanation. The grouping symbols tell us what to work on first. In this exercise, we have grouping symbols within grouping symbols, so any operation in there (the addition) should be executed first:

$$\begin{aligned} 30 - ((2+3) \cdot 2) &= 30 - (5 \cdot 2) \\ &= 30 - 10 \\ &= 20 \end{aligned}$$

1.4.2 Beyond Grouping Symbols

If all math expressions used grouping symbols for each and every arithmetic operation, we wouldn't need to say anything more here. In fact, some computer systems work that way, *requiring* the use of grouping symbols all the time. But it is much more common to permit math expressions with no grouping symbols at all, like $5 + 3 \cdot 2$. Should the addition $5 + 3$ be executed first, or should the multiplication $3 \cdot 2$? We need what's known formally as the **order of operations** to tell us what to do.

The **order of operations** is nothing more than an agreement that we all have made to prioritize the arithmetic operations in a certain order.

(P)arentheses and other grouping symbols Grouping symbols should always direct you to the highest priority arithmetic first.

(E)xponentiation After grouping symbols, exponentiation has the highest priority. Execute any exponentiation before other arithmetic operations.

(M)ultiplication, (D)ivision, and Negation After all exponentiation has been executed, start executing multiplications, divisions, and negations. These things all have equal priority. If there are more than one of them in your expression, the highest priority is the one that is leftmost (which comes first as you read it).

(A)ddition and (S)ubtraction After all other arithmetic has been executed, these are all that is left. Addition and subtraction have equal priority. If there are more than one of them in your expression, the highest priority is the one that is leftmost (which comes first as you read it).

List 1.4.5: Order of Operations

A common acronym to help you remember this order of operations is PEMDAS. There are a handful of mnemonic devices for remembering this ordering (such as Please Excuse My Dear Aunt Sally, People Eat

More Donuts After School, etc.).

We'll start with a few examples that only invoke a few operations each.

Example 1.4.6 Use the order of operations to simplify the following expressions.

a. $10 + 2 \cdot 3$. With this expression, we have the operations of addition and multiplication. The order of operations says the multiplication has higher priority, so execute that first:

$$10 + 2 \cdot 3 = 10 + \boxed{2 \cdot 3}$$
$$= 10 + 6$$
$$= 16$$

b. $4 + 10 \div 2 - 1$. With this expression, we have addition, division, and subtraction. According to the order of operations, the first thing we need to do is divide. After that, we'll apply the addition and subtraction, working left to right:

$$4 + 10 \div 2 - 1 = 4 + \boxed{10 \div 2} - 1$$
$$= \boxed{4 + 5} - 1$$
$$= 9 - 1$$
$$= 8$$

c. $7 - 10 + 4$. This example *only* has subtraction and addition. While the acronym PEMDAS may mislead you to do addition before subtraction, remember that these operations have the same priority, and so we work left to right when executing them:

$$7 - 10 + 4 = \boxed{7 - 10} + 4$$
$$= -3 + 4$$
$$= 1$$

d. $20 \div 4 \cdot 7$. This expression has only division and multiplication. Again, remember that although PEMDAS shows "MD," the operations of multiplication and division have the same priority, so we'll apply them left to right:

$$20 \div 4 \cdot 5 = \boxed{20 \div 4} \cdot 5$$
$$= 5 \cdot 5$$
$$= 25$$

e. $(6 + 7)^2$. With this expression, we have addition inside a set of parentheses, and an exponent of 2 outside of that. We must compute the operation inside the parentheses first, and after that we'll apply the exponent:

$$(6 + 7)^2 = (\boxed{6 + 7})^2$$
$$= 13^2$$
$$= 169$$

f. $4(2)^3$. This expression has multiplication and an exponent. There are parentheses too, but no operation inside them. Parentheses used in this manner make it clear that the 4 and 2 are separate numbers, not to be confused with 42. In other words, $4(2)^3$ and 42^3 mean very different things. Exponentiation has the higher priority, so we'll apply the exponent first, and then we'll multiply:

$$4(2)^3 = 4\boxed{(2)^3}$$
$$= 4(8)$$
$$= 32$$

Remark 1.4.7. There are many different ways that we write multiplication. We can use the symbols $\cdot$, $\times$, and $*$ to denote multiplication. We can also use parentheses to denote multiplication, as we've seen in Example 1.4.6, Item f. Once we start working with variables, there is even another way. No matter how multiplication is written, it does not change the priority that multiplication has in the order of operations.

Checkpoint 1.4.8 Practice with order of operations. Simplify this expression one step at a time, using the order of operations.

$5 - 3(7 - 4)^2 =$

$=$

$=$

$=$

Explanation.

$$5 - 3(\boxed{7-4})^2 = 5 - 3\boxed{(3)^2}$$
$$= 5 - \boxed{3(9)}$$
$$= 5 - 27$$
$$= -22$$

1.4.3 Absolute Value Bars, Radicals, and Fraction Bars are Grouping Symbols

When we first discussed grouping symbols, we only mentioned parentheses and brackets. Each of the following operations has an *implied* grouping symbol aside from parentheses and brackets.

Absolute Value Bars The absolute value bars, as in $|2 - 5|$, group the expression inside it just like a set of parentheses would.

Radicals The same is true of the radical symbol — everything inside the radical is grouped, as with $\sqrt{12 - 3}$.

Fraction Bars With a horizontal division bar, the numerator is treated as one group and the denominator as another, as with $\frac{2+3}{5-2}$.

We don't *need* parentheses for these three things since the absolute value bars, radical, and horizontal division bar each denote this grouping on their own. As far as priority in the order of operations goes, it's important to remember that these work just like our most familiar grouping symbols, parentheses.

With absolute value bars and radicals, these grouping symbols also *do* something to what's inside (but only *after* the operations inside have been executed). For example, $|-2| = 2$, and $\sqrt{9} = 3$.

Example 1.4.9 Use the order of operations to simplify the following expressions.

a. $4 - 3\,|5 - 7|$. For this expression, we'll treat the absolute value bars just like we treat parentheses. This implies we'll simplify what's inside the bars first, and then compute the absolute value. After that, we'll multiply and then finally subtract:

$$
\begin{aligned}
4 - 3\,|5 - 7| &= 4 - 3\,\boxed{|5 - 7|} \\
&= 4 - 3\,\boxed{|-2|} \\
&= 4 - \boxed{3(2)} \\
&= 4 - 6 \\
&= -2
\end{aligned}
$$

We may not do $4 - 3 = 1$ first, because 3 is connected to the absolute value bars by multiplication (although implicitly), which has a higher order than subtraction.

b. $8 - \sqrt{5^2 - 8 \cdot 2}$. This expression has an expression inside the radical of $5^2 - 8 \cdot 2$. We'll treat this radical like we would a set of parentheses, and simplify that internal expression first. We'll then apply the square root, and then our last step will be to subtract that expression from 8:

$$
\begin{aligned}
8 - \sqrt{5^2 - 8 \cdot 2} &= 8 - \sqrt{\boxed{5^2} - 8 \cdot 2} \\
&= 8 - \sqrt{25 - \boxed{8 \cdot 2}} \\
&= 8 - \sqrt{\boxed{25 - 16}} \\
&= 8 - \boxed{\sqrt{9}} \\
&= 8 - 3 \\
&= 5
\end{aligned}
$$

c. $\dfrac{2^4 + 3 \cdot 6}{5 - 18 \div 2}$. For this expression, the first thing we want to do is to recognize that the main fraction bar serves as a separator that groups the numerator and groups the denominator. Another way this expression could be written is $(2^4 + 3 \cdot 6) \div (15 - 18 \div 2)$. This implies we'll simplify the numerator and denominator separately according to the order of operations (since there are implicit parentheses around each of these). As a final step we'll simplify the resulting fraction (which is division).

$$
\begin{aligned}
\frac{2^4 + 3 \cdot 6}{5 - 18 \div 2} &= \frac{\boxed{2^4} + 3 \cdot 6}{5 - \boxed{18 \div 2}} \\
&= \frac{16 + \boxed{3 \cdot 6}}{\boxed{5 - 9}} \\
&= \frac{\boxed{16 + 18}}{-4} \\
&= \frac{34}{-4} \\
&= -\frac{17}{2}
\end{aligned}
$$

Checkpoint 1.4.10 More Practice with Order of Operations. Use the order of operations to evaluate

$$\frac{6 + 3\,|9 - 10|}{\sqrt{3 + 18 \div 3}}.$$

Explanation. We start by identifying the innermost, highest priority operations:

$$\frac{6 + 3\,|9 - 10|}{\sqrt{3 + 18 \div 3}} = \frac{6 + 3\,\boxed{|9 - 10|}}{\sqrt{3 + \boxed{18 \div 3}}}$$

$$= \frac{6 + 3\,\boxed{|-1|}}{\sqrt{\boxed{3 + 6}}}$$

$$= \frac{6 + \boxed{3(1)}}{\boxed{\sqrt{9}}}$$

$$= \frac{\boxed{6 + 3}}{3}$$

$$= \frac{9}{3} = 3$$

1.4.4 Negation and Distinguishing $(-a)^m$ from $-a^m$

We noted in the order of operations that using the negative sign to negate a number has the same priority as multiplication and division. To understand why this is, observe that $-1 \cdot 23 = -23$, just for one example. So negating 23 gives the same result as multiplying 23 by -1. For this reason, negation has the same priority in the order of operations as multiplication. This can be a source of misunderstandings.

How would you write a math expression that takes the number -4 and squares it?

$$-4^2? \qquad (-4)^2? \qquad \text{it doesn't matter?}$$

It *does* matter. The second option, $(-4)^2$ is squaring the number -4. The parentheses emphasize this.

But the expression -4^2 is different. There are two actions in this expression: a negation and and exponentiation. According to the order of operations, the exponentiation has higher priority than the negation, so the exponent of 2 in -4^2 applies to the 4 *before* the negative sign (multiplication by -1) is taken into account.

$$-4^2 = -\boxed{4^2}$$
$$= -16$$

and this is not the same as $(-4)^2$, which is *positive* 16.

Warning 1.4.11 Negative Numbers Raised to Powers. You may find yourself needing to raise a negative number to a power, and using a calculator to do the work for you. If you do not understand the issue described here, then you may get incorrect results.

- For example, entering -4^2 into a calculator will result in -16, the negative of 4^2.

- But entering (-4)^2 into a calculator will result in 16, the square of -4.

Go ahead and try entering these into your own calculator.

Checkpoint 1.4.12 Negating and Raising to Powers. Compute the following:

a. $-3^4 =$ [____________] and $(-3)^4 =$ [____________]

b. $-4^3 =$ [____________] and $(-4)^3 =$ [____________]

c. $-1.1^2 =$ [____________] and $(-1.1)^2 =$ [____________]

Explanation. In each part, the first expression asks you to exponentiate and then negate the result. The second expression has a negative number raised to a power. So the answers are:

a. $-3^4 = -81$ and $(-3)^4 = 81$

b. $-4^3 = -64$ and $(-4)^3 = -64$

c. $-1.1^2 = -1.21$ and $(-1.1)^2 = 1.21$

Remark 1.4.13. You might observe in the previous example that there is no difference between -4^3 and $(-4)^3$. It's true that the results are the same, -64, but the two expressions still do say different things. With -4^3, you raise to a power first, then negate. With $(-4)^3$, you negate first, then raise to a power.

As was discussed in Subsection 1.1.5, if the base of a power is negative, then whether or not the result is positive or negative depends on if the exponent is even or odd. It depends on whether or not the factors can all be paired up to "cancel" negative signs, or if there will be a lone factor left by itself.

1.4.5 More Examples

Here are some example exercises that involve applying the order of operations to more complicated expressions. Try these exercises and read the steps given in each solution.

Example 1.4.14 Simplify $10 - 4(5 - 7)^3$.

Explanation. For the expression $10 - 4(5 - 7)^3$, we have simplify what's inside parentheses first, then we'll apply the exponent, then multiply, and finally subtract:

$$\begin{aligned}
10 - 4(5 - 7)^3 &= 10 - 4(\boxed{5 - 7})^3 \\
&= 10 - 4\boxed{(-2)^3} \\
&= 10 - \boxed{4(-8)} \\
&= 10 - (-32) \\
&= 10 + 32 \\
&= 42
\end{aligned}$$

Checkpoint 1.4.15. Simplify $24 \div (15 \div 3 + 1) + 2$.

Explanation. With the expression $24 \div (15 \div 3 + 1) + 2$, we'll simplify what's inside the parentheses according to the order of operations, and then take 24 divided by that expression as our last step:

$$
\begin{aligned}
24 \div (15 \div 3 + 1) + 2 &= 24 \div (\boxed{15 \div 3} + 1) + 2 \\
&= 24 \div (\boxed{5 + 1}) + 2 \\
&= \boxed{24 \div 6} + 2 \\
&= 4 + 2 \\
&= 6
\end{aligned}
$$

Example 1.4.16 Simplify $6 - (-8)^2 \div 4 + 1$.

Explanation. To simplify $6 - (-8)^2 \div 4 + 1$, we'll first apply the exponent of 2 to -8, making sure to recall that $(-8)^2 = 64$. After this, we'll apply division. As a final step, we'll be have subtraction and addition, which we'll apply working left-to-right:

$$
\begin{aligned}
6 - (-8)^2 \div 4 + 1 &= 6 - \boxed{(-8)^2} \div 4 + 1 \\
&= 6 - \boxed{(64) \div 4} + 1 \\
&= \boxed{6 - 16} + 1 \\
&= -10 + 1 \\
&= -9
\end{aligned}
$$

Checkpoint 1.4.17. Simplify $(20 - 4^2) \div (4 - 6)^3$.

Explanation. The expression $(20 - 4^2) \div (4 - 6)^3$ has two sets of parentheses, so our first step will be to simplify what's inside each of those first according to the order of operations. Once we've done that, we'll apply the exponent and then finally divide:

$$
\begin{aligned}
(20 - 4^2) \div (4 - 6)^3 &= (20 - \boxed{4^2}) \div (4 - 6)^3 \\
&= (\boxed{20 - 16}) \div (4 - 6)^3 \\
&= 4 \div (\boxed{4 - 6})^3 \\
&= 4 \div \boxed{(-2)^3} \\
&= 4 \div (-8) \\
&= \frac{4}{-8} \\
&= \frac{1}{-2} \\
&= -\frac{1}{2}
\end{aligned}
$$

Checkpoint 1.4.18. Simplify $\dfrac{2\,|9 - 15| + 1}{\sqrt{(-5)^2 + 12^2}}$.

Explanation. To simplify this expression, the first thing we want to recognize is the role of the main fraction bar, which groups the numerator and denominator. This implies we'll simplify the numerator and denominator separately according to the order of operations, and then reduce the fraction that results:

$$\frac{2\,|9-15|+1}{\sqrt{(-5)^2+12^2}} = \frac{2\,\boxed{|9-15|}+1}{\sqrt{\boxed{(-5)^2}+12^2}}$$

$$= \frac{2\,\boxed{|-6|}+1}{\sqrt{25+\boxed{12^2}}}$$

$$= \frac{\boxed{2(6)}+1}{\sqrt{\boxed{25+144}}}$$

$$= \frac{\boxed{12+1}}{\boxed{\sqrt{169}}}$$

$$= \frac{13}{13}$$

$$= 1$$

Exercises

Review and Warmup

1. Multiply the following.

    a. $(-9)\cdot(-1) =$

    b. $(-4)\cdot 6 =$

    c. $8\cdot(-6) =$

    d. $(-4)\cdot 0 =$

2. Multiply the following.

    a. $(-8)\cdot(-3) =$

    b. $(-7)\cdot 4 =$

    c. $8\cdot(-3) =$

    d. $(-3)\cdot 0 =$

3. Multiply the following.

    a. $(-1)\cdot(-4)\cdot(-3) =$

    b. $2\cdot(-7)\cdot(-4) =$

    c. $(-81)\cdot(-56)\cdot 0 =$

4. Multiply the following.

    a. $(-1)\cdot(-5)\cdot(-5) =$

    b. $7\cdot(-7)\cdot(-2) =$

    c. $(-80)\cdot(-74)\cdot 0 =$

5. a. Compute -1^{96}.

    b. Calculate the power $(-3)^4$.

    c. Find $(-5)^3$.

    d. Calculate -6^3.

6. a. Compute -2^3.

    b. Calculate the power -8^2.

    c. Find $(-7)^2$.

    d. Calculate $(-3)^2$.

Order of Operations Skills Evaluate the following.

7. $4 + 6(3) =$

8. $7 + 3(4) =$

9. $5(3 + 2) =$

10. $4(4 + 4) =$

11. $(4 \cdot 2)^2 =$

12. $(4 \cdot 4)^2 =$

13. $5 \cdot 3^3 =$

14. $2 \cdot 2^2 =$

15. $(11 - 2) \cdot 4 =$

16. $(8 - 2) \cdot 2 =$

17. $11 - 3 \cdot 2 =$

18. $25 - 3 \cdot 5 =$

19. $6 + 3 \cdot 7 =$

20. $7 + 2 \cdot 9 =$

21. $4 - 4 \cdot 6 =$

22. $5 - 3 \cdot 8 =$

23. $1 - 2(-10) =$

24. $1 - 4(-8) =$

25. $-[2 - (1 - 5)^2] =$

26. $-[3 - (2 - 9)^2] =$

27. $7 - 5[7 - (5 + 3 \cdot 2)] =$

28. $7 - 3[1 - (2 + 3 \cdot 4)] =$

29. $4 + 4(139 - 5 \cdot 3^3) =$

30. $4 + 3(114 - 4 \cdot 3^3) =$

31. $-5[3 - (4 - 3 \cdot 3)^2] =$

32. $-2[9 - (7 - 4 \cdot 4)^2] =$

33. $9 - 5[2^2 - (4 - 1)] =$

34. $64 - 4[4^2 - (3 - 2)] =$

35. $(14 - 3)^2 + 5(14 - 3^2)$

36. $(12 - 3)^2 + 3(12 - 3^2)$

37. $(4 \cdot 2)^2 - 4 \cdot 2^2 =$

38. $(4 \cdot 5)^2 - 4 \cdot 5^2 =$

39. $9 \cdot 3^2 - 32 \div 4^2 \cdot 4 + 3 =$

40. $10 \cdot 2^2 - 75 \div 5^2 \cdot 6 + 3 =$

41. $4(7 - 2)^2 - 4(7 - 2^2) =$

42. $6(6 - 2)^2 - 6(6 - 2^2) =$

43. $\dfrac{1+3}{3-2} =$

44. $\dfrac{8+20}{7-3} =$

45. $\dfrac{8^2-3^2}{1+10} =$

46. $\dfrac{8^2-2^2}{5+5} =$

47. $\dfrac{27-(-4)^3}{10-17} =$

48. $\dfrac{27-(-2)^3}{5-10} =$

49. $\dfrac{(-2)\cdot(-1)-(-8)\cdot 3}{(-8)^2+(-66)} =$

50. $\dfrac{(-2)\cdot(-3)-(-8)\cdot 3}{(-8)^2+(-66)} =$

51. $-|1-4| =$

52. $-|2-10| =$

53. $2-7\,|3-5|+1 =$

54. $3-5\,|5-9|+1 =$

55. $-5^2-|8\cdot(-7)| =$

56. $-4^2-|6\cdot(-5)| =$

57. $8-2\,\bigl|-9+(2-5)^3\bigr| =$

58. $9-8\,\bigl|-5+(4-5)^3\bigr| =$

59. $\dfrac{\bigl|27+(-4)^3\bigr|}{-1} =$

60. $\dfrac{\bigl|1+(-2)^3\bigr|}{-1} =$

61. $\left|\dfrac{1+(-4)^3}{-3}\right| =$

62. $\left|\dfrac{1+(-2)^3}{-1}\right| =$

63. $\dfrac{-3\,|17-37|}{22-(-4)^2} =$

64. $\dfrac{-3\,|10-22|}{18-(-3)^2} =$

65. $\dfrac{8}{5}+6\cdot\dfrac{2}{5} =$

66. $\dfrac{8}{3}+10\cdot\dfrac{8}{15} =$

67. $\left(\dfrac{9}{8}-\dfrac{5}{24}\right)-5\left(\dfrac{5}{24}-\dfrac{9}{8}\right) =$

68. $\left(\dfrac{9}{4}-\dfrac{9}{8}\right)-5\left(\dfrac{9}{8}-\dfrac{9}{4}\right) =$

69. $\left|\dfrac{3}{2}-\dfrac{3}{10}\right|-5\left|\dfrac{3}{10}-\dfrac{3}{2}\right| =$

70. $\left|\dfrac{3}{8}-\dfrac{7}{32}\right|-5\left|\dfrac{7}{32}-\dfrac{3}{8}\right| =$

71. $\dfrac{2}{5}+2\left(\dfrac{3}{5}\right)^2 =$

72. $\dfrac{1}{5}+6\left(\dfrac{4}{5}\right)^2 =$

73. $\dfrac{3}{4}+\dfrac{5}{4}\div\dfrac{5}{2}-\dfrac{1}{2} =$

74. $\dfrac{4}{3}+\dfrac{2}{3}\div\dfrac{5}{4}-\dfrac{2}{5} =$

75. $5\sqrt{67+33} =$

76. $5\sqrt{9+27} =$

77. $5\sqrt{-27+6\cdot 6} =$

78. $2\sqrt{64+9\cdot 4} =$

79. $1-2\sqrt{34-9} =$

80. $8-3\sqrt{76-72} =$

81. $\sqrt{36}-2\sqrt{5+95} =$

82. $\sqrt{49} - 4\sqrt{9} + 7 = \boxed{}$

83. $\sqrt{40 + 3^2} = \boxed{}$

84. $\sqrt{-17 + 9^2} = \boxed{}$

85. $\sqrt{8^2 + 6^2} = \boxed{}$

86. $\sqrt{9^2 + 12^2} = \boxed{}$

87. $\dfrac{\sqrt{49} + 9}{\sqrt{49} - 9} = \boxed{}$

88. $\dfrac{\sqrt{9} + 5}{\sqrt{9} - 5} = \boxed{}$

89. $\dfrac{\sqrt{4 + 2 \cdot 6} + |-14 - 18|}{-4 - (-2)^3} = \boxed{}$

90. $\dfrac{\sqrt{94 + 2 \cdot 3} + |-17 - 45|}{-19 - (-3)^3} = \boxed{}$

91. $4[17 - 5(5 + 6)] = \boxed{}$

92. $4[15 - 7(3 + 6)] = \boxed{}$

93. $-8^2 - 7[2 - (2 - 3^3)] = \boxed{}$

94. $-10^2 - 5[6 - (7 - 3^3)] = \boxed{}$

Challenge

95. In this challenge, your job is to create expressions, using addition, subtraction, multiplication, and parentheses. You may use the numbers, $1, 2, 3,$ and 4 in your expression, using each number only once. For example, you could make the expression: $1 + 2 \cdot 3 - 4$.

    a. The greatest value that it is possible to create under these conditions is $\boxed{}$.

    b. The least value that it is possible to create under these conditions is $\boxed{}$.

1.5 Set Notation and Types of Numbers

When we talk about *how many* or *how much* of something we have, it often makes sense to use different types of numbers. For example, if we are counting dogs in a shelter, the possibilities are only $0, 1, 2, \ldots$. (It would be difficult to have $\frac{1}{2}$ of a dog.) On the other hand if you were weighing a dog in pounds, it doesn't make sense to only allow yourself to work with whole numbers. The dog might weigh something like 28.35 pounds. These examples highlight how certain kinds of numbers are appropriate for certain situations. We'll classify various types of numbers in this section.

1.5.1 Set Notation

What is the mathematical difference between these three "lists?"

$$28, 31, 30 \qquad \{28, 31, 30\} \qquad (28, 31, 30)$$

To a mathematician, the last one, $(28, 31, 30)$ is an *ordered* triple. What matters is not merely the three numbers, but *also* the order in which they come. The ordered triple $(28, 31, 30)$ is not the same as $(30, 31, 28)$; they have the same numbers in them, but the order has changed. For some context, February has 28 days; *then* March has 31 days; *then* April has 30 days. The order of the three numbers is meaningful in that context.

With curly braces and $\{28, 31, 30\}$, a mathematician sees a collection of numbers and does not particularly care in which order they are written. Such a collection is called a **set**. All that matters is that these numbers are part of a collection. They've been *written* in some particular order because that's necessary to write them down. But you might as well have put the three numbers in a bag and shaken up the bag. For some context, maybe your favorite three NBA players have jersey numbers 30, 31, and 28, and you like them all equally well. It doesn't really matter what order you use to list them.

So we can say:

$$\{28, 31, 30\} = \{30, 31, 28\} \qquad\qquad (28, 31, 30) \neq (30, 31, 28)$$

What about just writing $28, 31, 30$? This list of three numbers is ambiguous. Without the curly braces or parentheses, it's unclear to a reader if the order is important. **Set notation** is the use of curly braces to surround a list/collection of numbers, and we will use set notation frequently in this section.

Checkpoint 1.5.2 Set Notation. Practice using (and not using) set notation.

According to Google, the three most common error codes from visiting a web site are 403, 404, and 500.

 a. Without knowing which error code is most common, express this set mathematically.

 b. Error code 500 is the most common. Error code 403 is the least common of these three. And that leaves 404 in the middle. Express the error codes in a mathematical way that appreciates how frequently they happen, from most often to least often.

Explanation.

 a. Since we only have to describe a collection of three numbers and their order doesn't matter, we can write $\{403, 404, 500\}$.

 b. Now we must describe the same three numbers and we want readers to know that the order we are writing the numbers matters. We can write $(500, 404, 403)$.

1.5.2 Different Number Sets

In the introduction, we mentioned how different sets of numbers are appropriate for different situations. Here are the basic sets of numbers that are used in basic algebra.

Natural Numbers When we count, we begin: $1, 2, 3, \ldots$ and continue on in that pattern. These numbers are known as **natural numbers**.

$$\mathbb{N} = \{1, 2, 3, \ldots\}$$

Whole Numbers If we include zero, then we have the set of **whole numbers**.

$\{0, 1, 2, 3, \ldots\}$ has no standard symbol, but some options are $\mathbb{N}_0$, $\mathbb{N} \cup \{0\}$, and $\mathbb{Z}_{\geq 0}$.

Integers If we include the negatives of whole numbers, then we have the set of **integers**.

$$\mathbb{Z} = \{\ldots, -3, -2, -1, 0, 1, 2, 3, \ldots\}.$$

A $\mathbb{Z}$ is used because one word in German for "numbers" is "Zahlen."

Rational Numbers A **rational number** is any number that *can* be written as a fraction of integers, where the denominator is nonzero. Alternatively, a **rational number** is any number that *can* be written with a decimal that terminates or that repeats.

$$\mathbb{Q} = \left\{0, 1, -1, 2, \tfrac{1}{2}, -\tfrac{1}{2}, -2, 3, \tfrac{1}{3}, -\tfrac{1}{3}, -3, \tfrac{3}{2}, \tfrac{2}{3} \ldots\right\}$$

$$\mathbb{Q} = \left\{0, 1, -1, 2, 0.5, -0.5, -2, 3, 0.\overline{3}, -0.\overline{3}, -3, 1.5, 0.\overline{6} \ldots\right\}$$

A $\mathbb{Q}$ is used because fractions are *q*uotients of integers.

Irrational Numbers Any number that *cannot* be written as a fraction of integers belongs to the set of **irrational numbers**. Another way to say this is that any number whose decimal places goes on forever without repeating is an **irrational number**. Some examples include $\pi \approx 3.1415926\ldots$, $\sqrt{15} \approx 3.87298\ldots$, $e \approx 2.71828\ldots$

There is no standard symbol for the set of irrational numbers.

Real Numbers Any number that can be marked somewhere on a number line is a **real number**. Real numbers might be the only numbers you are familiar with. For a number to *not* be real, you have to start considering things called *complex numbers*, which are not our concern right now.

The set of real numbers can be denoted with $\mathbb{R}$ for short.

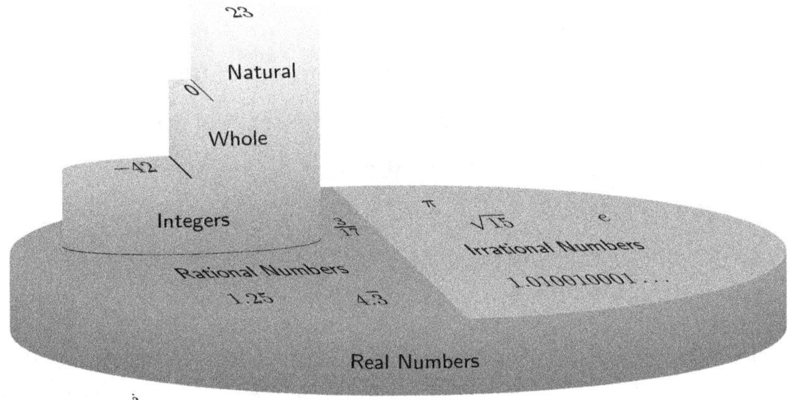

Figure 1.5.3: Types of Numbers

Warning 1.5.4 Rational Numbers in Other Forms. Any number that *can* be written as a ratio of integers is rational, even if it's not written that way at first. For example, these numbers might not look rational to you at first glance: -4, $\sqrt{9}$, 0π, and $\sqrt[3]{\sqrt{5}+2} - \sqrt[3]{\sqrt{5}-2}$. But they are all rational, because they can respectively be written as $\frac{-4}{1}$, $\frac{3}{1}$, $\frac{0}{1}$, and $\frac{1}{1}$.

> **Example 1.5.5 Determine If Numbers Are This Type or That Type.** Determine which numbers from the set $\left\{-102, -7.25, 0, \frac{\pi}{4}, 2, \frac{10}{3}, \sqrt{19}, \sqrt{25}, 10.\overline{7}\right\}$ are natural numbers, whole numbers, integers, rational numbers, irrational numbers, and real numbers.
>
> **Explanation.** All of these numbers are real numbers, because all of these numbers can be positioned on the real number line.
>
> Each real number is either rational or irrational, and not both. -102, -7.25, 0, and 2 are rational because we can see directly that their decimal expressions terminate. $10.\overline{7}$ is also rational, because its decimal expression repeats. $\frac{10}{3}$ is rational because it is a ratio of integers. And last but not least, $\sqrt{25}$ is rational, because that's the same thing as 5.
>
> This leaves only $\frac{\pi}{4}$ and $\sqrt{19}$ as irrational numbers. Their decimal expressions go on forever without entering a repetitive cycle.
>
> Only -102, 0, 2, and $\sqrt{25}$ (which is really 5) are integers.
>
> Of these, only 0, 2, and $\sqrt{25}$ are whole numbers, because whole numbers excludes the negative integers.
>
> Of these, only 2 and $\sqrt{25}$ are natural numbers, because the natural numbers exclude 0.

Checkpoint 1.5.6.

a. Give an example of a whole number that is not an integer.

b. Give an example of an integer that is not a whole number.

c. Give an example of a rational number that is not an integer.

d. Give an example of a irrational number.

e. Give an example of a irrational number that is also an integer.

Explanation.

a. Since all whole numbers belong to integers, we cannot write any whole number which is not an integer. Type DNE (does not exist) for this question.

b. Any negative integer, like -1, is not a whole number, but is an integer.

c. Any terminating decimal, like 1.2, is a rational number, but is not an integer.

d. π is the easiest number to remember as an irrational number. Another constant worth knowing is $e \approx 2.718$. Finally, the square root of most integers are irrational, like $\sqrt{2}$ and $\sqrt{3}$.

e. All irrational numbers are non-repeating and non-terminating decimals. No irrational numbers are integers.

Checkpoint 1.5.7. In the introduction, we mentioned that the different types of numbers are appropriate in different situation. Which number set do you think is most appropriate in each of the following situations?

a. The number of people in a math class that play the ukulele.

   This number is best considered as a   ($\square$ natural number $\square$ whole number $\square$ integer $\square$ rational number $\square$ irrational number $\square$ real number)   .

b. The hypotenuse's length in a given right triangle.

   This number is best considered as a   ($\square$ natural number $\square$ whole number $\square$ integer $\square$ rational number $\square$ irrational number $\square$ real number)   .

c. The proportion of people in a math class that have a cat.

   This number is best considered as a   ($\square$ natural number $\square$ whole number $\square$ integer $\square$ rational number $\square$ irrational number $\square$ real number)   .

d. The number of people in the room with you who have the same birthday as you.

   This number is best considered as a   ($\square$ natural number $\square$ whole number $\square$ integer $\square$ rational number $\square$ irrational number $\square$ real number)   .

e. The total revenue (in dollars) generated for ticket sales at a Timbers soccer game.

   This number is best considered as a   ($\square$ natural number $\square$ whole number $\square$ integer $\square$ rational number $\square$ irrational number $\square$ real number)   .

Explanation.

a. The number of people who play the ukulele could be $0, 1, 2, \ldots$, so the whole numbers are the appropriate set.

b. A hypotenuse's length could be 1, 1.2, $\sqrt{2}$ (which is irrational), or any other positive number. So the real numbers are the appropriate set.

c. This proportion will be a ratio of integers, as both the total number of people in the class and the number of people who have a cat are integers. So the rational numbers are the appropriate set.

d. We know that the number of people must be a counting number, and since *you* are in the room with yourself, there is at least one person in that room with your birthday. So the natural numbers are the appropriate set.

e. The total revenue will be some number of dollars and cents, such as $\$631{,}897.15$, which is a terminating decimal and thus a rational number. So the rational numbers are the appropriate set.

1.5.3 Converting Repeating Decimals to Fractions

We have learned that a terminating decimal number is a rational number. It's easy to convert a terminating decimal number into a fraction of integers: you just need to multiply and divide by one of the numbers in the set $\{10, 100, 1000, \ldots\}$. For example, when we say the number 0.123 out loud, we say "one hundred and twenty-three thousandths." While that's a lot to say, it makes it obvious that this number can be written as a ratio:

$$0.123 = \frac{123}{1000}.$$

Similarly,

$$21.28 = \frac{2128}{100} = \frac{532 \cdot 4}{25 \cdot 4} = \frac{532}{25},$$

demonstrating how *any* terminating decimal can be written as a fraction.

Repeating decimals can also be written as a fraction. To understand how, use a calculator to find the decimal for, say, $\frac{73}{99}$ and $\frac{189}{999}$ You will find that

$$\frac{73}{99} = 0.73737373\ldots = 0.\overline{73} \qquad \frac{189}{999} = 0.189189189\ldots = 0.\overline{189}.$$

The pattern is that diving a number by a number from $\{9, 99, 999, \ldots\}$ with the same number of digits will create a repeating decimal that starts as "0." and then repeats the numerator. We can use this observation to reverse engineer some fractions from repeating decimals.

Checkpoint 1.5.8.

a. Write the rational number $0.772772772\ldots$ as a fraction.

b. Write the rational number $0.69696969\ldots$ as a fraction.

Explanation.

a. The *three*-digit number 772 repeats after the decimal. So we will make use of the *three*-digit denominator 999. And we have $\frac{772}{999}$.

b. The *two*-digit number 69 repeats after the decimal. So we will make use of the *two*-digit denominator 99. And we have $\frac{69}{99}$. But this fraction can be reduced to $\frac{23}{33}$.

Converting a repeating decimal to a fraction is not always quite this straightforward. There are complications if the number takes a few digits before it begins repeating. For your interest, here is one example on how to do that.

Example 1.5.9 Can we convert the repeating decimal $9.134343434\ldots = 9.1\overline{34}$ to a fraction? The trick is to separate its terminating part from its repeating part, like this:

$$9.1 + 0.034343434\ldots.$$

Now note that the terminating part is $\frac{91}{10}$, and the repeating part is almost like our earlier examples, except it has an extra 0 right after the decimal. So we have:

$$\frac{91}{10} + \frac{1}{10} \cdot 0.34343434\ldots.$$

With what we learned in the earlier examples and basic fraction arithmetic, we can continue:

$$\begin{aligned}
9.134343434\ldots &= \frac{91}{10} + \frac{1}{10} \cdot 0.34343434\ldots \\
&= \frac{91}{10} + \frac{1}{10} \cdot \frac{34}{99} \\
&= \frac{91}{10} + \frac{34}{990} \\
&= \frac{91 \cdot 99}{10 \cdot 99} + \frac{34}{990} \\
&= \frac{9009}{990} + \frac{34}{990} = \frac{9043}{990}
\end{aligned}$$

Check that this is right by entering $\frac{9043}{990}$ into a calculator and seeing if it returns the decimal we started with, $9.134343434\ldots$.

Exercises

Review and Warmup

1. Write the decimal number as a fraction.

   $0.55 =$ ⬚

2. Write the decimal number as a fraction.

   $0.65 =$ ⬚

3. Write the decimal number as a fraction.

   $7.65 =$ ⬚

4. Write the decimal number as a fraction.

   $8.25 =$ ⬚

5. Write the decimal number as a fraction.

   $0.988 =$ ⬚

6. Write the decimal number as a fraction.

   $0.152 =$ ⬚

7. Write the fraction as a decimal number. Do not round your answers.

   a. $\frac{3}{5} =$ ⬚

   b. $\frac{7}{16} =$ ⬚

8. Write the fraction as a decimal number. Do not round your answers.

   a. $\frac{1}{5} =$ ⬚

   b. $\frac{3}{20} =$ ⬚

9. Write the mixed number as a decimal number. Do not round your answers.

   a. $5\frac{3}{8} =$ ⬚

   b. $3\frac{1}{4} =$ ⬚

10. Write the mixed number as a decimal number. Do not round your answers.

    a. $4\frac{13}{16} =$ ⬚

    b. $9\frac{3}{4} =$ ⬚

Set Notation

11. There are two numbers that you can square to get 36. Express this collection of two numbers using set notation.

12. There are four positive, even, one-digit numbers. Express this collection of four numbers using set notation.

13. There are six two-digit perfect square numbers. Express this collection of six numbers using set notation.

14. There is a set of three small positive integers where you can square all three numbers, then add the results, and get 61. Express this collection of three numbers using set notation.

Types of Numbers

15. Which of the following are whole numbers? There may be more than one correct answer.

    □ $-2.1\overline{97}$ □ $\sqrt{4}$ □ $\sqrt{3}$ □ 39
    □ -95159 □ -2 □ $5.101001000100001\ldots$
    □ 3.521

16. Which of the following are whole numbers? There may be more than one correct answer.

    □ $\frac{6}{31}$ □ 0   □ -2 □ $\sqrt{3}$ □ 4   □ -13126
    □ $-6.8\overline{71}$ □ $-4.101001000100001\ldots$

17. Which of the following are integers? There may be more than one correct answer.

    □ -6 □ $9.101001000100001\ldots$ □ 69
    □ $\sqrt{16}$ □ 4.097 □ $-\frac{1}{28}$ □ π
    □ -4233

18. Which of the following are integers? There may be more than one correct answer.

    □ 17956   □ -2 □ 6.267 □ $-\frac{1}{8}$ □ -95340
    □ 35   □ π □ $6.101001000100001\ldots$

19. Which of the following are rational numbers? There may be more than one correct answer.

    □ 8.157 □ $5.3\overline{85}$ □ -6 □ $2.101001000100001\ldots$
    □ 0   □ $\sqrt{3}$ □ -86447 □ 99

20. Which of the following are rational numbers? There may be more than one correct answer.

    □ -0.957000000000001 □ 0   □ $-\frac{4}{69}$
    □ $-7.101001000100001\ldots$ □ -77554
    □ $\sqrt{36}$ □ $\sqrt{6}$ □ -2

21. Which of the following are irrational numbers? There may be more than one correct answer.

    □ -68660 □ $\frac{7}{96}$ □ -6 □ π □ 9.077
    □ 43001 □ 0   □ $7.101001000100001\ldots$

22. Which of the following are irrational numbers? There may be more than one correct answer.

    □ -59767 □ $-\frac{5}{17}$ □ 4.245 □ -2
    □ π □ 95   □ $\sqrt{3}$ □ 0

23. Which of the following are real numbers? There may be more than one correct answer.

    □ 60   □ 0   □ π □ -86770 □ $-\frac{6}{95}$
    □ $0.5\overline{70}$ □ -5 □ $5.101001000100001\ldots$

24. Which of the following are real numbers? There may be more than one correct answer.

    □ 8.303 □ $\frac{7}{13}$ □ 0   □ $\sqrt{4}$
    □ -2 □ $\sqrt{11}$ □ π □ -41981

25. Determine the validity of each statement by selecting True or False.

   (a) The number $\frac{\pi}{2}$ is rational

   (b) The number 98 is an integer, but not a whole number

   (c) The number $-\frac{13}{102}$ is rational

   (d) The number $\sqrt{\frac{49}{25}}$ is an integer, but not a whole number

   (e) The number π is irrational

26. Determine the validity of each statement by selecting True or False.

   (a) The number $\frac{3}{2}$ is rational, but not an integer

   (b) The number $0.700700700700700\ldots$ is rational

   (c) The number -11 is an integer that is also a natural number

   (d) The number $\sqrt{2^2}$ is a real number, but not a rational number

   (e) The number $\sqrt{\frac{4}{36}}$ is rational, but not an integer

27.   In each situation, which number set do you think is most appropriate?

 a. The number of dogs a student has owned throughout their lifetime.

   This number is best considered as a   (□ natural number   □ whole number   □ integer   □ rational number   □ irrational number   □ real number)   .

 b. The difference between the projected annual expenditures and the actual annual expenditures for a given company.

   This number is best considered as a   (□ natural number   □ whole number   □ integer   □ rational number   □ irrational number   □ real number)   .

 c. The length around swimming pool in the shape of a half circle with radius 10 ft.

   This number is best considered as a   (□ natural number   □ whole number   □ integer   □ rational number   □ irrational number   □ real number)   .

 d. The proportion of students at a college who own a car.

   This number is best considered as a   (□ natural number   □ whole number   □ integer   □ rational number   □ irrational number   □ real number)   .

 e. The width of a sheet of paper, in inches.

   This number is best considered as a   (□ natural number   □ whole number   □ integer   □ rational number   □ irrational number   □ real number)   .

 f. The number of people eating in a non-empty restaurant.

   This number is best considered as a   (□ natural number   □ whole number   □ integer   □ rational number   □ irrational number   □ real number)   .

28.a. Give an example of a whole number that is not an integer.

 b. Give an example of an integer that is not a whole number.

 c. Give an example of a rational number that is not an integer.

 d. Give an example of a irrational number.

 e. Give an example of a irrational number that is also an integer.

Writing Decimals as Fractions

29. Write the rational number 6.35 as a fraction.

30. Write the rational number 77.162 as a fraction.

31. Write the rational number $0.\overline{78} = 0.7878\ldots$ as a fraction.

32. Write the rational number $0.\overline{955} = 0.955955\ldots$ as a fraction.

33. Write the rational number $4.4\overline{12} = 4.41212\ldots$ as a fraction.

34. Write the rational number $8.1\overline{238} = 8.1238238\ldots$ as a fraction.

Challenge

35. Imagine making up a number according to the following pattern. After the decimal point, write the natural numbers 1, 2, 3, 4, 5, etc. The decimal digits will extend infinitely according to my pattern.

$$0.12345\ldots$$

Is the number a rational number or an irrational number?

($\square$ rational $\square$ irrational)

1.6 Comparison Symbols and Notation for Intervals

As you know, 8 is larger than 3; that's a specific comparison between two numbers. We can also make a comparison between two less specific numbers, like if we say that average rent in Portland in 2016 is larger than it was in 2009. That makes a comparison using unspecified amounts. This section will go over the mathematical shorthand notation for making these kinds of comparisons.

In Oregon, only people who are 18 years old or older can vote in statewide elections.[1] Does that seem like a statement about the number 18? Maybe. But it's also a statement about numbers like 37 and 62: it says that people of these ages may vote as well. This section will also get into the mathematical notation for large collections of numbers like this.

1.6.1 Comparison Symbols

In everyday language you can say something like "8 is larger than 3." In mathematical writing, it's not convenient to write that out in English. Instead the symbol ">" has been adopted, and it's used like this:

$$8 > 3$$

and read out loud as "8 is greater than 3." The symbol ">" is called the **greater-than symbol.**

Checkpoint 1.6.2.

   a. Use mathematical notation to write "11.5 is greater than 4.2."

   b. Use mathematical notation to write "age is greater than 20."

Explanation.

   a. $11.5 > 4.2$

   b. We can just write the word age to represent age, and write age > 20. Or we could use an abbreviation like a for age, and write $a > 20$. Or, it is common to use x as a generic abbreviation, and we could write $x > 20$.

Remark 1.6.3. At some point in history, someone felt that $>$ was a good symbol for "is greater than." In "$8 > 3$," the tall side of the symbol is with the larger of the two numbers, and the small pointed side is with the smaller of the two numbers.

We have to be careful when negative numbers are part of the comparison though. Is -8 larger or smaller than -3? In some sense -8 is larger, because if you owe someone 8 dollars, that's *more* than owing them 3 dollars. But that is not how the $>$ symbol works. This symbol is meant to tell you which number is farther to the right on a number line. And if that's how it goes, then $-3 > -8$.

Alligator Jaws. Another visual was of thinking about the greater-than symbol ">" (and, as we will see later, the less-than symbol "<") is "the alligator wants to eat the larger number" as a way of remembering which direction to write the symbol.

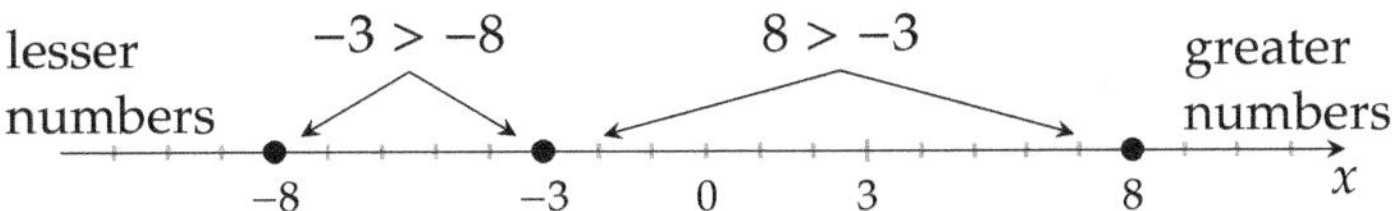

Figure 1.6.4: How the $>$ symbol works.

---

[1] Some other states like Washington allow 17-year-olds to vote in primary elections provided they will be 18 by the general election.

Checkpoint 1.6.5. Use the > symbol to arrange the following numbers in order from greatest to least. For example, your answer might look like 4>3>2>1>0.

$$-7.6 \quad 6 \quad -6 \quad 9.5 \quad 8$$

Explanation. We can order these numbers by placing these numbers on a number line.

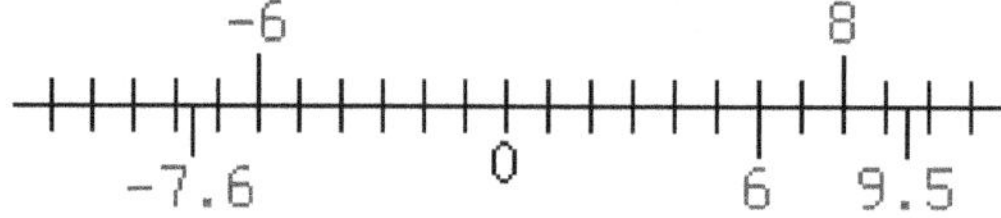

And so we see the answer is $9.5 > 8 > 6 > -6 > -7.6$.

Checkpoint 1.6.6. Use the > symbol to arrange the following numbers in order from greatest to least. For example, your answer might look like 4>3>2>1>0.

$$-5.2 \quad \pi \quad \frac{10}{3} \quad 4.6 \quad 8$$

Explanation. We can order these numbers by placing these numbers on a number line. Knowing or computing their decimals helps with this.

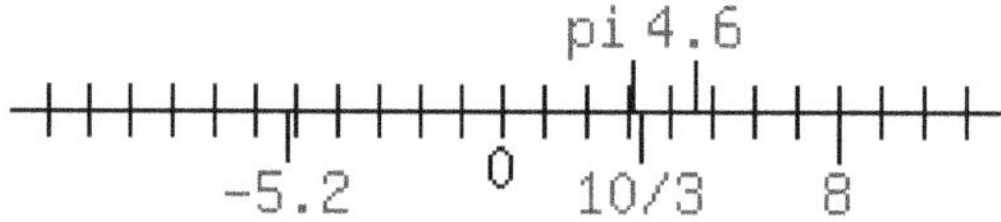

And so we see the answer is $8 > 4.6 > 3.33333 > 3.14159 > -5.2$.

The greater-than symbol has a close relative, the **greater-than-or-equal-to symbol**, "$\geq$." It means just like it sounds: the first number is either greater than, or equal to, the second number. These are all true statements:

$$8 \geq 3 \qquad\qquad 3 \geq -8 \qquad\qquad 3 \geq 3$$

but one of these three statements is false:

$$8 > 3 \qquad\qquad 3 > -8 \qquad\qquad 3 \overset{no}{>} 3$$

Remark 1.6.7. While it may not be that useful that we can write $3 \geq 3$, this symbol is quite useful when specific numbers aren't explicitly used on at least one side, like in these examples:

$$(\text{hourly pay rate}) \geq (\text{minimum wage})$$
$$(\text{age of a voter}) \geq 18$$

Sometimes you want to emphasize that one number is *less than* another number instead of emphasizing which number is greater. To do this, we have symbols that are reversed from > and $\geq$. The symbol "<" is the **less-than symbol** and it's used like this:

$$3 < 8$$

and read out loud as "3 is less than 8."

Table 1.6.8 gives the complete list of all six comparison symbols. Note that we've only discussed three in this section so far, but you already know the equals symbol and have likely also seen the symbol "$\neq$," which means "not equal to."

Symbol	Means		Examples	
$=$	equals	$13 = 13$	$\frac{5}{4} = 1.25$	$5 \overset{no}{=} 6$
$>$	is greater than	$13 > 11$	$\pi > 3$	$9 \overset{no}{>} 9$
$\geq$	is greater than or equal to	$13 \geq 11$	$3 \geq 3$	$11.2 \overset{no}{\geq} 10.2$
$<$	is less than	$-3 < 8$	$\frac{1}{2} < \frac{2}{3}$	$2 \overset{no}{<} -2$
$\leq$	is less than or equal to	$-3 \leq 8$	$3 \leq 3$	$\frac{4}{5} \overset{no}{\leq} \frac{3}{5}$
$\neq$	is not equal to	$10 \neq 20$	$\frac{1}{2} \neq 1.2$	$\frac{3}{8} \overset{no}{\neq} 0.375$

Table 1.6.8: Comparison Symbols

1.6.2 Set-Builder and Interval Notation

If you say
$$(\text{age of a voter}) \geq 18$$
and have a particular voter in mind, what is that person's age? There's no way to know for sure. *Maybe* they are 18, but maybe they are older. It's helpful to use a variable a to represent age (in years) and then to visualize the possibilities with a number line, as in Figure 1.6.9.

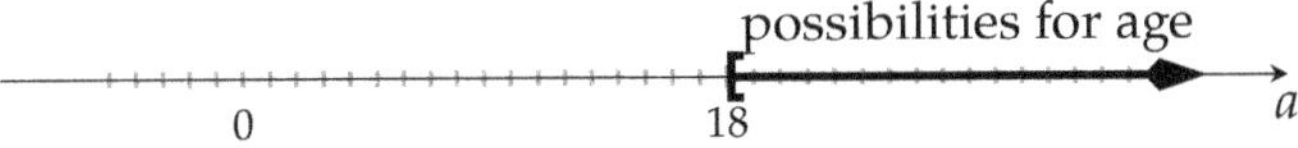

Figure 1.6.9: (age of a voter) ≥ 18

The shaded portion of the number line in Figure 1.6.9 is a mathematical **interval**. For now, that just means a collection of certain numbers. In this case, it's all the numbers 18 and above.

The number line in Figure 1.6.9 is a *graphical* representation of a collection of certain numbers. We have two notations, set-builder notation and interval notation, that we also use to represent such collections of numbers.

Definition 1.6.10 Set-Builder Notation. Set-builder notation attempts to directly say the condition that numbers in the interval satisfy. In general, we write set-builder notation like:
$$\{x \mid \text{condition on } x\}$$
and read it out loud as "the set of all x such that ...." For example,
$$\{x \mid x \geq 18\}$$
is read out loud as "the set of all x such that x is greater than or equal to 18." The breakdown is as follows.

$$\{x \mid x \geq 18\} \quad \text{the set of}$$
$$\{x \mid x \geq 18\} \quad \text{all } x$$
$$\{x \mid x \geq 18\} \quad \text{such that}$$
$$\{x \mid x \geq 18\} \quad x \text{ is greater than or equal to 18}$$

Definition 1.6.11 Interval Notation. Interval notation represents a collection of numbers by only stating where the collection starts and stops, using parentheses and square brackets to show if the end values are included (or not). For example, in Figure 1.6.9, the interval starts at 18. To the right, the interval extends forever and has no end, so we use the ∞ symbol (meaning "infinity"). This particular interval is denoted:

$$[18, \infty)$$

Why use "[" on one side and ")" on the other? The square bracket tells us that 18 *is* part of the interval and the round parenthesis tells us that ∞ is *not* part of the interval.[2]

In general there are four types of infinite intervals. Take note of the different uses of round parentheses and square brackets.

Figure 1.6.12: An **open, infinite** interval denoted by (a, ∞) means all numbers a or larger, *not including a.*

Figure 1.6.13: A **closed, infinite** interval denoted by $[a, \infty)$ means all numbers a or larger, *including a.*

Figure 1.6.14: An **open, infinite** interval denoted by $(-\infty, a)$ means all numbers a or smaller, *not including a.*

Figure 1.6.15: A **closed, infinite** interval denoted by $(-\infty, a]$ means all numbers a or smaller, *including a.*

Checkpoint 1.6.16 Interval and Set-Builder Notation from Number Lines. For each interval expressed in the number lines, give the interval notation and set-builder notation.

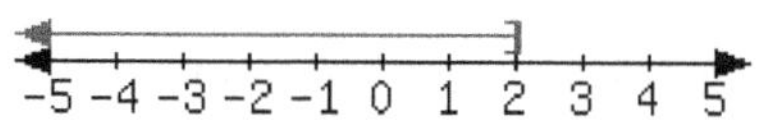 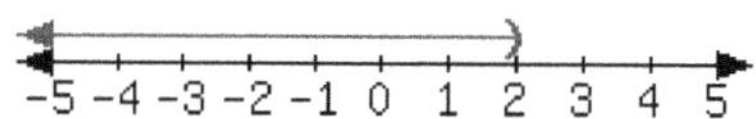 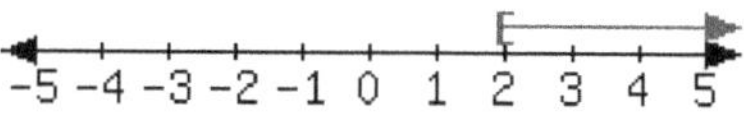

a. In set-builder notation: ☐

   In interval notation: ☐

b. In set-builder notation: ☐

   In interval notation: ☐

c. In set-builder notation: ☐

   In interval notation: ☐

Explanation.

a. Since all numbers less than or equal to 2 are shaded, the set-builder notation is { x | x <= 2 }. The shaded interval "starts" at $-\infty$ and ends at 2 (including 2) so the interval notation is (-infinity,2].

b. Since all numbers less than to 2 are shaded, the set-builder notation is { x | x < 2 }. The shaded interval "starts" at $-\infty$ and ends at 2 (excluding 2) so the interval notation is (-infinity,2)

c. Since all numbers greater than or equal to 2 are shaded, the set-builder notation is { x | x >= 2 }. The shaded interval starts at 2 (including 2) and "ends" at ∞, so the interval notation is [2,infinity)

---

[2]And how could it be, since ∞ is not even a number?

Exercises

Review and Warmup

1. Write the decimal number as a fraction.

    $0.85 = $ ⬚

2. Write the decimal number as a fraction.

    $0.95 = $ ⬚

3. Write the decimal number as a fraction.

    $1.95 = $ ⬚

4. Write the decimal number as a fraction.

    $2.65 = $ ⬚

5. Write the decimal number as a fraction.

    $0.332 = $ ⬚

6. Write the decimal number as a fraction.

    $0.494 = $ ⬚

7. Write the fraction as a decimal number. Do not round your answers.

    a. $\frac{13}{16} = $ ⬚

    b. $\frac{13}{25} = $ ⬚

8. Write the fraction as a decimal number. Do not round your answers.

    a. $\frac{5}{16} = $ ⬚

    b. $\frac{2}{5} = $ ⬚

9. Write the mixed number as a decimal number. Do not round your answers.

    a. $3\frac{17}{20} = $ ⬚

    b. $1\frac{18}{25} = $ ⬚

10. Write the mixed number as a decimal number. Do not round your answers.

    a. $8\frac{3}{25} = $ ⬚

    b. $3\frac{15}{16} = $ ⬚

Ordering Numbers   Use the $>$ symbol to arrange the following numbers in order from greatest to least. For example, your answer might look like 4>3>2>1>0.

11.
$$10 \quad -6 \quad 0 \quad 3 \quad 7$$

12.
$$-9 \quad 8 \quad 9 \quad -3 \quad 6$$

13.
$$-6.35 \quad 0.46 \quad -2.94 \quad -7.79 \quad 6.37$$

14.
$$-4.18 \quad -6.97 \quad 5.43 \quad 6.04 \quad -6.87$$

15.
$$-\frac{5}{4} \quad 7 \quad -7 \quad \frac{19}{7} \quad 6$$

16.
$$-\frac{19}{6} \quad -\frac{9}{5} \quad -\frac{11}{3} \quad -\frac{15}{2} \quad \frac{41}{8}$$

17.

$$3 \quad \frac{3}{7} \quad \frac{\pi}{2} \quad \frac{1}{2} \quad -8 \quad \sqrt{3}$$

18.

$$\frac{2}{3} \quad 6 \quad \sqrt{3} \quad \frac{4}{7} \quad \pi \quad 5$$

True/False

19. Decide if each comparison is true or false.

   a. $2 \neq -3$ ($\square$ True $\square$ False)

   b. $2 = -4$ ($\square$ True $\square$ False)

   c. $4 \neq 4$ ($\square$ True $\square$ False)

   d. $-3 \leq 7$ ($\square$ True $\square$ False)

   e. $-3 < -3$ ($\square$ True $\square$ False)

   f. $6 = 6$ ($\square$ True $\square$ False)

20. Decide if each comparison is true or false.

   a. $-7 < -7$ ($\square$ True $\square$ False)

   b. $-3 < 4$ ($\square$ True $\square$ False)

   c. $8 \geq -4$ ($\square$ True $\square$ False)

   d. $-5 \geq -5$ ($\square$ True $\square$ False)

   e. $-4 \neq -4$ ($\square$ True $\square$ False)

   f. $-6 \neq 4$ ($\square$ True $\square$ False)

21. Decide if each comparison is true or false.

   a. $\frac{4}{2} \neq \frac{12}{6}$ ($\square$ True $\square$ False)

   b. $\frac{5}{5} > \frac{15}{15}$ ($\square$ True $\square$ False)

   c. $-\frac{43}{8} \neq \frac{5}{2}$ ($\square$ True $\square$ False)

   d. $-\frac{46}{5} = \frac{25}{9}$ ($\square$ True $\square$ False)

   e. $\frac{2}{7} \leq \frac{4}{14}$ ($\square$ True $\square$ False)

   f. $-\frac{45}{7} < \frac{11}{2}$ ($\square$ True $\square$ False)

22. Decide if each comparison is true or false.

   a. $-\frac{7}{6} = -\frac{21}{18}$ ($\square$ True $\square$ False)

   b. $\frac{13}{2} \geq -\frac{74}{8}$ ($\square$ True $\square$ False)

   c. $\frac{3}{8} < \frac{3}{8}$ ($\square$ True $\square$ False)

   d. $\frac{14}{3} = \frac{19}{2}$ ($\square$ True $\square$ False)

   e. $-\frac{5}{5} \neq -\frac{5}{5}$ ($\square$ True $\square$ False)

   f. $-\frac{9}{4} \geq -\frac{9}{4}$ ($\square$ True $\square$ False)

Comparisons Choose $<$, $>$, or $=$ to make a true statement.

23. $-\frac{7}{2}$ ($\square <$ $\square >$ $\square =$) $-\frac{5}{4}$

24. $-\frac{6}{7}$ ($\square <$ $\square >$ $\square =$) $-\frac{1}{2}$

25. $\frac{3}{5} + \frac{4}{3}$ ($\square <$ $\square >$ $\square =$) $\frac{1}{2} \div \frac{5}{3}$

26. $\frac{4}{5} + \frac{3}{4}$ ($\square <$ $\square >$ $\square =$) $\frac{2}{5} \div \frac{4}{3}$

27. $\frac{14}{13} \div \frac{14}{13}$ ($\square <$ $\square >$ $\square =$) $\frac{8}{12} - \frac{2}{3}$

28. $\frac{17}{7} \div \frac{17}{7}$ ($\square <$ $\square >$ $\square =$) $\frac{12}{10} - \frac{6}{5}$

29. $-6\frac{1}{3}$ ($\square <$ $\square >$ $\square =$) -6

30. $-1\frac{2}{3}$ ($\square <$ $\square >$ $\square =$) -1

31. $-3\frac{1}{2}$ ($\square <$ $\square >$ $\square =$) 3

32. $-3\frac{2}{3}$ ($\square <$ $\square >$ $\square =$) 1

33. $\left|-\dfrac{3}{5}\right|$ ($\square <$ $\square >$ $\square =$) $|0.6|$

34. $\left|-\dfrac{3}{8}\right|$ ($\square <$ $\square >$ $\square =$) $|0.375|$

35. For each interval expressed in the number lines, give the interval notation and set-builder notation.

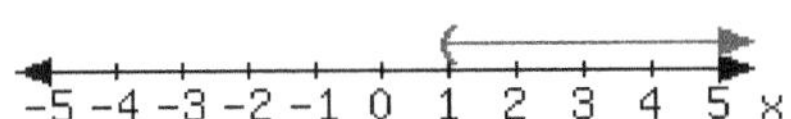 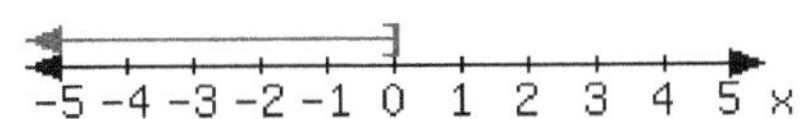 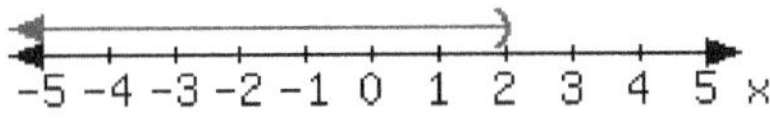

a. In set-builder notation: ☐ b. In set-builder notation: ☐ c. In set-builder notation: ☐

In interval notation: ☐ In interval notation: ☐ In interval notation: ☐

36. For each interval expressed in the number lines, give the interval notation and set-builder notation.

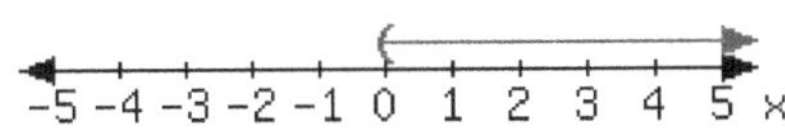 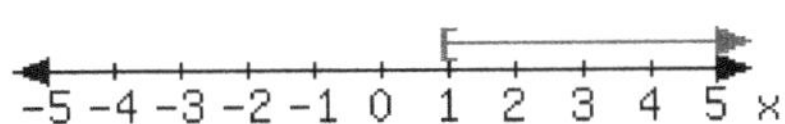 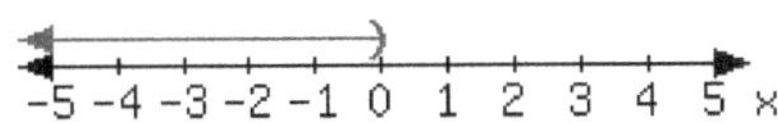

a. In set-builder notation: ☐ b. In set-builder notation: ☐ c. In set-builder notation: ☐

In interval notation: ☐ In interval notation: ☐ In interval notation: ☐

Set-builder and Interval Notation

37. Here is an interval:

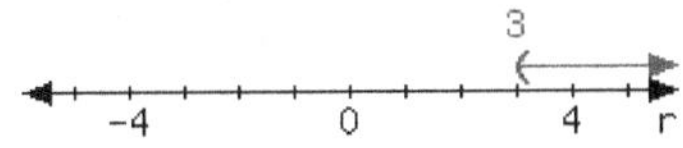

Write the interval using set-builder notation.

Write the interval using interval notation.

38. Here is an interval:

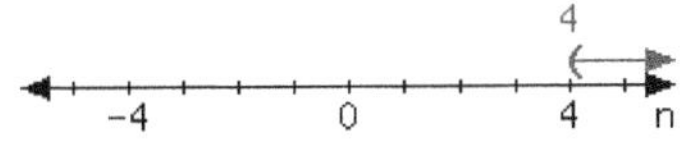

Write the interval using set-builder notation.

Write the interval using interval notation.

39. Here is an interval:

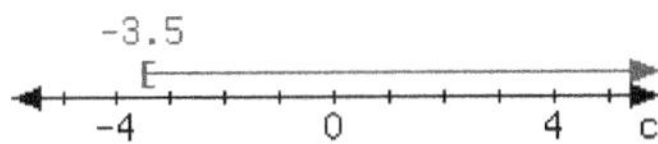

Write the interval using set-builder notation.

Write the interval using interval notation.

40. Here is an interval:

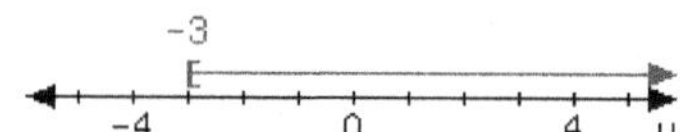

Write the interval using set-builder notation.

Write the interval using interval notation.

41. Here is an interval:

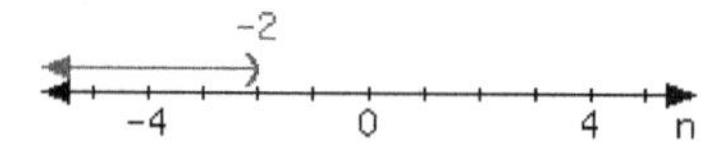

Write the interval using set-builder notation.

Write the interval using interval notation.

42. Here is an interval:

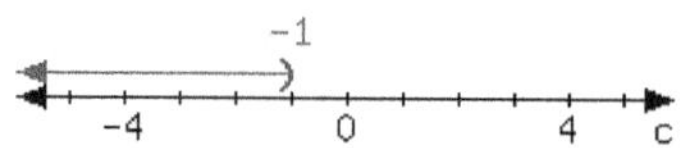

Write the interval using set-builder notation.

Write the interval using interval notation.

43. Here is an interval:

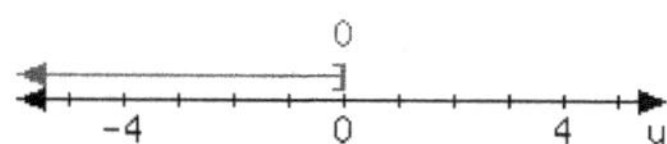

Write the interval using set-builder notation.

Write the interval using interval notation.

44. Here is an interval:

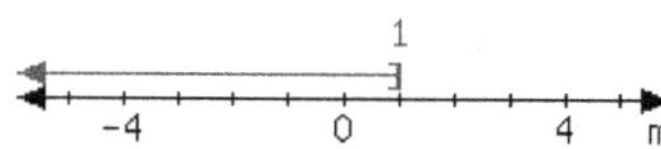

Write the interval using set-builder notation.

Write the interval using interval notation.

Convert to Interval Notation   A set is written using set-builder notation. Write it using interval notation.

45. $\{x \mid x \le 5\}$

46. $\{x \mid x \le 7\}$

47. $\{x \mid x \ge 9\}$

48. $\{x \mid x \ge -9\}$

49. $\{x \mid x < -7\}$

50. $\{x \mid x < -5\}$

51. $\{x \mid x > -2\}$

52. $\{x \mid x > 10\}$

53. $\{x \mid 2 > x\}$

54. $\{x \mid 5 > x\}$

55. $\{x \mid 7 \ge x\}$

56. $\{x \mid 9 \ge x\}$

57. $\{x \mid -9 \le x\}$

58. $\{x \mid -7 \le x\}$

59. $\{x \mid -5 < x\}$

60. $\{x \mid -2 < x\}$

61. $\left\{x \mid \dfrac{5}{9} < x\right\}$

62. $\left\{x \mid \dfrac{7}{6} < x\right\}$

63. $\left\{x \mid x \le -\dfrac{8}{3}\right\}$

64. $\left\{x \mid x \le -\dfrac{9}{7}\right\}$

65. $\{x \mid x \le 0\}$

66. $\{x \mid 0 < x\}$

1.7 Basic Math Chapter Review

1.7.1 Arithmetic with Negative Numbers

Adding Real Numbers with the Same Sign   When adding two numbers with the same sign, we can ignore the signs, and simply add the numbers as if they were both positive.

> **Example 1.7.1**
>
> a. $5 + 2 = 7$
> b. $-5 + (-2) = -7$

Adding Real Numbers with Opposite Signs   When adding two numbers with opposite signs, we find those two numbers' difference. The sum has the same sign as the number with the bigger value. If those two numbers have the same value, the sum is 0.

> **Example 1.7.2**
>
> a. $5 + (-2) = 3$
> b. $(-5) + 2 = -3$

Subtracting a Positive Number   When subtracting a positive number, we can change the problem to adding the *opposite* number, and then apply the methods of adding numbers.

> **Example 1.7.3**
>
> a. $5 - 2 = 5 + (-2)$
> $= 3$
> b. $2 - 5 = 2 + (-5)$
> $= -3$
> c. $-5 - 2 = -5 + (-2)$
> $= 3$

Subtracting a Negative Number   When subtracting a negative number, we can change those two negative signs to a positive sign, and then apply the methods of adding numbers.

> **Example 1.7.4**
>
> a. $5 - (-2) = 5 + 2$
> $= 7$
> b. $-5 - (-2) = -5 + 2$
> $= -3$
> c. $-2 - (-5) = -2 + 5$
> $= 3$

Multiplication and Division of Real Numbers   When multiplying and dividing real numbers, each pair of negative signs cancel out each other (becoming a positive sign). If there is still one negative sign left, the result is negative; otherwise the result is positive.

> **Example 1.7.5**
>
> a. $(6)(-2) = -12$
> b. $(-6)(2) = -12$
> c. $(-6)(-2) = 12$
> d. $(-6)(-2)(-1) = -12$
> e. $(-6)(-2)(-1)(-1) = 12$
> f. $\frac{12}{-2} = -6$
> g. $\frac{-12}{2} = -6$
> h. $\frac{-12}{-2} = 6$

Powers   When we raise a negative number to a certain power, apply the rules of multiplying real numbers: each pair of negative signs cancel out each other.

Example 1.7.6

a. $(-2)^2 = (-2)(-2)$
$\quad\quad\quad = 4$

b. $(-2)^3 = (-2)(-2)(-2)$
$\quad\quad\quad = -8$

c. $(-2)^4 = (-2)(-2)(-2)(-2)$
$\quad\quad\quad = 16$

Difference between $(-a)^n$ **and** $-a^n$ For the exponent expression 2^3, the number 2 is called the **base**, and the number 3 is called the **exponent**. The base of $(-a)^n$ is $-a$, while the base of $-a^n$ is a. This makes a difference in the result when the power is an even number.

Example 1.7.7

a. $(-4)^2 = (-4)(-4)$
$\quad\quad\quad = 16$

b. $-4^2 = -(4)(4)$
$\quad\quad\quad = -16$

c. $(-4)^3 = (-4)(-4)(-4)$
$\quad\quad\quad = -64$

d. $-4^3 = -(4)(4)(4)$
$\quad\quad\quad = -64$

1.7.2 Fraction Arithmetic

Example 1.7.8 Multiplying Fractions.

When multiplying two fractions, we simply multiply the numerators and denominators. To avoid big numbers, we should reduce fractions before multiplying. If one number is an integer, we can write it as a fraction with a denominator of 1. For example, $2 = \frac{2}{1}$.

$$\frac{1}{2} \cdot \frac{3}{4} = \frac{1 \cdot 3}{2 \cdot 4}$$
$$= \frac{3}{8}$$

Example 1.7.9 Dividing Fractions.

When dividing two fractions, we "flip" the second number, and then do multiplication.

$$\frac{1}{2} \div \frac{4}{3} = \frac{1}{2} \cdot \frac{3}{4}$$
$$= \frac{3}{8}$$

Example 1.7.10 Adding/Subtracting Fractions.

Before adding/subtracting fractions, we need to change each fraction's denominator to the same number, called the **common denominator**. Then, we add/subtract the numerators, and the denominator remains the same.

$$\frac{1}{2} - \frac{1}{3} = \frac{1}{2} \cdot \frac{3}{3} - \frac{1}{3} \cdot \frac{2}{2}$$
$$= \frac{3}{6} - \frac{2}{6}$$
$$= \frac{1}{6}$$

1.7.3 Absolute Value and Square Root

Example 1.7.11 Absolute Value.

The absolute value of a number is the distance from that number to 0 on the number line. An absolute value is always positive or 0.

a. $|2| = 2$

b. $\left|-\frac{1}{2}\right| = \frac{1}{2}$

c. $|0| = 0$

Example 1.7.12 Square Root.

The symbol $\sqrt{b}$ has meaning when $b \geq 0$. It means the positive number that can be squared to result in b.

a. $\sqrt{9} = 3$

b. $\sqrt{2} \approx 1.414$

c. $\sqrt{\frac{9}{16}} = \frac{3}{4}$

d. $\sqrt{-1}$ is undefined

1.7.4 Order of Operations

Example 1.7.13 Order of Operations.

When evaluating an expression with multiple operations, we must follow the order of operations:

1. (P)arentheses and other grouping symbols

2. (E)xponentiation

3. (M)ultiplication, (D)ivision, and Negation

4. (A)ddition and (S)ubtraction

$$4 - 2\left(3 - (2-4)^2\right) = 4 - 2\left(3 - \boxed{(2-4)}^2\right)$$
$$= 4 - 2\left(3 - \boxed{(-2)^2}\right)$$
$$= 4 - 2\left(\boxed{3-4}\right)$$
$$= 4 - \boxed{2(-1)}$$
$$= 4 - (-2)$$
$$= 6$$

1.7.5 Set Notation and Types of Numbers

A **set** is an unordered collection of items. Braces, {}, are used to show what items are in a set. For example, the set $\{1, 2, \pi\}$ is a set with three items that contains the numbers 1, 2, and π.

Types of Numbers   Real numbers are categorized into the following sets: natural numbers, whole numbers, integers, rational numbers and irrational numbers.

> **Example 1.7.14** Here are some examples of numbers from each set of numbers:
>
> **Natural Numbers** The natural numbers are all counting numbers larger 1 and larger.
>
> $$1, 251, 3462$$
>
> **Whole Numbers** The whole numbers are all counting numbers larger 0 and larger.
>
> $$0, 1, 42, 953$$
>
> **Integers** The integers are all counting numbers both negative and positive.
>
> $$-263, -10, 0, 1, 834$$
>
> **Rational Numbers** The rational numbers are all possible fractions of integers.
>
> $$\tfrac{1}{3}, -3, 1.1, 0, 0.\overline{73}$$
>
> **Irrational Numbers** The irrational numbers are all numbers that cannot be written as a fraction of integers.
>
> $$\pi, e, \sqrt{2}$$

1.7.6 Comparison Symbols and Notation for Intervals

The following are symbols used to compare numbers.

Symbol	Meaning	Examples	
$=$	equals	$13 = 13$	$\tfrac{5}{4} = 1.25$
$>$	is greater than	$13 > 11$	$\pi > 3$
$\geq$	is greater than or equal to	$13 \geq 11$	$3 \geq 3$
$<$	is less than	$-3 < 8$	$\tfrac{1}{2} < \tfrac{2}{3}$
$\leq$	is less than or equal to	$-3 \leq 8$	$3 \leq 3$
$\neq$	is not equal to	$10 \neq 20$	$\tfrac{1}{2} \neq 1.2$

Table 1.7.15: Comparison Symbols

The following are some examples of set-builder notation and interval notation.

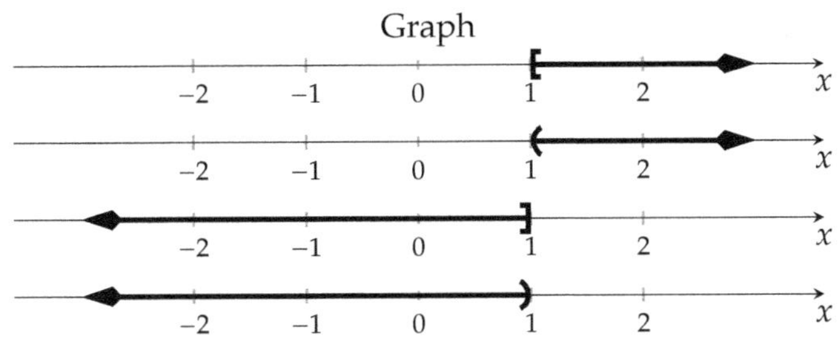

	Graph	Set-builder Notation	Interval Notation
		$\{x \mid x \geq 1\}$	$[1, \infty)$
		$\{x \mid x > 1\}$	$(1, \infty)$
		$\{x \mid x \leq 1\}$	$(-\infty, 1]$
		$\{x \mid x < 1\}$	$(-\infty, 1)$

Exercises

1. Perform the given addition and subtraction.

a. $-19 - 8 + (-2) = \boxed{}$

b. $2 - (-19) + (-14) = \boxed{}$

2. Perform the given addition and subtraction.

a. $-18 - 5 + (-8) = \boxed{}$

b. $9 - (-19) + (-19) = \boxed{}$

3. Multiply the following.

a. $(-2) \cdot (-6) \cdot (-3) = \boxed{}$

b. $5 \cdot (-9) \cdot (-2) = \boxed{}$

c. $(-99) \cdot (-60) \cdot 0 = \boxed{}$

4. Multiply the following.

a. $(-2) \cdot (-4) \cdot (-5) = \boxed{}$

b. $3 \cdot (-9) \cdot (-5) = \boxed{}$

c. $(-98) \cdot (-77) \cdot 0 = \boxed{}$

5. Evaluate the following.

a. $\dfrac{-25}{-5} = \boxed{}$

b. $\dfrac{10}{-5} = \boxed{}$

c. $\dfrac{-35}{5} = \boxed{}$

6. Evaluate the following.

a. $\dfrac{-8}{-4} = \boxed{}$

b. $\dfrac{32}{-4} = \boxed{}$

c. $\dfrac{-15}{5} = \boxed{}$

7. Evaluate the following.

a. $(-1)^2 = \boxed{}$

b. $-4^2 = \boxed{}$

8. Evaluate the following.

a. $(-1)^2 = \boxed{}$

b. $-8^2 = \boxed{}$

9. Evaluate the following.

a. $(-4)^3 = \boxed{}$

b. $-1^3 = \boxed{}$

10. Evaluate the following.

a. $(-4)^3 = \boxed{}$

b. $-3^3 = \boxed{}$

11. Add: $-\dfrac{9}{10} + \dfrac{5}{6}$

12. Add: $-\dfrac{1}{6} + \dfrac{7}{10}$

13. Subtract: $-\dfrac{5}{6} - \left(-\dfrac{9}{10}\right)$

14. Subtract: $-\dfrac{1}{10} - \left(-\dfrac{5}{6}\right)$

15. Subtract: $2 - \dfrac{28}{9}$

16. Subtract: $4 - \dfrac{25}{6}$

17. Multiply: $-\dfrac{12}{13} \cdot \dfrac{7}{22}$

18. Multiply: $-\dfrac{2}{13} \cdot \dfrac{5}{26}$

19. Multiply: $-4 \cdot \dfrac{5}{6}$

20. Multiply: $-5 \cdot \dfrac{9}{20}$

21. Divide: $\dfrac{7}{15} \div \left(-\dfrac{5}{12}\right)$

22. Divide: $\dfrac{1}{9} \div \left(-\dfrac{5}{12}\right)$

23. Divide: $27 \div \dfrac{9}{4}$

24. Divide: $9 \div \dfrac{9}{4}$

25. Evaluate the following.

a. $-|3 - 10| = \rule{2cm}{0.4pt}$

b. $|-3 - 10| = \rule{2cm}{0.4pt}$

c. $-2|10 - 3| = \rule{2cm}{0.4pt}$

26. Evaluate the following.

a. $-|1 - 7| = \rule{2cm}{0.4pt}$

b. $|-1 - 7| = \rule{2cm}{0.4pt}$

c. $-2|7 - 1| = \rule{2cm}{0.4pt}$

27. Evaluate the following.

a. $\sqrt{1} = \rule{2cm}{0.4pt}$

b. $\sqrt{81} = \rule{2cm}{0.4pt}$

c. $\sqrt{100} = \rule{2cm}{0.4pt}$

28. Evaluate the following.

a. $\sqrt{4} = \rule{2cm}{0.4pt}$

b. $\sqrt{25} = \rule{2cm}{0.4pt}$

c. $\sqrt{9} = \rule{2cm}{0.4pt}$

29. Evaluate the following.

a. $\sqrt{\dfrac{16}{49}} = \rule{2cm}{0.4pt}$

b. $\sqrt{-\dfrac{25}{64}} = \rule{2cm}{0.4pt}$

30. Evaluate the following.

a. $\sqrt{\dfrac{25}{81}} = \rule{2cm}{0.4pt}$

b. $\sqrt{-\dfrac{144}{49}} = \rule{2cm}{0.4pt}$

31. Evaluate the following.

$-6^2 - 5[4 - (6 - 4^3)] = \rule{1.5cm}{0.4pt}$

32. Evaluate the following.

$-6^2 - 9[8 - (4 - 4^3)] = \rule{1.5cm}{0.4pt}$

33. Evaluate the following.

$\dfrac{27 - (-4)^3}{3 - 10} = \rule{1.5cm}{0.4pt}$

34. Evaluate the following.

$\dfrac{27 - (-2)^3}{7 - 12} = \rule{1.5cm}{0.4pt}$

35. Evaluate the following.

$10 - 8\left|-9 + (4 - 7)^3\right| = \rule{1.5cm}{0.4pt}$

36. Evaluate the following.

$1 - 6\left|-5 + (3 - 6)^3\right| = \rule{1.5cm}{0.4pt}$

Compare the following integers:

37. a. 2 ($\square <$ $\square >$ $\square =$) -7

b. -2 ($\square <$ $\square >$ $\square =$) -7

c. -7 ($\square <$ $\square >$ $\square =$) 0

38. a. 3 ($\square <$ $\square >$ $\square =$) -6

b. -1 ($\square <$ $\square >$ $\square =$) -6

c. -6 ($\square <$ $\square >$ $\square =$) 0

Determine the validity of each statement by selecting True or False.

39.
(a) The number $\sqrt{(-60)^2}$ is irrational

(b) The number $\sqrt{\frac{9}{16}}$ is an integer, but not a whole number

(c) The number $\sqrt{23}$ is rational

(d) The number 60 is an integer, but not a whole number

(e) The number 0 is a natural number

40.
(a) The number $\sqrt{\frac{25}{81}}$ is rational, but not an integer

(b) The number $\frac{19}{43}$ is rational, but not an integer

(c) The number $\sqrt{11}$ is a real number, but not an irrational number

(d) The number 0.14404004000400004... is rational

(e) The number $\sqrt{4}$ is a real number, but not a rational number

A set is written using set-builder notation. Write it using interval notation.

41. $\{x \mid x > 2\}$

42. $\{x \mid x > 4\}$

43. For each interval expressed in the number lines, give the interval notation and set-builder notation.

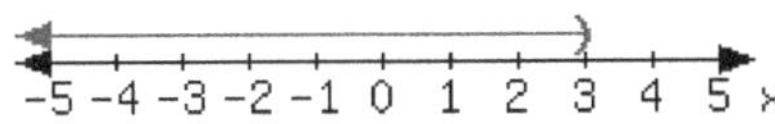 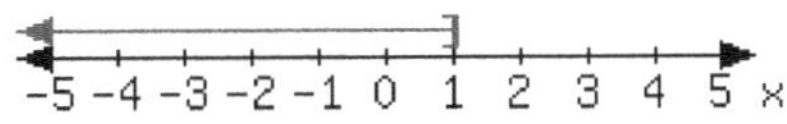 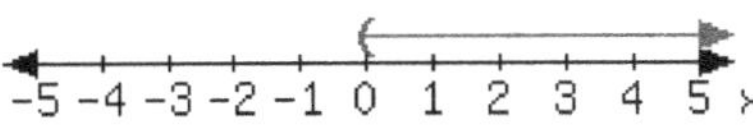

a. In set-builder notation: ☐

   In interval notation: ☐

b. In set-builder notation: ☐

   In interval notation: ☐

c. In set-builder notation: ☐

   In interval notation: ☐

44. For each interval expressed in the number lines, give the interval notation and set-builder notation.

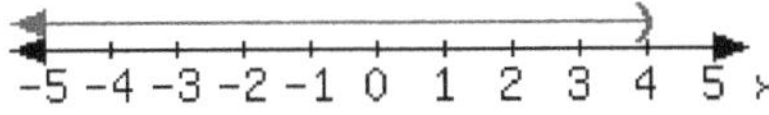 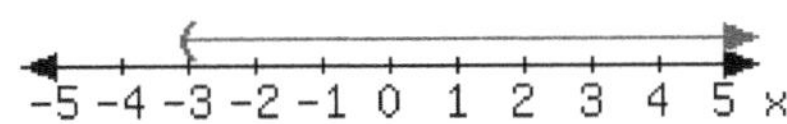 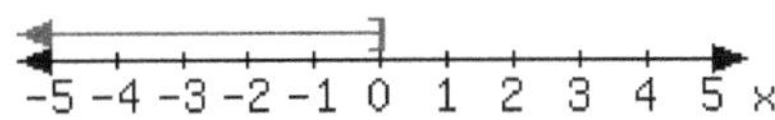

a. In set-builder notation: ☐

   In interval notation: ☐

b. In set-builder notation: ☐

   In interval notation: ☐

c. In set-builder notation: ☐

   In interval notation: ☐

CHAPTER 2

Variables, Expressions, and Equations

2.1 Variables and Evaluating Expressions

To move past *arithmetic* to *algebra*, we begin working with **variables**. Any combination of numbers and variables using mathematical operations is called a mathematical **expression**. Some expressions are simple, and some are complicated. Some expressions are abstract, whereas some have context and meaning. One example of a simple mathematical expression with context is "$220 - a$," which has one variable, a, and is the expression for the maximum heart rate of a person who is a years old.

In this section, we'll focus on variables and expressions. In Section 2.2 we'll continue with a focus on geometry formulas. In the remainder of this chapter, we'll focus on mathematical **equations** and **inequalities** which are also very important in algebra.

2.1.1 Introduction to Variables

When we want to represent an unknown or changing numerical quantity, we use a **variable** to do so. For example, if you'd like to discuss the gas mileage of various cars, you could let the symbol "g" represent a car's gas mileage. The mileage might be 25 mpg, 30 mpg, or some other quantity. If we agree to use mpg for the units of measure, g might be a place holder for 25, 30, or some other number. Since we are using a variable and not a specific number, we can discuss gas mileage for Honda Civics at the same time we discuss gas mileage for Ford Explorers.

When variables stand for actual physical quantities, it's good to use letters that clearly correspond to the quantity they represent. For example, it's wise to use g to represent *g*as mileage. This helps the people who might read your work in the future to understand it better.

It is also important to identify what unit of measurement goes along with each variable you use, and clearly tell your reader this. For example, suppose you are working with $g = 25$. A car whose gas mileage is 25 mpg is very different from a car whose gas mileage is 25 kpg (kilometers per gallon). So it would be important to tell readers that g represents gas mileage *in miles per gallon*.

Checkpoint 2.1.2. Identify a variable you might use to represent each quantity. And identify what units would be most appropriate.

a. Let ☐ be the age of a student, measured in ☐.

b. Let ☐ be the amount of time passed since a driver left Portland, Oregon, bound for Boise, Idaho, measured in ☐.

c. Let ☐ be the area of a three-bedroom apartment, measured in ☐.

Explanation.

a. The unknown quantity is age, which we generally measure in years. So we could define this variable as:

"Let a be the age of a student, measured in years."

b. The amount of time passed is the unknown quantity. Since this is a drive from Portland to Boise, it would make sense to measure this in hours. So we could define this variable as:

"Let t be the amount of time passed since a driver left Portland, Oregon, bound for Boise, Idaho, measured in hours."

c. The unknown quantity is area. Apartment area is usually measured in square feet. So we'll define this variable as:

"Let A be the area of a three-bedroom apartment, measured in ft^2."

Unless an algebra problem specifies which letter(s) to use, we may *choose* which letter(s) to use for our variable(s). However *without* any context to a problem, x, y, and z are the most common letters used as variables, and you may see these variables (especially x) a lot.

Also note that the units we use are often determined indirectly by other information given in an algebra problem. For example, if we're told that a car has used so many gallons of gas after traveling so many miles, then it suggests we should measure gas mileage in mpg.

2.1.2 Mathematical Expressions

A mathematical **expression** is any combination of variables and numbers using arithmetic operations. The following are all examples of mathematical expressions:

$$x + 1 \qquad 2\ell + 2w \qquad \frac{\sqrt{x}}{y + 1} \qquad nRT$$

Note that this definition of "mathematical expression" does *not* include anything with signs like these in them: $=$, $<$, $\leq$, etc.

Example 2.1.3 The expression:

$$\frac{5}{9}(F - 32)$$

can be used to convert from degrees Fahrenheit to degrees Celsius. To do this, we need a Fahrenheit temperature, F. Then we can **evaluate** the expression. This means replacing its variable(s) with specific numbers and calculating the result. In this case, we can replace F with a specific number.

Let's convert the temperature $89\,°$F to the Celsius scale by evaluating the expression.

$$\frac{5}{9}(F - 32) = \frac{5}{9}(89 - 32)$$
$$= \frac{5}{9}(57)$$
$$= \frac{285}{9} \approx 31.67$$

This shows us that $89\,°$F is equivalent to approximately $31.67\,°$C.

Warning 2.1.4 Evaluating Versus Solving. The steps in Example 2.1.3 are not considered "solving" anything. "Solving" is a word you might be tempted to use, because in some sense the steps from Example 2.1.3

are "finding an answer." There is a special meaning in algebra for words like "solve" and "solution" that will come soon. Instead, when we substitute a value and compute the result, the proper vocabulary is "evaluating an expression."

Checkpoint 2.1.5. Try evaluating the temperature expression for yourself.

Use the expression $\frac{5}{9}(F - 32)$ to evaluate some Celsius temperatures.

    a. If a temperature is 50°F what is that temperature measured in Celsius?

    b. If a temperature is −20°F what is that temperature measured in Celsius?

Explanation.

a.
$$\begin{aligned}
\frac{5}{9}(F - 32) &= \frac{5}{9}(50 - 32) \\
&= \frac{5}{9}(18) \\
&= \frac{5}{1}(2) \\
&= 10
\end{aligned}$$

So 50°F is equavalent to 10°C.

b.
$$\begin{aligned}
\frac{5}{9}(F - 32) &= \frac{5}{9}(-20 - 32) \\
&= \frac{5}{9}(-52) \\
&= -\frac{260}{9} \\
&\approx -28.89
\end{aligned}$$

So −20°F is equivalent to about −28.89°C.

Example 2.1.6 Target heart rate. According to the American Heart Association, a person's maximum heart rate, in beats per minute (bpm), is given by $220 - a$, where a is their age in years.

    a. Determine the maximum heart rate for someone who is 31 years old.

    b. A person's *target* heart rate for moderate exercise is 50% to 70% of their maximum heart rate. If they want to reach 65% of their maximum heart rate during moderate exercise, we'd use the expression $0.65(220 - a)$, where a is their age in years. Determine the target heart rate at this 65% level for someone who is 31 years old.

Explanation. Both of these parts ask us to evaluate an expression.

    a. Since a is defined to be age in years, we will evaluate this expression by substituting a with 31:

$$\begin{aligned}
220 - a &= 220 - 31 \\
&= 189
\end{aligned}$$

This tells us that the maximum heart rate for someone who is 31 years old is 189 bpm.

    b. We'll again substitute a with 31, but this time using the target heart rate expression:

$$\begin{aligned}
0.65(220 - a) &= 0.65(220 - 31) \\
&= 0.65(189) \\
&= 122.85
\end{aligned}$$

This tells us that the target heart rate for someone who is 31 years old undertaking moderate exercise is 122.85 bpm.

Checkpoint 2.1.7. The target heart rate for moderate exercise is 50% to 70% of maximum heart rate. We can use the expression $\frac{p}{100}(220 - a)$ to represent a person's target heart rate when their target rate is p% of their maximum heart rate, and they are a years old.

Determine the target heart rate at the 53% level for moderate exercise for someone who is 56 years old.

At the 53% level, the target heart rate for moderate exercise for someone who is 56 years old is ⬚ beats per minute.

Explanation.

$$\frac{p}{100}(220 - a) = \frac{53}{100}(220 - 56)$$
$$= \frac{53}{100}(164)$$
$$= \frac{53}{25}(41)$$
$$= 86.92$$

At the 53% level, the target heart rate for moderate exercise for someone who is 56 years old is 86.92 beats per minute.

Checkpoint 2.1.8 Rising Rents. An expression estimating the average rent of a one-bedroom apartment in Portland, Oregon, from January, 2011 to October, 2016, is given by $10.173x + 974.78$, where x is the number of months since January, 2011.

    a. According to this model, what was the average rent of a one-bedroom apartment in Portland in January, 2011?

    b. According to this model, what was the average rent of a one-bedroom apartment in Portland in January, 2016?

Explanation.

    a. This model uses x as the number of months after January, 2011. So in January, 2011, x is 0:

$$10.173x + 974.48 = 10.173(0) + 974.48$$
$$\approx 974.48$$

According to this model, the average monthly rent for a one-bedroom apartment in Portland, Oregon, in January, 2011, was \$974.78.

    b. The date we are given is January, 2016, which is 5 years after January, 2011. Recall that x is the number of *months* since January, 2011. So we need to use $x = 60$:

$$10.173x + 974.48 = 10.173(60) + 974.48$$
$$\approx 1584.86$$

According to this model, the average monthly rent for a one-bedroom apartment in Portland, Oregon, was \$1584.86 in January 2016.

2.1.3 Evaluating Expressions with Exponents, Absolute Value, and Radicals

Mathematical expressions will often have exponents, absolute value bars, and radicals. This does not change the basic approach to evaluating them.

Example 2.1.9 Tsunami Speed. The speed of a tsunami (in meters per second) can be modeled by $\sqrt{9.8d}$, where d is the depth of the tsunami (in meters). Determine the speed of a tsunami that has a depth of 30 m to four significant digits.

Explanation. Using $d = 30$, we find:

$$\sqrt{9.8d} = \sqrt{9.8(30)}$$
$$= \sqrt{294}$$
$$\approx 17.14\,\overset{\text{four}}{\overbrace{6428}}\ldots$$

The speed of tsunami with a depth of 30 m is about $17.15\ \tfrac{\text{m}}{\text{s}}$.

Up to now, we have been evaluating expressions, but we can evaluate formulas in the same way. A **formula** usually has a single variable that represents the output of an expression. For example, the expression for a person's maximum heart rate in beats per minute, $220 - a$, can be written as the formula, $H = 220 - a$. When we substitute a value for a we are evaluating the formula. Even though we have an equation, we are not solving yet. That will come soon.

Checkpoint 2.1.10 Tent Height. While camping, the height (in feet) inside a tent when you are d ft from the north side of the tent is given by the formula $h = -2\,|d - 3| + 6$.

a. When you are 5 ft from the north side, the height will be [   ].

b. When you are 2.5 ft from the north side, the height will be [   ].

Explanation.

a. When $d = 5$, we have:

$$h = -2\,|d - 3| + 6$$
$$= -2\,|5 - 3| + 6$$
$$= -2\,|2| + 6$$
$$= -2(2) + 6$$
$$= -4 + 6$$
$$= 2$$

Thus when you are 5 ft from the north side, the height in the tent is 2 ft.

b. When $d = 2.5$, we have:

$$h = -2\,|d - 3| + 6$$
$$= -2\,|2.5 - 3| + 6$$
$$= -2\,|-0.5| + 6$$
$$= -2(0.5) + 6$$
$$= -1 + 6$$
$$= 5$$

Thus when you are 2.5 ft from the north side, the height in the tent is 5 ft.

Checkpoint 2.1.11 Mortgage Payments. If we borrow L dollars for a home mortgage loan at an annual interest rate r, and intend to pay off the loan after n months, then the amount we should pay each month M, in dollars, is given by the formula

$$M = \frac{rL\left(1 + \frac{r}{12}\right)^{n}}{12\left(\left(1 + \frac{r}{12}\right)^{n} - 1\right)}$$

If we borrow \$200,000 at an interest rate of 6% with the intent to pay off the loan in 30 years, what should our monthly payment be? (Using a calculator is appropriate here.)

Explanation. We must use $L = 200000$. Because the interest rate is a percentage, $r = 0.06$ (not 6). The variable n is supposed to be a number on months, but we will pay off the loan in 30 years. Therefore we take $n = 360$.

$$M = \frac{rL\left(1 + \frac{r}{12}\right)^n}{12\left(\left(1 + \frac{r}{12}\right)^n - 1\right)} = \frac{(0.06)(200000)\left(1 + \frac{0.06}{12}\right)^{360}}{12\left(\left(1 + \frac{0.06}{12}\right)^{360} - 1\right)}$$

$$= \frac{(0.06)(200000)(1 + 0.005)^{360}}{12\left((1 + 0.005)^{360} - 1\right)}$$

$$\approx \frac{(0.06)(200000)(6.022575\ldots)}{12\left(6.022575\ldots - 1\right)}$$

$$\approx \frac{(0.06)(200000)(6.022575\ldots)}{12(5.022575\ldots)}$$

$$\approx \frac{72270.90\ldots}{60.2709\ldots}$$

$$\approx 1199.10$$

Our monthly payment should be \$1,199.10.

Warning 2.1.12 Evaluating Expressions with Negative Numbers. When we substitute negative numbers into an expression, it's important to use parentheses around them or else it's easy to forget that a *negative* number is being raised to a power. Let's look at some examples.

Example 2.1.13 Evaluate x^2 if $x = -2$.

We substitute:

$$x^2 = (-2)^2$$
$$= 4$$

If we don't use parentheses, we would have:

$$x^2 = -2^2 \qquad\qquad \text{incorrect!}$$
$$= -4$$

The original expression takes x and squares it. With $-2^2 = -4$, the number -2 is not being squared. Since the exponent has higher priority than the negation, it's just the number 2 that is being squared. With $(-2)^2 = 4$ the number -2 *is* being squared, which is what we would want given the expression x^2.

So it is wise to always use some parentheses when substituting in any negative number.

Checkpoint 2.1.14. Evaluate and simplify the following expressions for $x = -5$ and $y = -2$:

a. $x^3 y^2 = \boxed{}$
b. $(-2x)^3 = \boxed{}$
c. $-3x^2 y = \boxed{}$

Explanation.

a. $x^3 y^2 = (-5)^3(-2)^2$
$ = (-125)(4)$
$ = -500$

b. $(-2x)^3 = (-2(-5))^3$
$ = (10)^3$
$ = 1000$

c. $-3x^2 y = -3(-5)^2(-2)$
$ = -3(25)(-2)$
$ = 150$

Exercises

Evaluating Expressions

1. Evaluate $x - 10$ for $x = 6$.

2. Evaluate $x + 4$ for $x = 9$.

3. Evaluate $-4 - x$ for $x = -10$.

4. Evaluate $10 - x$ for $x = -8$.

5. Evaluate $3x + 6$ for $x = -5$.

6. Evaluate $-5x - 6$ for $x = -3$.

7. Evaluate $-7c$ for $c = 9$.

8. Evaluate $-2B$ for $B = 2$.

9. Evaluate the expression r^2:

   a. When $r = 3$, $r^2 = \boxed{}$

   b. When $r = -5$, $r^2 = \boxed{}$

10. Evaluate the expression t^2:

   a. When $t = 9$, $t^2 = \boxed{}$

   b. When $t = -9$, $t^2 = \boxed{}$

11. Evaluate the expression t^3:

   a. When $t = 4$, $t^3 = \boxed{}$

   b. When $t = -3$, $t^3 = \boxed{}$

12. Evaluate the expression x^3:

   a. When $x = 2$, $x^3 = \boxed{}$

   b. When $x = -4$, $x^3 = \boxed{}$

13. Evaluate the following expressions.

   a. Evaluate $5x^2$ when $x = 2$. $5x^2 = \boxed{}$

   b. Evaluate $(5x)^2$ when $x = 2$. $(5x)^2 = \boxed{}$

14. Evaluate the following expressions.

   a. Evaluate $3x^2$ when $x = 2$. $3x^2 = \boxed{}$

   b. Evaluate $(3x)^2$ when $x = 2$. $(3x)^2 = \boxed{}$

15. Evaluate $-(y + 2)$ for $y = -7$.

16. Evaluate $-7(y + 9)$ for $y = 6$.

17. Evaluate $\dfrac{9r - 5}{3r}$ for $r = -1$.

18. Evaluate $\dfrac{3r - 7}{3r}$ for $r = -8$.

19. Evaluate $-9B + 9A$ for $B = -10$ and $A = -6$.

20. Evaluate $-2C - c$ for $C = -9$ and $c = 5$.

21. Evaluate $\dfrac{-5}{x} - \dfrac{4}{a}$ for $x = 7$ and $a = -3$.

22. Evaluate $\dfrac{-5}{y} - \dfrac{8}{B}$ for $y = 9$ and $B = -5$.

23. Evaluate $\dfrac{-7t + 9c - 3}{7t - 2c}$ for $t = -4$ and $c = 9$.

24. Evaluate $\dfrac{-2a - B + 9}{-8a + 7B}$ for $a = 3$ and $B = -7$.

25. Evaluate the expression $\dfrac{1}{7}(x + 2)^2 - 7$ when $x = -9$.

26. Evaluate the expression $\dfrac{1}{4}(x + 3)^2 - 4$ when $x = -7$.

27. Evaluate the expression $\dfrac{1}{2}h(B + b)$ when $h = 10$, $B = 8$, $b = 7$.

28. Evaluate the expression $\dfrac{1}{2}h(B + b)$ when $h = 12$, $B = 6$, $b = 5$.

29. Evaluate the expression $-16t^2 + 64t + 128$ when $t = 3$.

30. Evaluate the expression $-16t^2 + 64t + 128$ when $t = -5$.

31. Evaluate the following expressions.

    a. Evaluate $x^2 r^3$ when $x = -3$ and $r = -1$.

    $x^2 r^3 = $ ____________

    b. Evaluate $x^3 r^2$ when $x = -3$ and $r = -1$.

    $x^3 r^2 = $ ____________

32. Evaluate the following expressions.

    a. Evaluate $x^2 y^3$ when $x = -1$ and $y = -2$.

    $x^2 y^3 = $ ____________

    b. Evaluate $x^3 y^2$ when $x = -1$ and $y = -2$.

    $x^3 y^2 = $ ____________

33. Evaluate the following expressions.

    a. Evaluate $(-3y)^2$ when $y = -1$.

    $(-3y)^2 = $ ____________

    b. Evaluate $(-3y)^3$ when $y = -1$.

    $(-3y)^3 = $ ____________

34. Evaluate the following expressions.

    a. Evaluate $(-y)^2$ when $y = -2$.

    $(-y)^2 = $ ____________

    b. Evaluate $(-y)^3$ when $y = -2$.

    $(-y)^3 = $ ____________

35. Evaluate each algebraic expression for the given value(s):

$\dfrac{y^3 + \sqrt{x - 4}}{|2x - y|}$, for $x = 104$ and $y = -4$: ____________

36. Evaluate each algebraic expression for the given value(s):

$\dfrac{y^3 + \sqrt{x - 4}}{|5x - y|}$, for $x = 29$ and $y = 7$: ____________

37. Evaluate each algebraic expression for the given value(s):

$\dfrac{\sqrt{x}}{y} - \dfrac{y}{x}$, for $x = 81$ and $y = 3$: ____________

38. Evaluate each algebraic expression for the given value(s):

$\dfrac{\sqrt{x}}{y} - \dfrac{y}{x}$, for $x = 100$ and $y = -6$: ____________

39. Evaluate

$$\frac{y_2 - y_1}{x_2 - x_1}$$

for $x_1 = -20$, $x_2 = 13$, $y_1 = 19$, and $y_2 = -1$:

40. Evaluate

$$\frac{y_2 - y_1}{x_2 - x_1}$$

for $x_1 = -15$, $x_2 = -2$, $y_1 = -5$, and $y_2 = -12$:

41. Evaluate

$$\sqrt{(x_2 - x_1)^2 + (y_2 - y_1)^2}$$

for $x_1 = 8$, $x_2 = 4$, $y_1 = 8$, and $y_2 = 11$:

42. Evaluate

$$\sqrt{(x_2 - x_1)^2 + (y_2 - y_1)^2}$$

for $x_1 = -2$, $x_2 = 4$, $y_1 = -8$, and $y_2 = -16$:

43. Evaluate the algebraic expression $3a + b$ for $a = \frac{5}{7}$ and $b = \frac{4}{9}$.

44. Evaluate the algebraic expression $-8a + b$ for $a = \frac{6}{5}$ and $b = \frac{1}{2}$.

45. Evaluate each algebraic expression for the given value(s):

$$\frac{5 + 5|y - x|}{x + 5y}, \text{ for } x = 6 \text{ and } y = -3:$$

46. Evaluate each algebraic expression for the given value(s):

$$\frac{4 + 5|y - x|}{x + 3y}, \text{ for } x = 11 \text{ and } y = -3:$$

To convert a temperature measured in degrees Fahrenheit to degrees Celsius, there is a formula:

$$C = \frac{5}{9}(F - 32)$$

where C represents the temperature in degrees Celsius and F represents the temperature in degrees Fahrenheit.

47. If a temperature is $113°F$, what is that temperature measured in Celsius?

48. If a temperature is $5°F$, what is that temperature measured in Celsius?

49. If a temperature is $14°F$, what is that temperature measured in Celsius?

50. If a temperature is $122°F$, what is that temperature measured in Celsius?

A formula for converting meters into feet is

$$F = 3.28M$$

where M is a number of meters, and F is the corresponding number of feet.

51. Use the formula to find the number of feet that corresponds to fourteen meters.

feet corresponds to fourteen meters.

52. Use the formula to find the number of feet that corresponds to eight meters.

feet corresponds to eight meters.

The formula

$$y = \frac{1}{2} a\, t^2 + v_0\, t + y_0$$

gives the vertical position of an object, at time t, thrown with an initial velocity v_0, from an initial position y_0 in a place where the acceleration of gravity is a. The acceleration of gravity on earth is $-9.8\,\frac{m}{s^2}$. It is negative, because we consider the upward direction as positive in this situation, and gravity pulls down.

53. What is the height of a baseball thrown with an initial velocity of $v_0 = 78\,\frac{m}{s}$, from an initial position of $y_0 = 97$ m, and at time $t = 14$ s?

Fourteen seconds after the baseball was thrown, it was ⬚ high in the air.

54. What is the height of a baseball thrown with an initial velocity of $v_0 = 84\,\frac{m}{s}$, from an initial position of $y_0 = 80$ m, and at time $t = 5$ s?

Five seconds after the baseball was thrown, it was ⬚ high in the air.

The percentage of births in the U.S. delivered via C-section can be given by the following formula for the years since 1996:

$$p = 0.8(y - 1996) + 21$$

In this formula y is a year after 1996 and p is the percentage of births delivered via C-section for that year.

55. What percentage of births in the U.S. were delivered via C-section in the year 2010?

⬚ of births in the U.S. were delivered via C-section in the year 2010.

56. What percentage of births in the U.S. were delivered via C-section in the year 2012?

⬚ of births in the U.S. were delivered via C-section in the year 2012.

Target heart rate for moderate exercise is 50% to 70% of maximum heart rate. If we want to represent a certain percent of an individual's maximum heart rate, we'd use the formula

$$\text{rate} = p(220 - a)$$

where p is the percent, and a is age in years.

57. Determine the target heart rate at 51% level for someone who is 43 years old. Round your answer to an integer.

The target heart rate at 51% level for someone who is 43 years old is ⬚ beats per minute.

58. Determine the target heart rate at 53% level for someone who is 22 years old. Round your answer to an integer.

The target heart rate at 53% level for someone who is 22 years old is ⬚ beats per minute.

The diagonal length (D) of a rectangle with side lengths L and W is given by:

$$D = \sqrt{L^2 + W^2}$$

59. Determine the diagonal length of rectangles with $L = 6$ ft and $W = 8$ ft.

The diagonal length of rectangles with $L = 6$ ft and $W = 8$ ft is ⬚.

60. Determine the diagonal length of rectangles with $L = 9$ ft and $W = 12$ ft.

The diagonal length of rectangles with $L = 9$ ft and $W = 12$ ft is ⬚.

61. The height inside a camping tent when you are d feet from the edge of the tent is given by

$$h = -1.1|d - 5.4| + 6$$

where h stands for height in feet.

Determine the height when you are:

a. 6.5 ft from the edge.

The height inside a camping tent when you 6.5 ft from the edge of the tent is ☐

b. 4.4 ft from the edge.

The height inside a camping tent when you 4.4 ft from the edge of the tent is ☐

62. The height inside a camping tent when you are d feet from the edge of the tent is given by

$$h = -0.6|d - 5.8| + 5.5$$

where h stands for height in feet.

Determine the height when you are:

a. 9.7 ft from the edge.

The height inside a camping tent when you 9.7 ft from the edge of the tent is ☐

b. 2.9 ft from the edge.

The height inside a camping tent when you 2.9 ft from the edge of the tent is ☐

2.2 Geometry Formulas

Two- and three- dimensional shapes provide some formulas with variables that we can evaluate.

2.2.1 Evaluating Perimeter and Area Formulas

Rectangles   The rectangle in Figure 2.2.2 has a length (as measured by the edges on the top and bottom) and a width (as measured by the edges on the left and right).

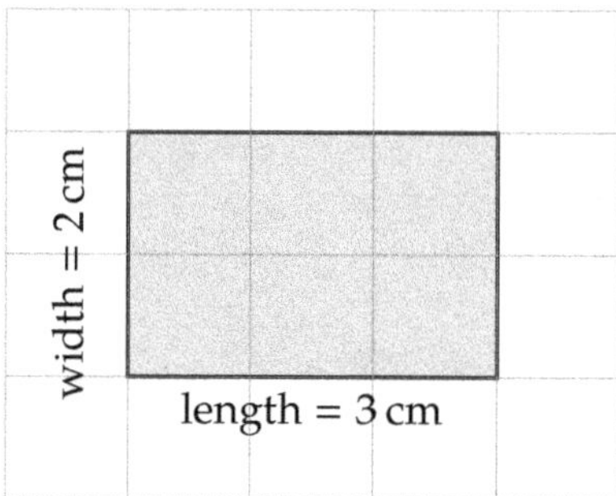

Figure 2.2.2: A Rectangle

Perimeter is the distance around the edge(s) of a two-dimensional shape. To calculate perimeter, start from a point on the shape (usually a corner), travel around the shape, and add up the total distance traveled. For the rectangle in the Figure 2.2.2, if we travel around it, the total distance would be:

$$\text{rectangle perimeter} = 3\,\text{cm} + 2\,\text{cm} + 3\,\text{cm} + 2\,\text{cm}$$
$$= 10\,\text{cm}.$$

Another way to compute a rectangle's perimeter would be to start at one corner, add up the edge length half-way around, and then double that. So we could have calculated the perimeter this way:

$$\text{rectangle perimeter} = 2(3\,\text{cm} + 2\,\text{cm})$$
$$= 2(5\,\text{cm})$$
$$= 10\,\text{cm}.$$

There is nothing special about this rectangle having length $3\,\text{cm}$ and width $2\,\text{cm}$. With a generic rectangle, it has some length we can represent with the variable ℓ and some width we can represent with the variable w. We can use P to represent its perimeter, and then the perimeter of the rectangle will be given by:

$$P = 2(\ell + w).$$

Area is the number of 1×1 squares that fit inside a two-dimensional shape (possibly after morphing them into non-square shapes). If the edges of the squares are, say, $1\,\text{cm}$ long, then the area is measured in "square cm," written cm^2. In Figure 2.2.2, the rectangle has six $1\,\text{cm} \times 1\,\text{cm}$ squares, so its area is 6 square centimeters.

Note that we can find that area by multiplying the length and the width:

$$\text{rectangle area} = (3\,\text{cm}) \cdot (2\,\text{cm})$$
$$= 6\,\text{cm}^2$$

Again, there is nothing special about this rectangle having length 3 cm and width 2 cm. With a generic rectangle, it has some length we can represent with the variable ℓ and some width we can represent with the variable w. We can represent its area with the variable A, and then the area of the rectangle will be given by:

$$A = \ell \cdot w.$$

Checkpoint 2.2.3. Find the perimeter and area of the rectangle.

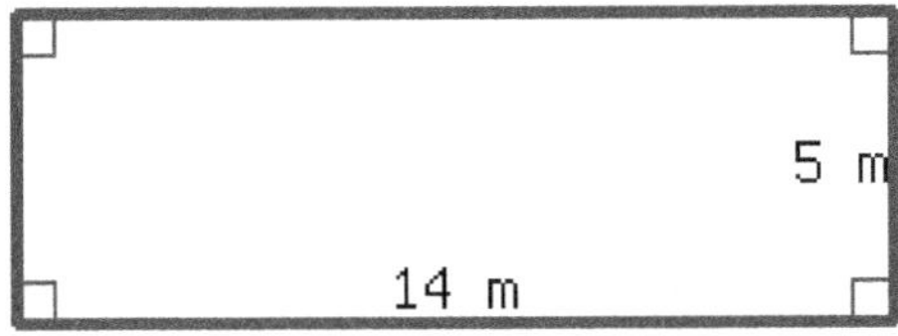

Its perimeter is [____________] and its area is [____________].

Explanation. Using the perimeter and area formulas for a rectangle, we have:

$$\begin{aligned} P &= 2(\ell + w) & A &= \ell \cdot w \\ &= 2(14 + 10) & &= 14 \cdot 10 \\ &= 2(24) & &= 140 \\ &= 48 \end{aligned}$$

Since length and width were measured in meters, we find that the perimeter is 28 meters and the area is 140 square meters.

Triangles The perimeter of a general triangle has no special formula — all that is needed is to add the lengths of its three sides. The *area* of a triangle is a bit more interesting. In Figure 2.2.4, there are three triangles. From left to right, there is an acute triangle, a right triangle, and an obtuse triangle. Each triangle is drawn so that there is a "bottom" horizontal edge. This edge is referred to as the "base" of the triangle. With each triangle, a "height" that is perpendicular to the base is also illustrated.

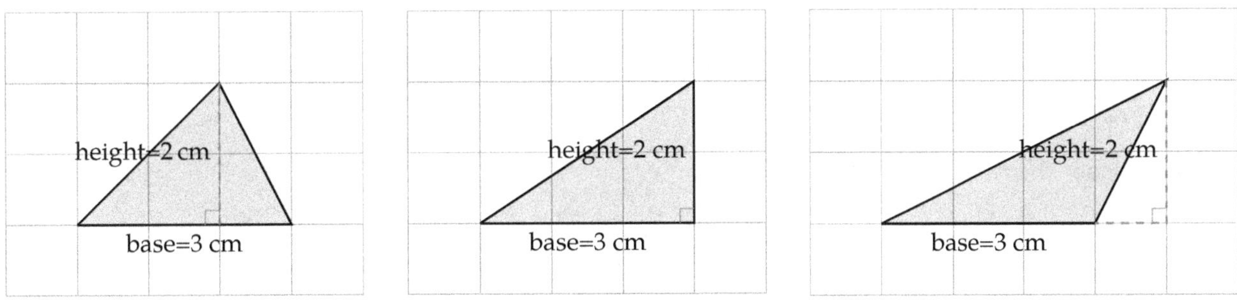

Figure 2.2.4: Triangles

Each of these triangles has the same base width, 3 cm, and the same height, 2 cm. Note that they each have the same area as well. Figure 2.2.5 illustrates how they each have an area of $3 \, \text{cm}^2$.

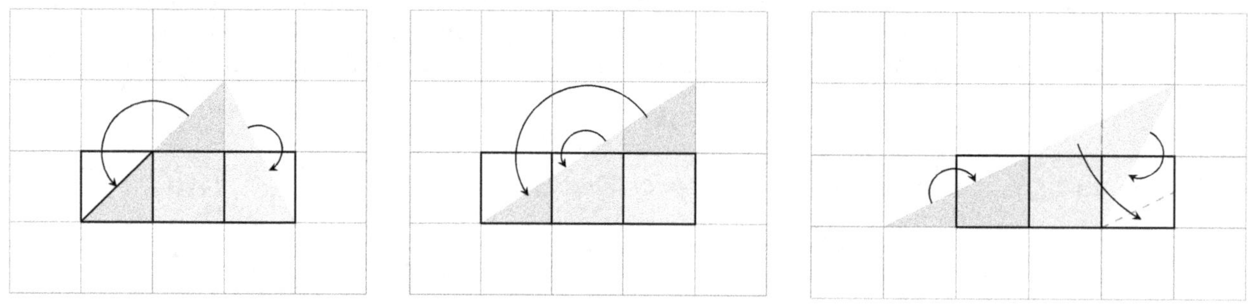

Figure 2.2.5: Triangles

As with the triangles in Figure 2.2.5, you can always rearrange little pieces of a triangle so that the resulting shape is a rectangle with the same base width, but with a height that's one-half of the triangle's height. With a generic rectangle, it has some base width we can represent with the variable b and some height we can represent with the variable h. We can represent its area with the variable A, and then the area of the triangle will be given by $A = b \cdot \left(\frac{1}{2}h\right)$, or more conventionally:

$$A = \frac{1}{2}bh.$$

Checkpoint 2.2.6. Find the perimeter and area of the triangle.

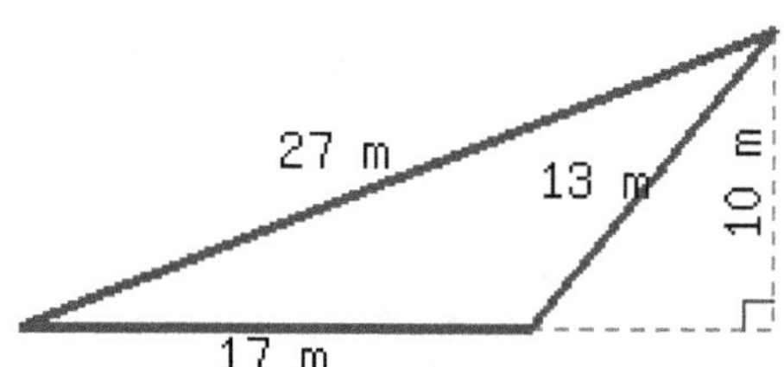

Its perimeter is [   ] and its area is [   ].

Explanation. For perimeter, we just add the three side lengths:

$$P = 13 + 25 + 33$$
$$= 71$$

For area, we use the triangle area formula:

$$A = \frac{1}{2}bh$$
$$= \frac{1}{2}(13)(22)$$
$$= 13(11)$$
$$= 143$$

Since length and width were measured in meters, we find that the perimeter is 71 meters and the area is 143 square meters.

Circles   To find formulas for the perimeter and area of a circle, it helps to first know that there is a special number called π (spelled "pi" and pronounced like "pie") that appears in many places in mathematics. The decimal value of π is about $3.14159265\ldots$, and it helps to memorize some of these digits. It also helps to understand that π is a little larger than 3. There are many definitions for π that can explain where it comes from and how you can find all its decimal places, but here we are just going to accept that it is a special number, and it is roughly $3.14159265\ldots$.

The perimeter of a circle is the distance around its edge. For circles, the perimeter has a special name: the **circumference**. Imagine wrapping a string around the circle and cutting it so that it makes one complete loop. If we straighten out that piece of string, we have a length that is just as long as the circle's circumference.

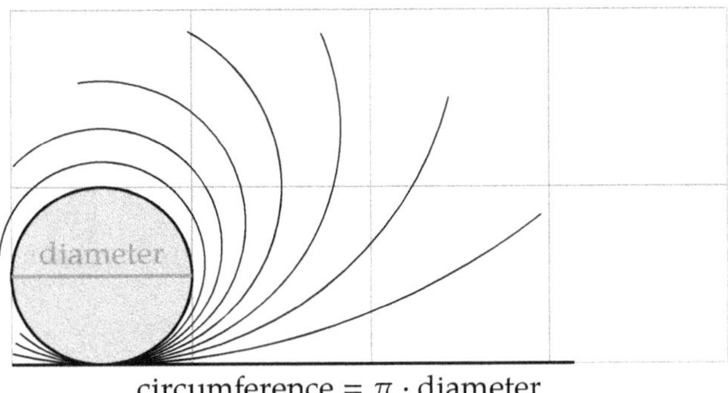

circumference $= \pi \cdot$ diameter

Figure 2.2.7: Circle Diameter and Circumference

As we can see in Figure 2.2.7, the circumference of a circle is a little more than three times as long as its diameter. (The diameter of a circle is the length of a straight line running from a point on the edge through the center to the opposite edge.) In fact, the circumference is actually exactly π times the length of the diameter. With a generic circle, it has some diameter we can represent with the variable d. We can represent its circumference with the variable c, and then the circumference of the circle will be given by:

$$c = \pi d.$$

Alternatively, we often prefer to work with a circle's **radius** instead of its diameter. The radius is the distance from any point on the circle's edge to its center. (Note that the radius is half the diameter.) From this perspective, we can see in Figure 2.2.8 that the circumference is a little more than 6 times the radius.

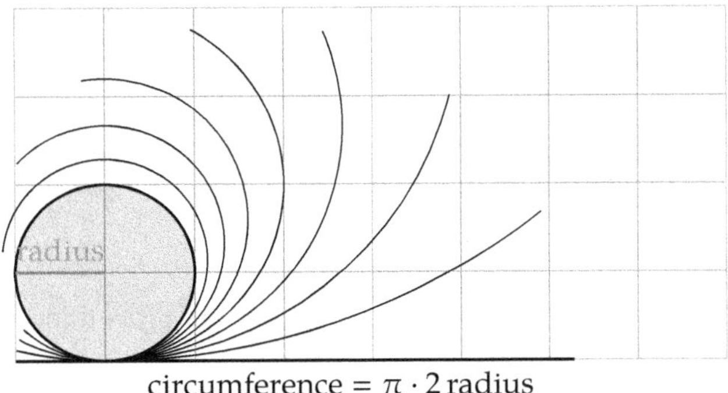

circumference $= \pi \cdot 2$ radius

Figure 2.2.8: Circle Diameter and Circumference

This gives us another formula for a circle's circumerence that uses the variable r for its radius: $c = \pi \cdot 2r$.

Or more conventionally,

$$c = 2\pi r.$$

There is also a formula for the *area* of a circle based on its radius. Figure 2.2.9 shows how three squares can be cut up and rearranged to fit inside a circle. This shows how the area of a circle of radius r is just a little larger than $3r^2$. Since π is just a little larger than 3, could it be that the area of a circle is given by πr^2?

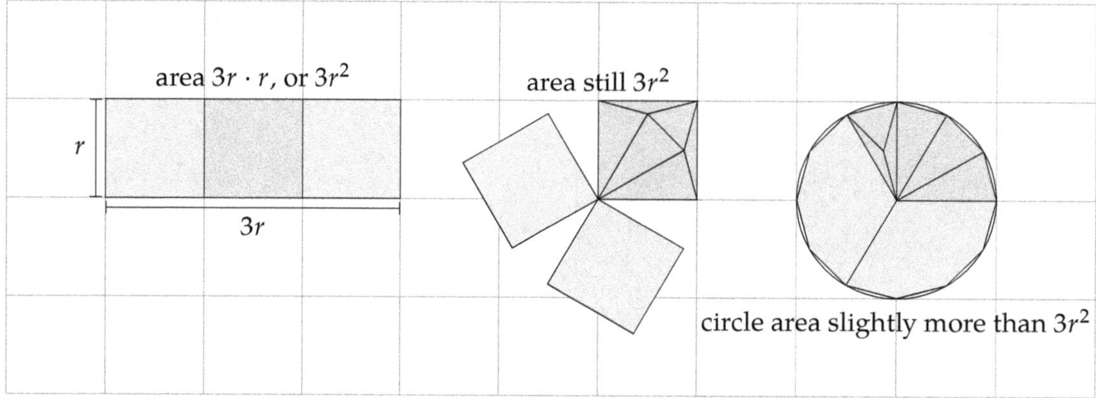

Figure 2.2.9: Circle area is slightly larger than $3r^2$.

One way to establish this formula is to imagine slicing up the circle into many pie slices as in Figure 2.2.10. Then you can rearrange the slices into a strange shape that is *almost* a rectangle with height equal to the radius of the original circle, and width equal to half the circumference of the original circle.

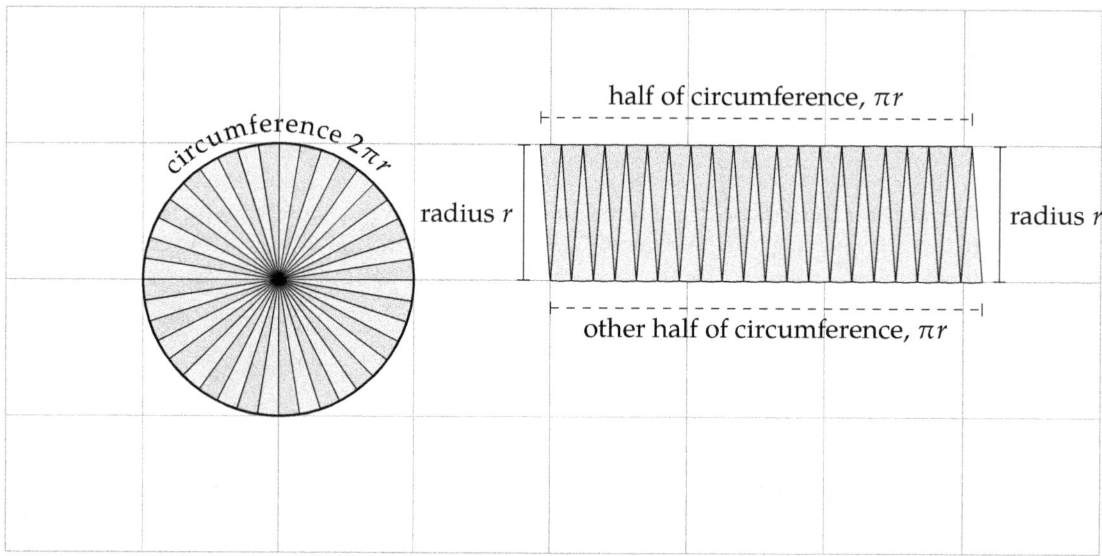

Figure 2.2.10: Reasoning the circle area formula.

Since the area of the circle is equal to the area of the almost-rectangular shape in Figure 2.2.10, we have the circle area formula:

$$A = \pi r^2.$$

Checkpoint 2.2.11. A circle's diameter is 6 m.

    a. This circle's circumference, in terms of π, is [________].

    b. This circle's circumference, rounded to the hundredth place, is [________].

    c. This circle's area, in terms of π, is [________].

    d. This circle's area, rounded to the hundredth place, is [________].

Explanation. We use r to represent radius and d to represent diameter. In this problem, it's given that the diameter is 6 m. A circle's radius is half as long as its diameter, so the radius is 3 m.

Throughout these computations, all quantities have units attached, but we only show them in the final step.

a. $c = \pi d$

$\quad = \pi \cdot 6$

$\quad = 6\pi$ m

b. $c = \pi d$

$\quad \approx 3.1415926 \cdot 6$

$\quad \approx 18.85$ m

c. $A = \pi r^2$

$\quad = \pi \cdot 3^2$

$\quad = \pi \cdot 9$

$\quad = 9\pi$ m^2

d. $A = \pi r^2$

$\quad \approx 3.1415926 \cdot 3^2$

$\quad \approx 3.1415926 \cdot 9$

$\quad \approx 28.27$ m^2

2.2.2 Volume

The **volume** of a three-dimensional object is the number of $1 \times 1 \times 1$ cubes that fit inside the object (possibly after morphing them into non-cube shapes). If the edges of the cubes are, say, 1 cm long, then the volume is measured in "cubic cm," written cm^3.

Rectangular Prisms The 3D shape in Figure 2.2.12 is called a rectangular prism.

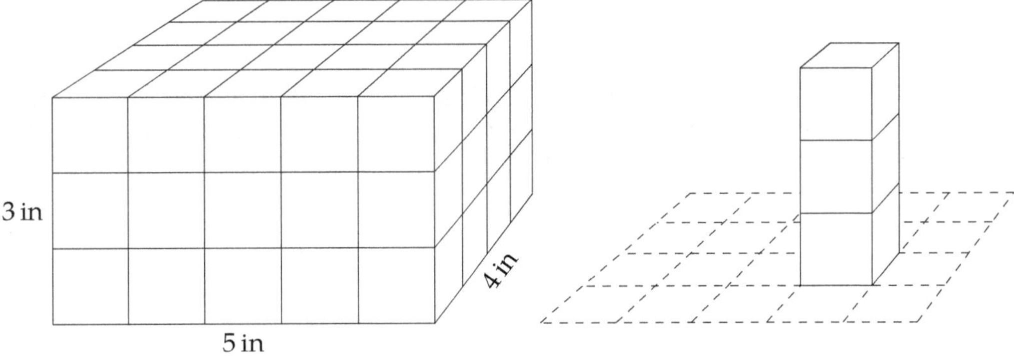

Figure 2.2.12: Volume of a Rectangular Prism

The rectangular prism in Figure 2.2.12 is composed of 1 in $\times$ 1 in $\times$ 1 in unit cubes, with each cube's volume being 1 cubic inch (or in^3). The shape's volume is the number of such unit cubes. The bottom face has

$5 \cdot 4 = 20$ unit squares. Since there are 3 layers of cubes, the shape has a total of $3 \cdot 20 = 60$ unit cubes. In other words, the shape's volume is $60\,\text{in}^3$ because it has sixty $1\,\text{in} \times 1\,\text{in} \times 1\,\text{in}$ cubes inside it.

We found the number of unit squares in the bottom face by multiplying $5 \cdot 4 = 20$. Then to find the volume, we multiplied by 3 because there are three layers of cubes. So one formula for a prism's volume is

$$V = wdh$$

where V stands for volume, w for width, d for depth, and h for height.

Checkpoint 2.2.13. A masonry brick is in the shape of a rectangular prism and is 8 inches wide, 3.5 inches deep, and 2.25 inches high. What is its volume?

Explanation.   Using the formula for the volume of a rectangular prism:

$$\begin{aligned}
V &= wdh \\
&= 8(3.5)(2.25) \\
&= 63
\end{aligned}$$

So the brick's volume is 63 cubic inches.

Cylinders   A cylinder is not a prism, but it has some similarities. Instead of a square base, the base is a circle. Its volume can also be calculated in a similar way to how prism volume is calculated. Let's look at an example.

Example 2.2.14 Find the volume of a cylinder with a radius of 3 meters and a height of 2 meters.

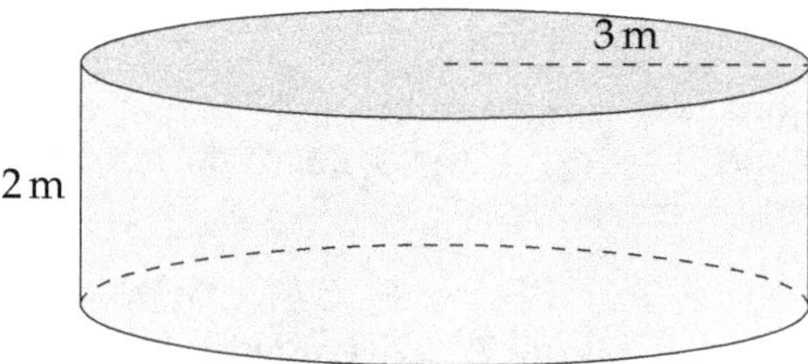

Figure 2.2.15: A Cylinder

Explanation.   The base of the cylinder is a circle. We know the area of a circle is given by the formula $A = \pi r^2$, so the base area is $9\pi\,\text{m}^2$, or about $28.27\,\text{m}^2$. That means about 28.27 unit squares can fit into the base. One of them is drawn in Figure 2.2.16 along with two unit cubes above it.

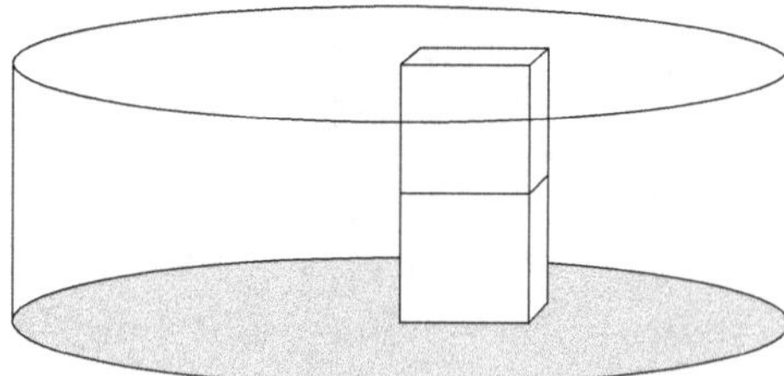

Figure 2.2.16: Finding Cylinder Volume

For each unit square in the base circle, there are two unit cubes of volume. So the volume is the base area times the height: $9\pi\,\text{m}^2 \cdot 2\,\text{m}$, which equals $18\pi\,\text{m}^3$. Approximating π with a decimal value, this is about $56.55\,\text{m}^3$.

Example 2.2.14 demonstrates that the volume of a cylinder can be calculated with the formula

$$V = \pi r^2 h$$

where r is the radius and h is the height.

Checkpoint 2.2.17. A soda can is basically in the shape of a cylinder with radius 1.3 inches and height 4.8 inches. What is its volume?

Its exact volume in terms of π is: ____________ .

As a decimal approximation rounded to four significant digits, its volume is: ____________ .

Explanation. Using the formula for the volume of a cylinder:

$$\begin{aligned}
V &= \pi r^2 h \\
&= \pi (1.3)^2 (4.8) \\
&= 8.112\pi \\
&\approx 25.48
\end{aligned}$$

So the can's volume is 8.112π cubic inches, which is about 25.48 cubic inches.

Note that the volume formulas for a rectangular prism and a cylinder have something in common: both formulas first find the area of the base (which is a rectangle for a prism and a circle for a cylinder) and then multiply by the height. So there is another formula

$$V = Bh$$

that works for both shapes. Here, B stands for the base area (which is wd for a prism and πr^2 for a cylinder.)

2.2.3 Summary

You may be required to memorize the geometry formulas that have been cataloged in this section. Check that you have each of them memorized here.

Checkpoint 2.2.18. Fill out the table with various formulas as they were given in this section.

Rectangle Perimeter	__________________________
Rectangle Area	__________________________
Triangle Area	__________________________
Circle Circumference	__________________________
Circle Area	__________________________
Rectangular Prism Volume	__________________________
Cylinder Volume	__________________________
Volume of either Rectangular Prism or Cylinder	__________________________

Exercises

Perimeter and Area

1. Find the perimeter and area of the rectangle.

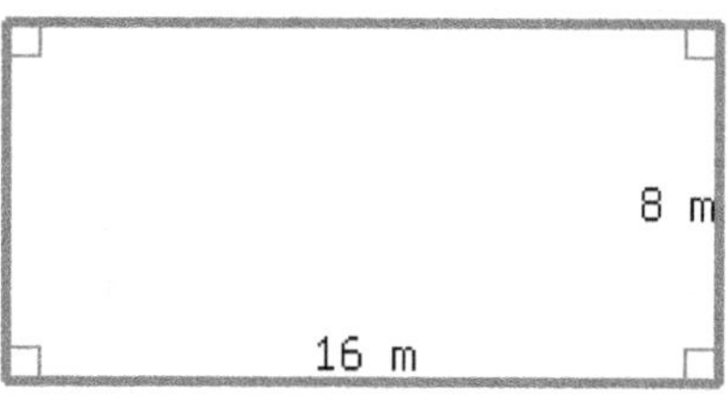

Its perimeter is [______________] and its area is [______________].

2. Find the perimeter and area of the rectangle.

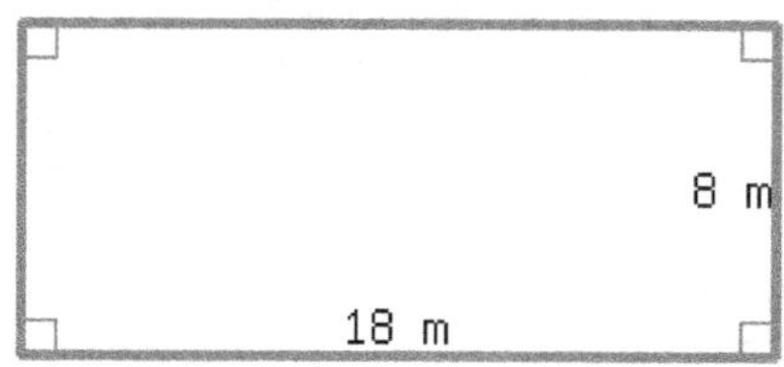

Its perimeter is [______________] and its area is [______________].

3. Find the perimeter of the rectangle below.

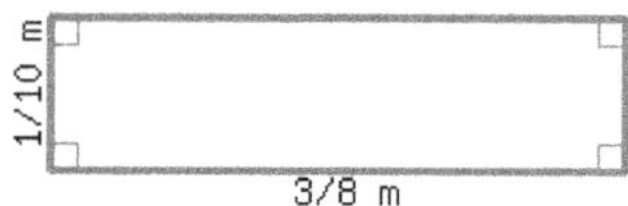

4. Find the perimeter of the rectangle below.

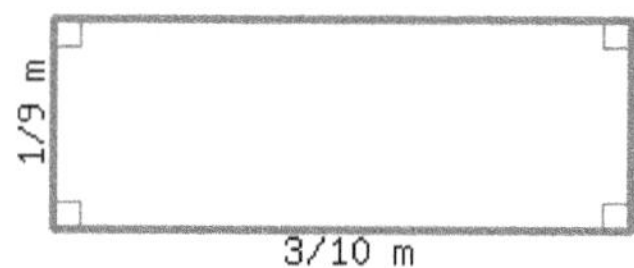

5. Find the area of the rectangle below.

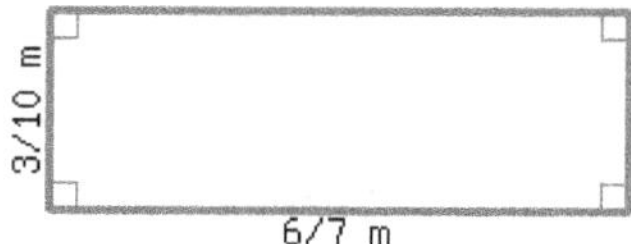

6. Find the area of the rectangle below.

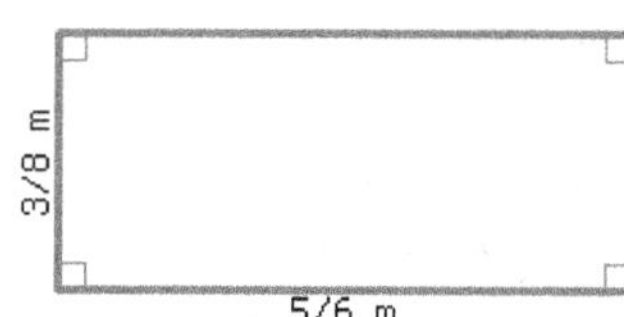

7. Find the perimeter and area of a rectangular table top with a length of 4 ft and a width of 27 in.

Its perimeter is [______________] and its area is [______________].

8. Find the perimeter and area of a rectangular table top with a length of 4.2 ft and a width of 36 in.

Its perimeter is [______________] and its area is [______________].

9. Find the perimeter and area of the square.

a. The square's perimeter is [   ].

b. The square's area is [   ].

10. Find the perimeter and area of the square.

a. The square's perimeter is [   ].

b. The square's area is [   ].

11. Find the perimeter and area of the triangle.

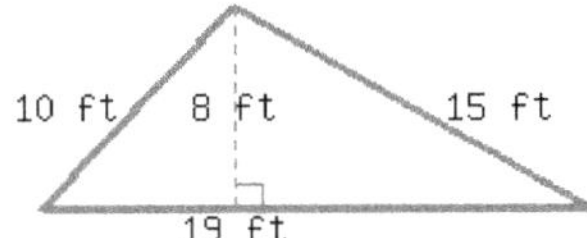

Its perimeter is [   ] and its area is [   ].

12. Find the perimeter and area of the triangle.

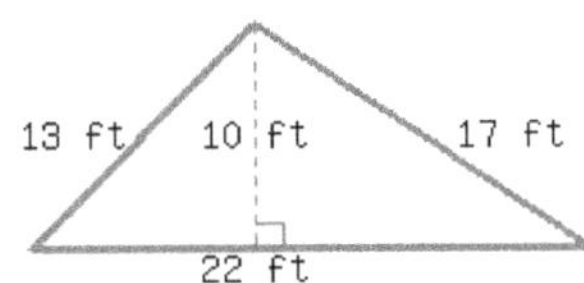

Its perimeter is [   ] and its area is [   ].

13. Find the perimeter and area of the right triangle.

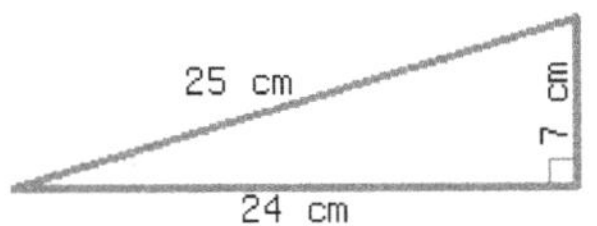

Its perimeter is [   ] and its area is [   ].

14. Find the perimeter and area of the right triangle.

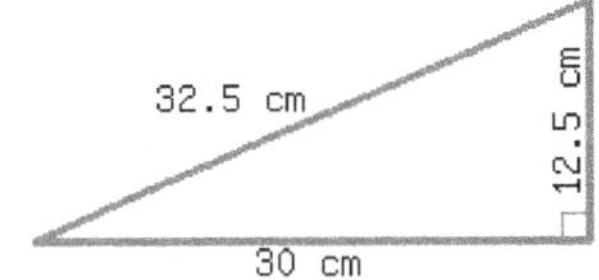

Its perimeter is [   ] and its area is [   ].

15. Find the perimeter and area of the triangle.

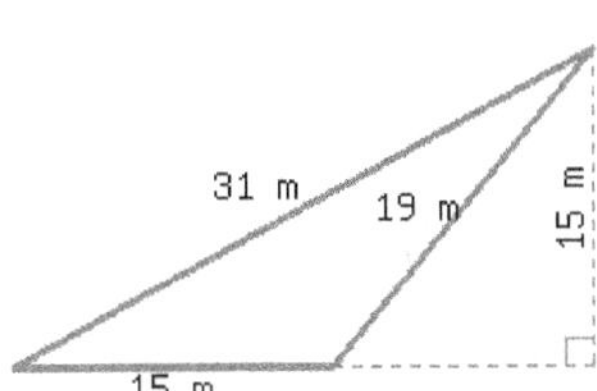

Its perimeter is [   ] and its area is [   ].

16. Find the perimeter and area of the triangle.

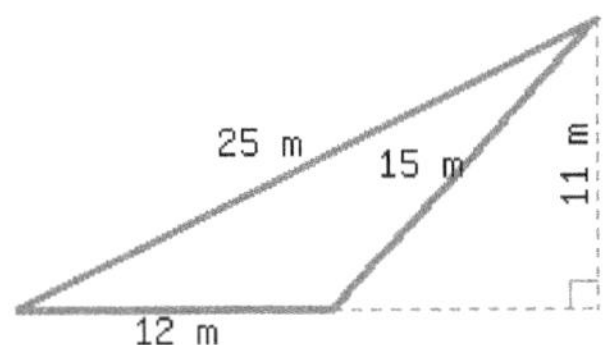

Its perimeter is [   ] and its area is [   ].

17. The area of the triangle below is [   ] square feet.

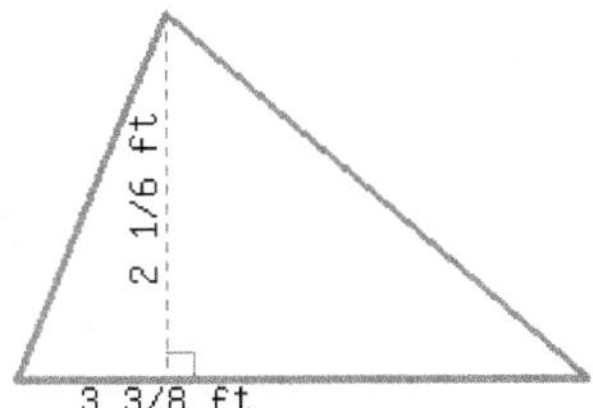

18. The area of the triangle below is [____] square feet.

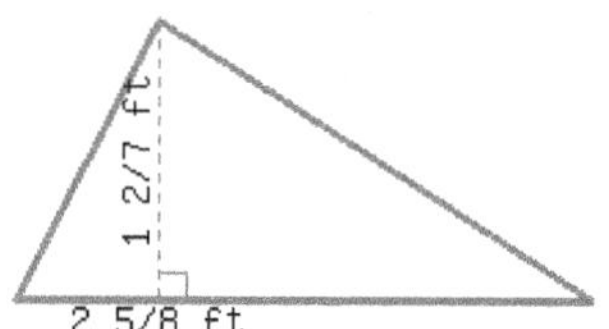

19. Find the area of a triangular flag with a base of 2.6 m and a height of 60 cm.

Its area is [____].

20. Find the area of a triangular flag with a base of 2.9 m and a height of 130 cm.

Its area is [____].

21. A circle's radius is 6 m.

a. This circle's circumference, in terms of π, is [____].

b. This circle's circumference, rounded to the hundredth place, is [____].

c. This circle's area, in terms of π, is [____].

d. This circle's area, rounded to the hundredth place, is [____].

22. A circle's radius is 7 m.

a. This circle's circumference, in terms of π, is [____].

b. This circle's circumference, rounded to the hundredth place, is [____].

c. This circle's area, in terms of π, is [____].

d. This circle's area, rounded to the hundredth place, is [____].

23. A circle's diameter is 16 m.

a. This circle's circumference, in terms of π, is [____].

b. This circle's circumference, rounded to the hundredth place, is [____].

c. This circle's area, in terms of π, is [____].

d. This circle's area, rounded to the hundredth place, is [____].

24. A circle's diameter is 18 m.

a. This circle's circumference, in terms of π, is [____].

b. This circle's circumference, rounded to the hundredth place, is [____].

c. This circle's area, in terms of π, is [____].

d. This circle's area, rounded to the hundredth place, is [____].

25. Find the perimeter and area of this shape, which is a semicircle on top of a rectangle.

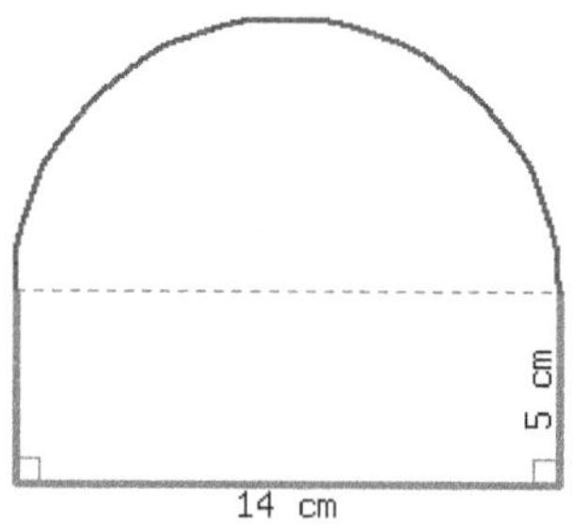

Its perimeter is [____] and its area is [____].

26. Find the perimeter and area of this shape, which is a semicircle on top of a rectangle.

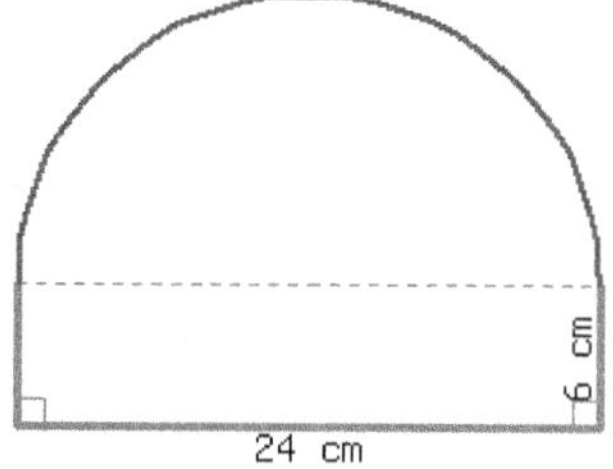

Its perimeter is [____] and its area is [____].

27. Find the perimeter and area of this polygon.

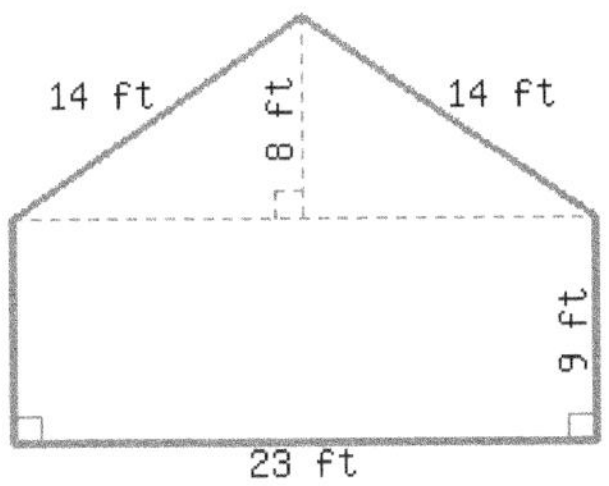

Its perimeter is ☐ and its area is ☐.

28. Find the perimeter and area of this polygon.

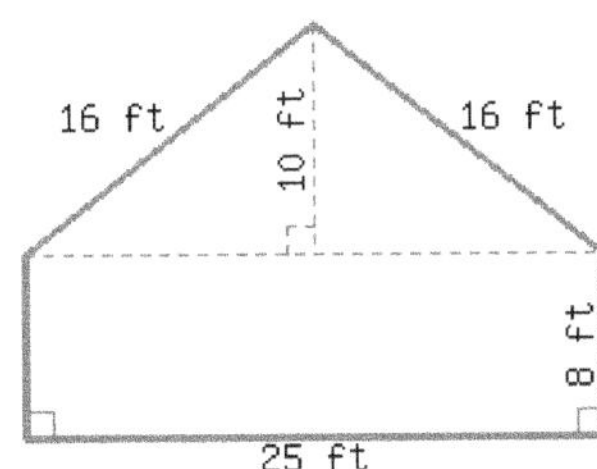

Its perimeter is ☐ and its area is ☐.

29. Find the perimeter and area of this shape.

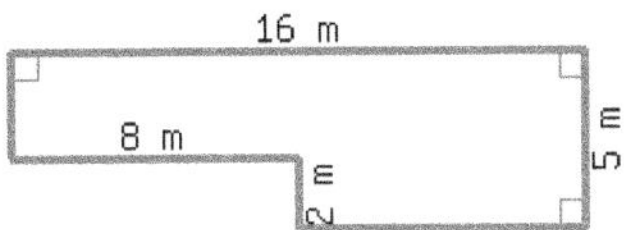

Its perimeter is ☐ and its area is ☐.

30. Find the perimeter and area of this shape.

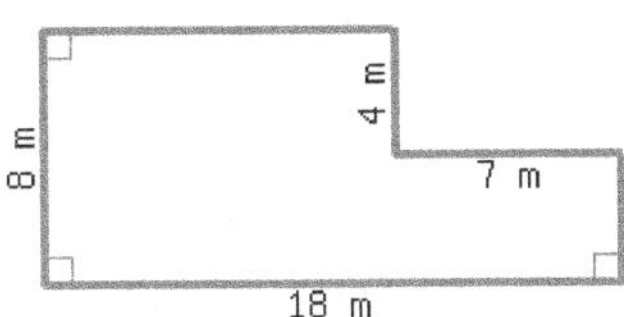

Its perimeter is ☐ and its area is ☐.

31. Find the perimeter and area of this polygon.

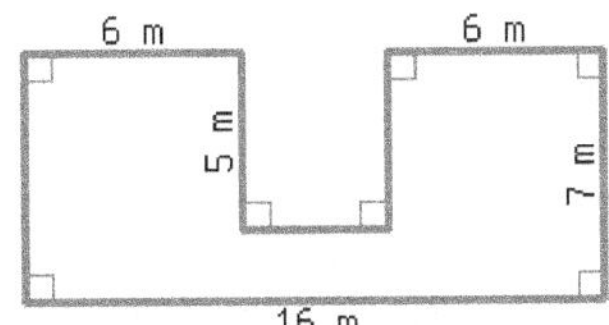

Its perimeter is ☐ and its area is ☐.

32. Find the perimeter and area of this polygon.

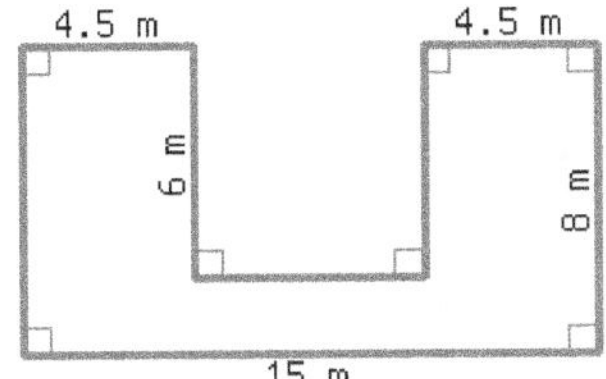

Its perimeter is ☐ and its area is ☐.

The formula

$$A = \frac{1}{2}\,r\,n\,s$$

gives the area of a regular polygon with side length s, number of sides n and, apothem r. (The *apothem* is the distance from the center of the polygon to one of its sides.)

33. What is the area of a regular pentagon with $s = 90$ in and $r = 82$ in?

The area is ☐.

34. What is the area of a regular 28-gon with $s = 99$ in and $r = 90$ in?

The area is ☐.

A trapezoid's area can be calculated by the formula $A = \frac{1}{2}(b_1 + b_2)h$, where A stands for area, b_1 for the first base's length, b_2 for the second base's length, and h for height.

35. Find the area of the trapeoid below.

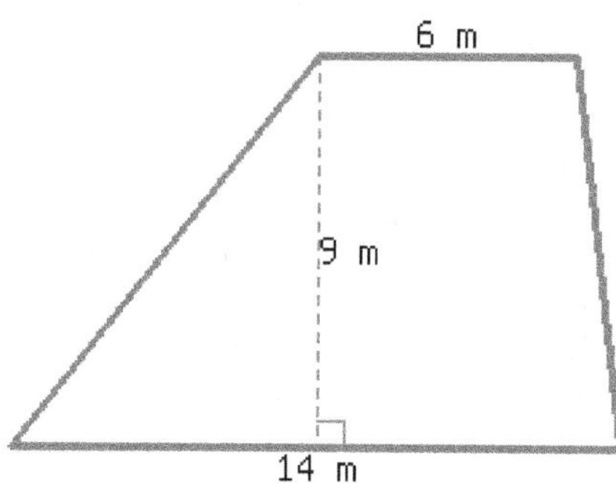

36. Find the area of the trapeoid below.

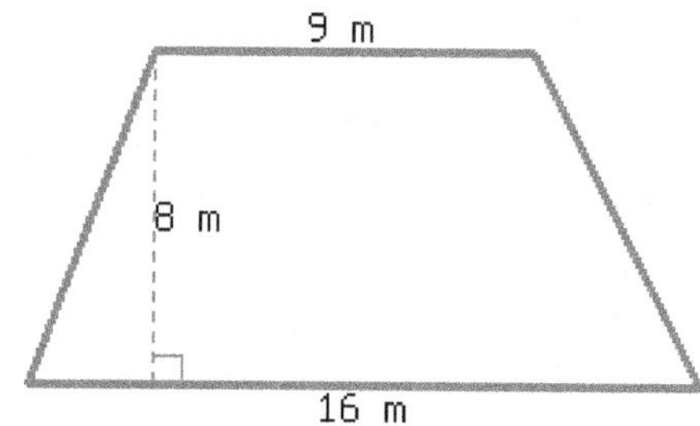

Volume

37. Find the volume of this rectangular prism.

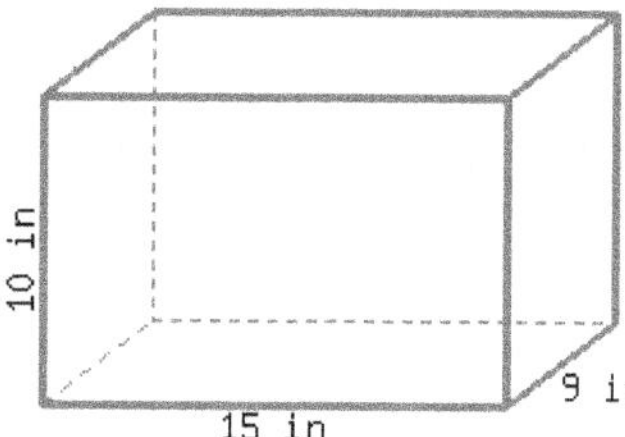

38. Find the volume of this rectangular prism.

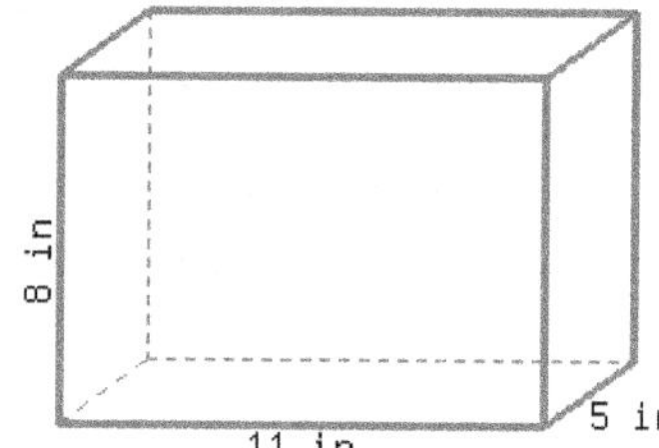

39. Find the volume of this rectangular prism.

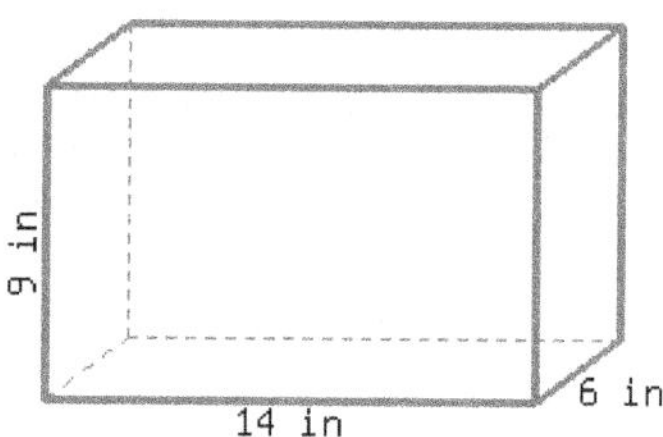

40. Find the volume of this rectangular prism.

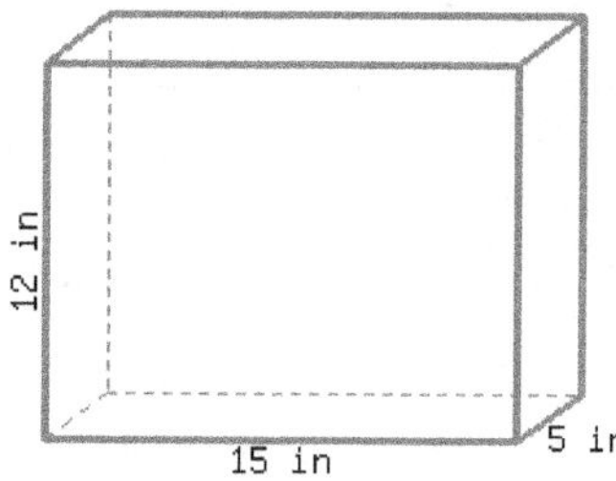

41. A cube's side length is 8 cm.

Its volume is ____________.

42. A cube's side length is 9 cm.

Its volume is ____________.

43. Find the volume of this cylinder.

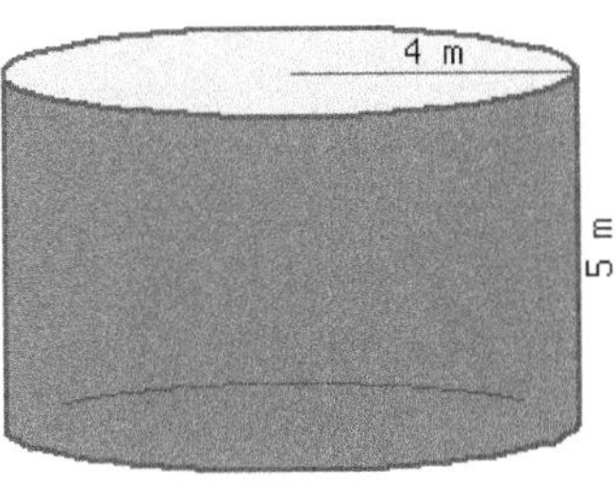

   a. This cylinder's volume, in terms of π, is ☐.

   b. This cylinder's volume, rounded to the hundredth place, is ☐.

44. Find the volume of this cylinder.

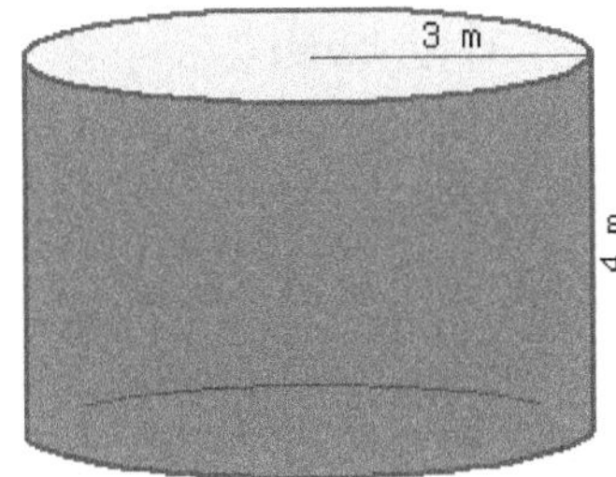

   a. This cylinder's volume, in terms of π, is ☐.

   b. This cylinder's volume, rounded to the hundredth place, is ☐.

45. Find the volume of this cylinder.

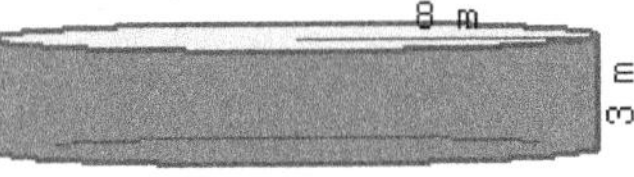

   a. This cylinder's volume, in terms of π, is ☐.

   b. This cylinder's volume, rounded to the hundredth place, is ☐.

46. Find the volume of this cylinder.

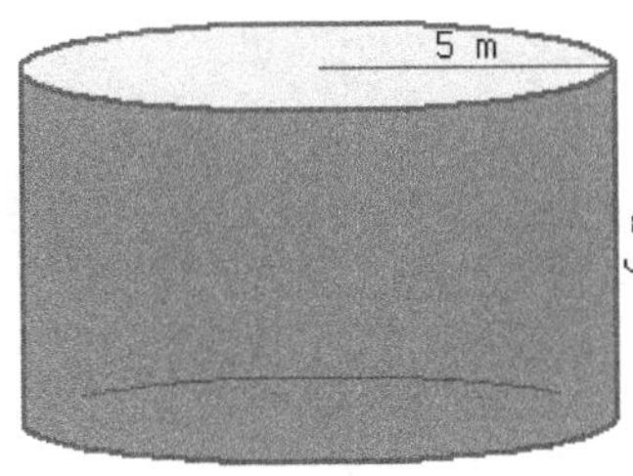

   a. This cylinder's volume, in terms of π, is ☐.

   b. This cylinder's volume, rounded to the hundredth place, is ☐.

47. A cylinder's base's diameter is 10 ft, and its height is 5 ft.

   a. This cylinder's volume, in terms of π, is ☐.

   b. This cylinder's volume, rounded to the hundredth place, is ☐.

48. A cylinder's base's diameter is 4 ft, and its height is 6 ft.

   a. This cylinder's volume, in terms of π, is ☐.

   b. This cylinder's volume, rounded to the hundredth place, is ☐.

The formula $V = \frac{1}{3} \cdot s^2 \cdot h$ gives the volume of a right square pyramid.

49. What is the volume of a right square pyramid with $s = 66$ in and $h = 69$ in?

50. What is the volume of a right square pyramid with $s = 78$ in and $h = 34$ in?

2.3 Combining Like Terms

In the last section we worked with algebraic expressions. In order to simplify algebraic expressions, it is useful to identify which quantities we can combine.

2.3.1 Identifying Terms

In an algebraic expression, the **terms** are the quantities that are added. For example, the expression $3x + 2y$ has two terms, which are $3x$ and $2y$. Let's look at some more examples.

> **Example 2.3.2** List the terms in the expression $2l + 2w$.
>
> The expression has two terms that are being added, $2l$ and $2w$.

If there is any subtraction, we will rewrite the expression using addition. Here is an example of that.

> **Example 2.3.3** List the terms in the expression $-3x^2 + 5x - 4$.
>
> We can rewrite this expression as $-3x^2 + 5x + (-4)$ to see that the terms are $-3x^2$, $5x$, and -4. The last term is negative because subtracting is the same as adding the opposite.

> **Example 2.3.4** List the terms in the expression $3\,\text{cm} + 2\,\text{cm} + 3\,\text{cm} + 2\,\text{cm}$.
>
> This expression has four terms: $3\,\text{cm}$, $2\,\text{cm}$, $3\,\text{cm}$, and $2\,\text{cm}$.

Checkpoint 2.3.5. List the terms in the expression $5x - 4x + 10z$.

Explanation. The terms are $5x$, $-4x$, and $10z$.

2.3.2 Combining Like Terms

In the examples above, you may have wanted to combine some of the terms. Look at the quantities below to see which ones you can add or subtract.

a. $5\,\text{in} + 20\,\text{in}$

b. $16\,\text{ft} - 4\,\text{ft}^2$

c. $2\,🍎 + 5\,🍎$

d. $5\,\text{min} + 50\,\text{ft}$

e. $5\,🐶 - 2\,🐱$

f. $20\,\text{m} - 6\,\text{m}$

The terms that we can combine are called **like terms**. We can combine terms with the same units, but we cannot combine units such as minutes and feet or cats and dogs. Here are the answers:

a. $5\,\text{in} + 20\,\text{in} = 25\,\text{in}$

b. $16\,\text{ft} - 4\,\text{ft}^2$ cannot be simplified

c. $2\,🍎 + 5\,🍎 = 7\,🍎$

d. $5\,\text{min} + 50\,\text{ft}$ cannot be simplified

e. $5\,🐶 - 2\,🐱$ cannot be simplified

f. $20\,\text{m} - 6\,\text{m} = 14\,\text{min}$

Now let's look at some examples that have variables in them.

Checkpoint 2.3.6. Which expressions have like terms that you can combine?

a. $10x + 3y$ ($\square$ can $\square$ cannot) be combined.

b. $4x - 8x$ ($\square$ can $\square$ cannot) be combined.

c. $9y - 4y$ ($\square$ can $\square$ cannot) be combined.

d. $-6x + 17z$ ($\square$ can $\square$ cannot) be combined.

e. $-3x - 7x$ ($\square$ can $\square$ cannot) be combined.

f. $5t + 8t^2$ ($\square$ can $\square$ cannot) be combined.

Explanation. The terms that we can combine have the same variable part, including any exponents.

a. $10x + 3y$ cannot be combined.

b. $4x - 8x = -4x$

c. $9y - 4y = 5y$

d. $-6x + 17z$ cannot be combined.

e. $-3x - 7x = -10x$

f. $5t + 8t^2$ cannot be combined.

Example 2.3.7 Simplify the expression $20x - 16x + 4y$, if possible, by combining like terms.

This expression has two like terms, $20x$ and $-16x$, which we can combine.

$$20x - 16x + 4y = 4x + 4y$$

Note that we cannot combine $4x$ and $4y$ because x and y represent different quantities.

Example 2.3.8 Simplify the expression $100x + 100x^2$, if possible, by combining like terms.

This expression cannot be simplified because the variable parts are not the same. We cannot add x and x^2 just like we cannot add feet, a measure of length, and square feet, a measure of area.

Example 2.3.9 Simplify the expression $-10r + 2s - 5t$, if possible, by combining like terms.

This expression cannot be simplified because there are not any like terms.

Example 2.3.10 Simplify the expression $y + 5y$, if possible, by combining like terms.

This expression can be thought of as $1y + 5y$. When we have a single y, the coefficient of 1 is not usually written. Now we have two like terms, $1y$ and $5y$. We will add those together:

$$x + 5x = 1x + 5x$$
$$= 6x$$

So far we have combined terms with whole numbers and integers, but we can also combine like terms when the coefficients are decimals (or fractions).

Example 2.3.11 Simplify the expression $x - 0.15x$, if possible, by combining like terms.

Note that this expression can be rewritten as $1.00x - 0.15x$, and combined like this:

$$x - 0.15x = 1.00x - 0.15x$$
$$= 0.85x$$

Checkpoint 2.3.12. Simplify each expression, if possible, by combining like terms.

a. $4x - 7y + 10x$

b. $y - 8y + 2x^2$

c. $x + 0.25x$

d. $4x + 1.5y - 9z$

Explanation.

a. This expression has two like terms that can be combined to get $14x - 7y$.

b. In this expression we can combine the y terms to get $-7y + 2x^2$.

c. Rewrite this expression as $1.00x + 0.25x$ and simplify to get $1.25x$.

d. This expression cannot be simplified further because there are not any like terms.

Remark 2.3.13 The Difference Between Terms and Factors. We have learned that terms are quantities that are added, such as $3x$ and $-2x$ in $3x - 2x$. These are different than **factors**, which are parts that are multiplied together. For example, the term $2x$ has two factors: 2 and x (with the multiplication symbol implied between them). The term $2\pi r$ has three factors: 2, π, and r.

Exercises

Review and Warmup

1. Add the following.

a. $2 + (-6) = \boxed{}$

b. $8 + (-3) = \boxed{}$

c. $9 + (-9) = \boxed{}$

2. Add the following.

a. $2 + (-7) = \boxed{}$

b. $10 + (-4) = \boxed{}$

c. $9 + (-9) = \boxed{}$

3. Add the following.

a. $-9 + 3 = \boxed{}$

b. $-1 + 7 = \boxed{}$

c. $-2 + 2 = \boxed{}$

4. Add the following.

a. $-6 + 3 = \boxed{}$

b. $-2 + 9 = \boxed{}$

c. $-2 + 2 = \boxed{}$

5. Subtract the following.

a. $4 - 8 = \boxed{}$

b. $6 - 2 = \boxed{}$

c. $6 - 12 = \boxed{}$

6. Subtract the following.

a. $4 - 7 = \boxed{}$

b. $8 - 1 = \boxed{}$

c. $6 - 17 = \boxed{}$

7. Subtract the following.

a. $-1 - 5 = \boxed{}$

b. $-10 - 5 = \boxed{}$

c. $-4 - 4 = \boxed{}$

8. Subtract the following.

a. $-1 - 3 = \boxed{}$

b. $-8 - 3 = \boxed{}$

c. $-4 - 4 = \boxed{}$

9. Subtract the following.

a. $-1 - (-6) = \boxed{}$

b. $-10 - (-1) = \boxed{}$

c. $-5 - (-5) = \boxed{}$

10. Subtract the following.

    a. $-2 - (-8) =$ ☐

    b. $-6 - (-2) =$ ☐

    c. $-5 - (-5) =$ ☐

Identifying Terms

11. Count the number of terms in each expression.

    a. $8t^2 - 6t$

    b. $4x - 4 - 9y$

    c. $-8z^2 - z^2 + 2y^2 + 6t$

    d. $-3s + 2t + 2z$

12. Count the number of terms in each expression.

    a. $-7t - 5x$

    b. $-4s^2$

    c. $2x^2 - 6x^2$

    d. $-2s^2 + 8z + 4t^2$

13. Count the number of terms in each expression.

    a. $1.8t + 0.7z^2 + x^2$

    b. $8.1s - 8.9s + 5.2 + 0.6y$

    c. $3s^2 + 5.9z$

    d. $0.1s$

14. Count the number of terms in each expression.

    a. $-3.2t - 6.6x^2 + 4.2$

    b. $-8.5t + 2.5s$

    c. $-0.4x - 6.2x - 6.8 + 2s^2$

    d. $-3.6x^2 + 7.6t + 2.7 + 3.6s$

15. List the terms in each expression.

    a. $-2t + 6x + 3x - 2x^2$

    b. $2t$

    c. $-9s + 4 + 6x$

    d. $s^2 - 3z + 2t^2$

16. List the terms in each expression.

    a. $6t - 1 + 4x + 8t^2$

    b. $-7s - 3z^2 + 8y^2$

    c. $-6s - x - 3y + t^2$

    d. $9z^2 - 8s$

17. List the terms in each expression.

    a. $-5t - 8t + 4.4x - 3.6y^2$

    b. $8.3s^2 - 5.8z$

    c. $6.7t - 4.9s$

    d. $3.9z$

18. List the terms in each expression.

    a. $3.3t$

    b. $3.4t^2 + 5.2x$

    c. $-7.9s^2 - 4.4t^2 + 0.7 - 7.5x$

    d. $-2.4t + 4.5y + 7.5t + 0.8t$

19. List the terms in each expression.

    a. $-1.8t^2 + 0.3y^2$

    b. $7.8t + 3.2t - 3 + 8.9y$

    c. $-6.6z^2$

    d. $-2.2s^2 - 8.1x + 5.2x$

20. List the terms in each expression.

    a. $2.9t + 6.9s^2$

    b. $-3.7x + 4.1y + 1.1z$

    c. $-6.6z - 5.3z^2 - 4t$

    d. $-3.9t$

Combining Like Terms   Simplify each expression, if possible, by combining like terms.

21. a. $7t^2 + 2s^2$

    b. $8s^2 + 5z$

    c. $8t^2 + t^2$

    d. $-5y^2 - 5z$

22. a. $-9z + 5z$

    b. $7y^2 - 4s^2$

    c. $-s^2 + 8x$

    d. $-2z + 5t$

23. a. $z + 8z^2 + 8z$

    b. $-3z^2 + 7y^2$

    c. $-9y - y - 7t^2 + s^2$

    d. $3s^2 + 6y$

24. a. $-5z + 3z^2 + 2z^2 - 3z^2$

    b. $-2z^2 - 9 + 3y^2 + 9z^2$

    c. $-9s - 8z$

    d. $7z^2 + y^2$

25. a. $-5z - 79z - 76 - 99z$

    b. $-79x - 51t - 53s$

    c. $-89x - 44x - 16z$

    d. $22x + 38x^2 + 50x^2 + 90$

26. a. $-21z + 14s - 18z - 56z$

    b. $90z - 32$

    c. $30y^2 - 61t^2 + 89t^2 + 94t^2$

    d. $23y^2 + 53y^2 - 90y^2$

27. a. $-3.8z - 2.5t$

    b. $-5.2s - 6.7s - 5.4y - z^2$

    c. $-1.5t - 4.5s^2 + 1.3t^2$

    d. $8.8y + 2.3y$

28. a. $1.3z - 0.4z$

    b. $-5.7s^2 - 7.6s^2 + 4.5y^2$

    c. $0.8y - 8.8 - 2.7t - 5.1t$

    d. $5t^2 + 2.5t$

29. a. $4z - \frac{5}{8}s$

    b. $\frac{9}{5}y - \frac{2}{3}y$

    c. $\frac{3}{4}y - 9y + y$

    d. $\frac{4}{7}y + z^2 - z - \frac{5}{2}z^2$

30. a. $\frac{5}{4}z + \frac{5}{9}x + \frac{4}{3}x$

    b. $-\frac{4}{9}t^2 + \frac{7}{6}z^2 + 1 - \frac{6}{7}y^2$

    c. $\frac{1}{3}y + \frac{8}{3}s^2 - \frac{8}{7}t$

    d. $\frac{7}{3}x - \frac{3}{7}y$

31. a. $\frac{8}{9}z^2 + \frac{4}{7}t^2 - \frac{7}{2}z$

    b. $-\frac{2}{5}x^2 + \frac{8}{9}y^2 - \frac{7}{3}$

    c. $6t^2 - \frac{2}{9}t^2$

    d. $-\frac{7}{2}t^2 + \frac{5}{2}t$

32. a. $-\frac{8}{3}z + \frac{4}{5}z + \frac{3}{4}z$

    b. $-\frac{9}{4}y + y^2$

    c. $\frac{4}{5}y + \frac{3}{2}t + z^2$

    d. $\frac{9}{5}t + 6t + \frac{1}{3}t + 7x$

2.4 Equations and Inequalities as True/False Statements

This section introduces the concepts of algebraic **equations** and **inequalities**, and what it means for a number to be a **solution** to an equation or inequality.

2.4.1 Equations, Inequalities, and Solutions

An **equation** is two mathematical expressions with an equals sign between them. The two expressions can be relatively simple or more complicated:

A relatively simple equation:

$$x + 1 = 2$$

A more complicated equation:

$$\left(x^2 + y^2 - 1\right)^3 = x^2 y^3$$

An **inequality** is quite similar, but the sign between the expressions is one of these: $<, \leq, >, \geq,$ or $\neq$.

A relatively simple inequality:

$$x \geq 15$$

A more complicated inequality:

$$x^2 + y^2 < 1$$

A **linear equation** in one variable can be written in the form $ax + b = 0$, where a, b are real numbers, and $a \neq 0$. The variable doesn't have to be x. The variable cannot have an exponent other than 1 ($x = x^1$), and the variable cannot be inside a root symbol (square root, cube root, etc.) or in a denominator.

The following are some linear equations in one variable:

$$4 - y = 5 \qquad 4 - z = 5z \qquad 0 = \frac{1}{2}p$$

$$3 - 2(q + 2) = 10 \qquad \sqrt{2} \cdot r + 3 = 10 \qquad \frac{s}{2} + 3 = 5$$

(Note that r is outside the square root symbol.) We will see in later sections that all equations above can be converted into the form $ax + b = 0$.

The following are some non-linear equations:

$$1 + 2 = 3 \qquad \text{(There is no variable.)}$$
$$4 - 2y^2 = 5 \qquad \text{(The exponent of } y \text{ is not 1.)}$$
$$\sqrt{2r} + 3 = 10 \qquad (r \text{ is inside the square root.)}$$
$$\frac{2}{s} + 3 = 5 \qquad (s \text{ is in a denominator.)}$$

This chapter focuses on linear equations in *one* variable. We will study other types of equations in later chapters.

The simplest equations and inequalities have numbers and no variables. When this happens, the equation is either *true* or *false*. The following equations and inequalities are *true* statements:

$$2 = 2 \qquad -4 = -4 \qquad 2 > 1 \qquad -2 < -1 \qquad 3 \geq 3$$

The following equations and inequalities are *false* statements:

$$2 = 1 \qquad -4 = 4 \qquad 2 < 1 \qquad -2 \geq -1 \qquad 0 \neq 0$$

When equations and inequalities have variables, we can consider substituting values in for the variables. If replacing a variable with a number makes an equation or inequality *true*, then that number is called a **solution** to the equation.

Example 2.4.2 A Solution. Consider the equation $y + 2 = 3$, which has only one variable, y. If we substitute in 1 for y and then simplify:

$$y + 2 = 3$$
$$1 + 2 \overset{?}{=} 3$$
$$3 \overset{\checkmark}{=} 3$$

we get a true equation. So we say that 1 is a solution to $y + 2 = 3$. Notice that we used a question mark at first because we are unsure if the equation is true or false until the end.

If replacing a variable with a value makes a false equation or inequality, that number is not a solution.

Example 2.4.3 Not a Solution. Consider the inequality $x + 4 > 5$, which has only one variable, x. If we substitute in 0 for x and then simplify:

$$x + 4 > 5$$
$$0 + 4 \overset{?}{>} 5$$
$$4 \overset{no}{>} 5$$

we get a false equation. So we say that 0 is *not* a solution to $x + 4 > 5$.

2.4.2 Checking Possible Solutions

Given an equation or an inequality (with one variable), checking if some particular number is a solution is just a matter of replacing the value of the variable with the specified number and determining if the resulting equation/inequality is true or false. This may involve some amount of arithmetic simplification.

Example 2.4.4 Is 8 a solution to $x^2 - 5x = \sqrt{2x} + 20$?

To find out, substitute in 8 for x and see what happens.

$$x^2 - 5x = \sqrt{2x} + 20$$
$$8^2 - 5(8) \overset{?}{=} \sqrt{2(8)} + 20$$
$$64 - 5(8) \overset{?}{=} \sqrt{16} + 20$$
$$64 - 40 \overset{?}{=} 4 + 20$$
$$24 \overset{\checkmark}{=} 24$$

So yes, 8 is a solution to $x^2 - 5x = \sqrt{2x} + 20$.

Example 2.4.5 Is -5 a solution to $\sqrt{169 - y^2} = y^2 - 2y$?

To find out, substitute in -5 for y and see what happens.

$$\sqrt{169 - y^2} = y^2 - 2y$$
$$\sqrt{169 - (-5)^2} \overset{?}{=} (-5)^2 - 2(-5)$$
$$\sqrt{169 - 25} \overset{?}{=} 25 - 2(-5)$$
$$\sqrt{144} \overset{?}{=} 25 - (-10)$$
$$12 \overset{\text{no}}{=} 35$$

So no, -5 is not a solution to $\sqrt{169 - y^2} = y^2 - 2y$.

But is -5 a solution to the *inequality* $\sqrt{169 - y^2} \leq y^2 - 2y$? Yes, because substituting -5 in for y would give you

$$12 \leq 35,$$

which is true.

Checkpoint 2.4.6. Is -3 a solution for x in the equation $2x - 3 = 5 - (4 + x)$? Evaluating the left and right sides gives:

$$
\begin{array}{ccc}
2x - 3 & = & 5 - (4 + x) \\
& \overset{?}{=} & \\
\underline{} & & \underline{}
\end{array}
$$

So -3 ($\square$ is $\square$ is not)   a solution to $2x - 3 = 5 - (4 + x)$.

Explanation.   We will substitute x with -3 in the equation and simplify each side of the equation to determine if the statement is true or false:

$$2x - 3 = 5 - (4 + x)$$
$$2(-3) - 3 \overset{?}{=} 5 - (4 + (-3))$$
$$-6 - 3 \overset{?}{=} 5 - (1)$$
$$-9 \overset{\times}{=} 4$$

Since $-9 = 4$ is not true, -3 is not a solution for x in the equation $2x - 3 = 5 - (4 + x)$.

Checkpoint 2.4.7. Is $\frac{1}{3}$ a solution for t in the equation $2t = 4\left(t - \frac{1}{2}\right)$? Evaluating the left and right sides gives:

$$
\begin{array}{ccc}
2t & = & 4\left(t - \frac{1}{2}\right) \\
& \overset{?}{=} & \\
\underline{} & & \underline{}
\end{array}
$$

So $\frac{1}{3}$ ($\square$ is $\square$ is not)   a solution to $2t = 4\left(t - \frac{1}{2}\right)$.

Checkpoint 2.4.8. Is -2 a solution to $y^2 + y - 5 \leq y - 1$? Evaluating the left and right sides gives:

$$y^2 + y - 5 \quad \overset{?}{\underset{\leq}{\leq}} \quad y - 1$$

$$\underline{} \qquad \underline{}$$

So -3 ($\square$ is $\square$ is not)   a solution to $y^2 + y - 5 \leq y - 1$.

Checkpoint 2.4.9. Is 2 a solution to $\frac{z+3}{z-1} = \sqrt{18z}$? Evaluating the left and right sides gives:

$$\frac{z+3}{z-1} \quad \overset{?}{\underset{=}{=}} \quad \sqrt{18z}$$

$$\underline{} \qquad \underline{}$$

So 2 ($\square$ is $\square$ is not)   a solution to $\frac{z+3}{z-1} = \sqrt{18z}$.

Checkpoint 2.4.10. Is -3 a solution to $x^2 + x + 1 \leq \frac{3x+2}{x+2}$? Evaluating the left and right sides gives:

$$x^2 + x + 1 \quad \overset{?}{\underset{\leq}{\leq}} \quad \frac{3x + 2}{x + 2}$$

$$\underline{} \qquad \underline{}$$

So -3 ($\square$ is $\square$ is not)   a solution to $x^2 + x + 1 \leq \frac{3x+2}{x+2}$.

A cylinder's volume is related to its radius and its height by:

$$V = \pi r^2 h,$$

where V is the volume, r is the base's radius, and h is the height. If we know the volume is 96π cm^3 and the radius is 4 cm, then we have:

$$96\pi = 16\pi h$$

Is 4 cm the height of the cylinder? In other words, is 4 a solution to $96\pi = 16\pi h$? We will substitute h in the equation with 4 to check:

$$96\pi = 16\pi h$$

$$96\pi \overset{?}{=} 16\pi \cdot 4$$

$$96\pi \overset{\text{no}}{=} 64\pi$$

Since $96\pi = 64\pi$ is false, $h = 4$ does *not* satisfy the equation $96\pi = 16\pi h$.

Figure 2.4.12: A cylinder

Example 2.4.11 Cylinder Volume. Next, we will try $h = 6$:

$$96\pi = 16\pi h$$

$$96\pi \overset{?}{=} 16\pi \cdot 6$$

$$96\pi \overset{\checkmark}{=} 96\pi$$

When $h = 6$, the equation $96\pi = 16\pi h$ is true. This tells us that 6 *is* a solution to $96\pi = 16\pi h$.

Remark 2.4.13. Note that we did not approximate π with 3.14 or any other approximation. We often leave π as π throughout our calculations. If we need to round, we do so as a final step.

Example 2.4.14 Jaylen has budgeted a maximum of \$300 for an appliance repair. The total cost of the repair can be modeled by $89 + 110(h - 0.25)$, where \$89 is the initial cost and \$110 is the hourly labor charge after the first quarter hour. Is 2 hours a solution for h in the inequality $89 + 110(h - 0.25) \le 300$?

To determine if $h = 2$ satisfies the inequality, we will replace h with 2 and check if the statement is true:

$$89 + 110(h - 0.25) \le 300$$

$$89 + 110(2 - 0.25) \overset{?}{\le} 300$$

$$89 + 110(1.75) \overset{?}{\le} 300$$

$$89 + 192.5 \overset{?}{\le} 300$$

$$281.5 \overset{\checkmark}{\le} 300$$

Thus, 2 hours is a solution for h in the inequality $89 + 110(h - 0.25) \le 300$. In context, this means that Jaylen would stay within their \$300 budget if 2 hours of labor were performed.

Exercises

Review and Warmup

1. Evaluate $-9 - x$ for $x = -4$.

2. Evaluate $5 - x$ for $x = -2$.

3. Evaluate $-2x + 7$ for $x = 1$.

4. Evaluate $-10x - 5$ for $x = 3$.

5. Evaluate $-4(r + 2)$ for $r = 4$.

6. Evaluate $-10(t + 9)$ for $t = -3$.

7. Evaluate the expression $\frac{1}{3}(x + 4)^2 - 2$ when $x = -7$.

8. Evaluate the expression $\frac{1}{4}(x + 1)^2 - 7$ when $x = -5$.

9. Evaluate the expression $-16t^2 + 64t + 128$ when $t = -3$.

10. Evaluate the expression $-16t^2 + 64t + 128$ when $t = -5$.

Identifying Linear Equations and Inequalities

11. Are the equations below linear equations in one variable?

    a. $\sqrt{9 - 5.5r} = 7$ ($\square$ is $\square$ is not) a linear equation in one variable.

    b. $2 - 3r^2 = 18$ ($\square$ is $\square$ is not) a linear equation in one variable.

    c. $x - 6q^2 = 27$ ($\square$ is $\square$ is not) a linear equation in one variable.

    d. $2\pi r = 2\pi$ ($\square$ is $\square$ is not) a linear equation in one variable.

    e. $8.59y = -9$ ($\square$ is $\square$ is not) a linear equation in one variable.

    f. $12 - 3z = 4$ ($\square$ is $\square$ is not) a linear equation in one variable.

12. Are the equations below linear equations in one variable?

    a. $-5 - 4y^2 = -21$ ($\square$ is $\square$ is not) a linear equation in one variable.

    b. $\sqrt{1 - 4.7x} = 2$ ($\square$ is $\square$ is not) a linear equation in one variable.

    c. $9q - y^2 = 1$ ($\square$ is $\square$ is not) a linear equation in one variable.

    d. $2\pi r = 12\pi$ ($\square$ is $\square$ is not) a linear equation in one variable.

    e. $-r - 11 = 3$ ($\square$ is $\square$ is not) a linear equation in one variable.

    f. $6.6y = 6$ ($\square$ is $\square$ is not) a linear equation in one variable.

13. Are the equations below linear equations in one variable?

    a. $4.61q = 14$ ($\square$ is $\square$ is not) a linear equation in one variable.

    b. $5q\sqrt{x} = -42$ ($\square$ is $\square$ is not) a linear equation in one variable.

    c. $\sqrt{2} + y^2 = -68$ ($\square$ is $\square$ is not) a linear equation in one variable.

    d. $\pi r^2 = 48\pi$ ($\square$ is $\square$ is not) a linear equation in one variable.

    e. $7 - 2r = 17$ ($\square$ is $\square$ is not) a linear equation in one variable.

    f. $r\sqrt{12} = -84$ ($\square$ is $\square$ is not) a linear equation in one variable.

14. Are the equations below linear equations in one variable?

    a. $q\sqrt{24} = 63$ ($\square$ is $\square$ is not) a linear equation in one variable.

    b. $\pi r^2 = 35\pi$ ($\square$ is $\square$ is not) a linear equation in one variable.

    c. $-2.96x = -22$ ($\square$ is $\square$ is not) a linear equation in one variable.

    d. $-8prz = 52$ ($\square$ is $\square$ is not) a linear equation in one variable.

    e. $15x - 4 = 28$ ($\square$ is $\square$ is not) a linear equation in one variable.

    f. $q^2 + y^2 = 4$ ($\square$ is $\square$ is not) a linear equation in one variable.

15. Are the inequalities below linear inequalities in one variable?

    a. $4p^2 - y > -51$ ($\square$ is $\square$ is not) a linear inequality in one variable.

    b. $-1 > 1 - 5p$ ($\square$ is $\square$ is not) a linear inequality in one variable.

    c. $-3z^2 - 7y^2 < 1$ ($\square$ is $\square$ is not) a linear inequality in one variable.

16. Are the inequalities below linear inequalities in one variable?

    a. $3x^2 + 6z^2 > 1$ ($\square$ is $\square$ is not) a linear inequality in one variable.

    b. $9V^2 + 6p > 51$ ($\square$ is $\square$ is not) a linear inequality in one variable.

    c. $4 \geq -5 - 3y$ ($\square$ is $\square$ is not) a linear inequality in one variable.

17. Are the inequalities below linear inequalities in one variable?

   a. $-5.8x > 96$ ($\square$ is $\square$ is not)   a linear inequality in one variable.

   b. $-1 > -4878r - 2301p$ ($\square$ is $\square$ is not) a linear inequality in one variable.

   c. $\sqrt{9p} - 9 \le -6$ ($\square$ is $\square$ is not)   a linear inequality in one variable.

18. Are the inequalities below linear inequalities in one variable?

   a. $-6y < -43$ ($\square$ is $\square$ is not)   a linear inequality in one variable.

   b. $\sqrt{9r} + 6 < -6$ ($\square$ is $\square$ is not)   a linear inequality in one variable.

   c. $198 \le 4182y - 9693x$ ($\square$ is $\square$ is not)   a linear inequality in one variable.

Checking a Solution for an Equation

19. Is -3 a solution for x in the equation $x + 7 = 3$? ($\square$ Yes $\square$ No)

20. Is 2 a solution for x in the equation $x + 9 = 7$? ($\square$ Yes $\square$ No)

21. Is -9 a solution for r in the equation $-3 - r = 5$?   ($\square$ Yes $\square$ No)

22. Is 6 a solution for r in the equation $5 - r = -1$? ($\square$ Yes $\square$ No)

23. Is -1 a solution for r in the equation $-7r - 8 = -1$?   ($\square$ Yes $\square$ No)

24. Is -9 a solution for t in the equation $2t + 8 = -10$?   ($\square$ Yes $\square$ No)

25. Is 5 a solution for t in the equation $-10t + 2 = -9t - 3$?   ($\square$ Yes $\square$ No)

26. Is 2 a solution for x in the equation $-2x - 4 = -9x - 18$?   ($\square$ Yes $\square$ No)

27. Is -9 a solution for x in the equation $9(x - 11) = 20x$?   ($\square$ Yes $\square$ No)

28. Is 3 a solution for y in the equation $2(y + 14) = 9y$?   ($\square$ Yes $\square$ No)

29. Is -3 a solution for y in the equation $-3(y - 13) = 16(y + 6)$?   ($\square$ Yes $\square$ No)

30. Is -10 a solution for r in the equation $19(r + 1) = 9(r - 9)$?   ($\square$ Yes $\square$ No)

31. Is $\frac{7}{10}$ a solution for x in the equation $10x - 1 = -8$?   ($\square$ Yes $\square$ No)

32. Is $\frac{14}{5}$ a solution for x in the equation $5x - 4 = 9$?   ($\square$ Yes $\square$ No)

33. Is 4 a solution for t in the equation $-\frac{10}{3}t - \frac{3}{10} = -\frac{17}{15}$?   ($\square$ Yes $\square$ No)

34. Is $-\frac{9}{2}$ a solution for t in the equation $\frac{2}{5}t - \frac{9}{10} = -\frac{27}{10}$?   ($\square$ Yes $\square$ No)

Checking a Solution for an Inequality   Decide whether each value is a solution to the given inequality.

35. $-2x + 9 > 7$

    a. $x = -5$ ($\square$ is $\square$ is not)   a solution.

    b. $x = 6$ ($\square$ is $\square$ is not)   a solution.

    c. $x = 0$ ($\square$ is $\square$ is not)   a solution.

    d. $x = 1$ ($\square$ is $\square$ is not)   a solution.

36. $2x - 5 > 1$

    a. $x = 3$ ($\square$ is $\square$ is not)   a solution.

    b. $x = 0$ ($\square$ is $\square$ is not)   a solution.

    c. $x = 13$ ($\square$ is $\square$ is not)   a solution.

    d. $x = 2$ ($\square$ is $\square$ is not)   a solution.

37. $3x - 8 \geq -5$

    a. $x = 0$ ($\square$ is $\square$ is not)   a solution.

    b. $x = 8$ ($\square$ is $\square$ is not)   a solution.

    c. $x = 1$ ($\square$ is $\square$ is not)   a solution.

    d. $x = -5$ ($\square$ is $\square$ is not)   a solution.

38. $-3x + 19 \geq 10$

    a. $x = 2$ ($\square$ is $\square$ is not)   a solution.

    b. $x = 0$ ($\square$ is $\square$ is not)   a solution.

    c. $x = 3$ ($\square$ is $\square$ is not)   a solution.

    d. $x = 13$ ($\square$ is $\square$ is not)   a solution.

39. $4x - 16 \leq 4$

    a. $x = 0$ ($\square$ is $\square$ is not)   a solution.

    b. $x = 10$ ($\square$ is $\square$ is not)   a solution.

    c. $x = 5$ ($\square$ is $\square$ is not)   a solution.

    d. $x = 3$ ($\square$ is $\square$ is not)   a solution.

40. $4x - 10 \leq -2$

    a. $x = 1$ ($\square$ is $\square$ is not)   a solution.

    b. $x = 2$ ($\square$ is $\square$ is not)   a solution.

    c. $x = 12$ ($\square$ is $\square$ is not)   a solution.

    d. $x = 0$ ($\square$ is $\square$ is not)   a solution.

Checking Solutions for Application Problems

41. A triangle's area is 99 square meters. Its height is 11 meters. Suppose we wanted to find how long is the triangle's base. A triangle's area formula is

$$A = \frac{1}{2}bh$$

where A stands for area, b for base and h for height. If we let b be the triangle's base, in meters, we can solve this problem using the equation:

$$99 = \frac{1}{2}(b)(11)$$

Check whether 36 is a solution for b of this equation.   ($\square$ Yes $\square$ No)

42. A triangle's area is 95 square meters. Its height is 10 meters. Suppose we wanted to find how long is the triangle's base. A triangle's area formula is

$$A = \frac{1}{2}bh$$

where A stands for area, b for base and h for height. If we let b be the triangle's base, in meters, we can solve this problem using the equation:

$$95 = \frac{1}{2}(b)(10)$$

Check whether 38 is a solution for b of this equation.   ($\square$ Yes $\square$ No)

43. When a plant was purchased, it was 2 inches tall. It grows 1 inches per day. How many days later will the plant be 17 inches tall?

Assume the plant will be 17 inches tall d days later. We can solve this problem using the equation:

$$1d + 2 = 17$$

Check whether 17 is a solution for d of this equation.   ($\square$ Yes $\square$ No)

44. When a plant was purchased, it was 1.2 inches tall. It grows 0.2 inches per day. How many days later will the plant be 5.2 inches tall?

Assume the plant will be 5.2 inches tall d days later. We can solve this problem using the equation:

$$0.2d + 1.2 = 5.2$$

Check whether 21 is a solution for d of this equation.   ($\square$ Yes $\square$ No)

45. A water tank has 163 gallons of water in it, and it is being drained at the rate of 7 gallons per minute. After how many minutes will there be 37 gallons of water left?

Assume the tank will have 37 gallons of water after m minutes. We can solve this problem using the equation:

$$163 - 7m = 37$$

Check whether 19 is a solution for m of this equation.   ($\square$ Yes $\square$ No)

46. A water tank has 171 gallons of water in it, and it is being drained at the rate of 9 gallons per minute. After how many minutes will there be 45 gallons of water left?

Assume the tank will have 45 gallons of water after m minutes. We can solve this problem using the equation:

$$171 - 9m = 45$$

Check whether 17 is a solution for m of this equation.   ($\square$ Yes $\square$ No)

47. A cylinder's volume is 126π cubic centimeters. Its height is 14 centimeters. Suppose we wanted to find how long is the cylinder's radius. A cylinder's volume formula is

$$V = \pi r^2 h$$

where V stands for volume, r for radius and h for height. Let r represent the cylinder's radius, in centimeters. We can solve this problem using the equation:

$$126\pi = \pi r^2(14)$$

Check whether 3 is a solution for r of this equation.   ($\square$ Yes $\square$ No)

48. A cylinder's volume is 960π cubic centimeters. Its height is 15 centimeters. Suppose we wanted to find how long is the cylinder's radius. A cylinder's volume formula is

$$V = \pi r^2 h$$

where V stands for volume, r for radius and h for height. Let r represent the cylinder's radius, in centimeters. We can solve this problem using the equation:

$$960\pi = \pi r^2(15)$$

Check whether 64 is a solution for r of this equation.   ($\square$ Yes $\square$ No)

49. A country's national debt was 140 million dollars in 2010. The debt increased at 50 million dollars per year. If this trend continues, when will the country's national debt increase to 640 million dollars?

Assume the country's national debt will become 640 million dollars y years after 2010. We can solve this problem using the equation:

$$50y + 140 = 640$$

Check whether 10 is a solution for y of this equation.　(□ Yes　□ No)

50. A country's national debt was 100 million dollars in 2010. The debt increased at 60 million dollars per year. If this trend continues, when will the country's national debt increase to 1180 million dollars?

Assume the country's national debt will become 1180 million dollars y years after 2010. We can solve this problem using the equation:

$$60y + 100 = 1180$$

Check whether 20 is a solution for y of this equation.　(□ Yes　□ No)

51. A school district has a reserve fund worth 41.1 million dollars. It plans to spend 2.9 million dollars per year. After how many years, will there be 15 million dollars left?

Assume there will be 15 million dollars left after y years. We can solve this problem using the equation:

$$41.1 - 2.9y = 15$$

Check whether 11 is a solution for y of this equation.　(□ Yes　□ No)

52. A school district has a reserve fund worth 32 million dollars. It plans to spend 3 million dollars per year. After how many years, will there be 11 million dollars left?

Assume there will be 11 million dollars left after y years. We can solve this problem using the equation:

$$32 - 3y = 11$$

Check whether 8 is a solution for y of this equation.　(□ Yes　□ No)

53. A rectangular frame's perimeter is 6 feet. If its length is 2 feet, suppose we want to find how long is its width. A rectangle's perimeter formula is

$$P = 2(l + w)$$

where P stands for perimeter, l for length and w for width. We can solve this problem using the equation:

$$6 = 2(2 + w)$$

Check whether 4 is a solution for w of this equation.　(□ Yes　□ No)

54. A rectangular frame's perimeter is 7 feet. If its length is 2.1 feet, suppose we want to find how long is its width. A rectangle's perimeter formula is

$$P = 2(l + w)$$

where P stands for perimeter, l for length and w for width. We can solve this problem using the equation:

$$7 = 2(2.1 + w)$$

Check whether 1.4 is a solution for w of this equation.　(□ Yes　□ No)

2.5 Solving One-Step Equations

We have learned how to check whether a specific number is a solution to an equation or inequality. In this section, we will begin learning how to *find* the solution(s) to basic equations ourselves.

2.5.1 Imagine Filling in the Blanks

Let's start with a very simple situation — so simple, that you might have success entirely in your head without writing much down. It's not exactly the algebra we want you to learn, but the example may serve as a good warm-up.

> **Example 2.5.2** A number plus 2 is 6. What is that number?
>
> You may be so familiar with basic arithmetic that you know the answer already. The *algebra* approach would be to start by translating "A number plus 2 is 6" into a math statement — in this case, an equation:
>
> $$x + 2 = 6$$
>
> where x is the number we are trying to find. In other words, what should be substituted in for x...
>
> $$x + 2 = 6$$
>
> ... to make the equation true?
>
> Now, how do you determine what x is? One valid option is to just *imagine* what number you could put in place of x that would result in a true equation. Would 0 work? No, that would mean $0 + 2 \stackrel{no}{=} 6$. Would 17 work? No, that would mean $17 + 2 \stackrel{no}{=} 6$. Would 4 work? Yes, because $4 + 2 = 6$ is a true equation.
>
> So one solution to $x + 2 = 6$ is 4. No other numbers are going to be solutions, because when you add 2 to something smaller or larger than 4, the result is going to be smaller or larger than 6.

This approach might work for you to solve *very basic equations*, but in general equations are going to be too complicated to solve in your head this way. So we move on to more systematic approaches.

2.5.2 The Basic Principle of Algebra

2.5.2.1 Opposite Operations

Let's revisit Example 2.5.2, thinking it through differently.

> **Example 2.5.3** If a number plus 2 is 6, what is the number?
>
> One way to solve this riddle is to use the opposite operation. If a number *plus* 2 is 6, we should be able to *subtract* 2 from 6 and get that unknown number. So the unknown number is $6 - 2 = 4$.

Let's try this strategy with another riddle.

> **Example 2.5.4** If a number minus 2 is 6, what is the number? This time, we should be able to *add* 2 to 6 to get the unknown number. So the unknown number is $6 + 2 = 8$.

Does this strategy work with multiplication and division?

Example 2.5.5 If a number times 2 is 6, what is the number? This time, we should be able to *divide* 6 by 2 to get the unknown number. So the unknown number is $6 \div 2 = 3$.

Example 2.5.6 If a number divided by 2 equals 6, what is the number? This time, we can *multiply* 6 by 2, and the unknown number is $6 \cdot 2 = 12$.

These examples explore an important principle for solving an equation — applying an opposite arithmetic operation.

2.5.2.2 Balancing Equations Like a Scale

We can revisit Example 2.5.2 with yet another strategy. If a number plus 2 is 6, what is the number? As is common in algebra, we can use x to represent the unknown number. The question translates into the math equation

$$x + 2 = 6.$$

Try to envision the equals sign as the middle of a balanced scale. The left side has 2 one-pound objects and a block with unknown weight x lb. Together, the weight on the left is $x + 2$. The right side has 6 one-pound objects. Figure 2.5.7 shows the scale.

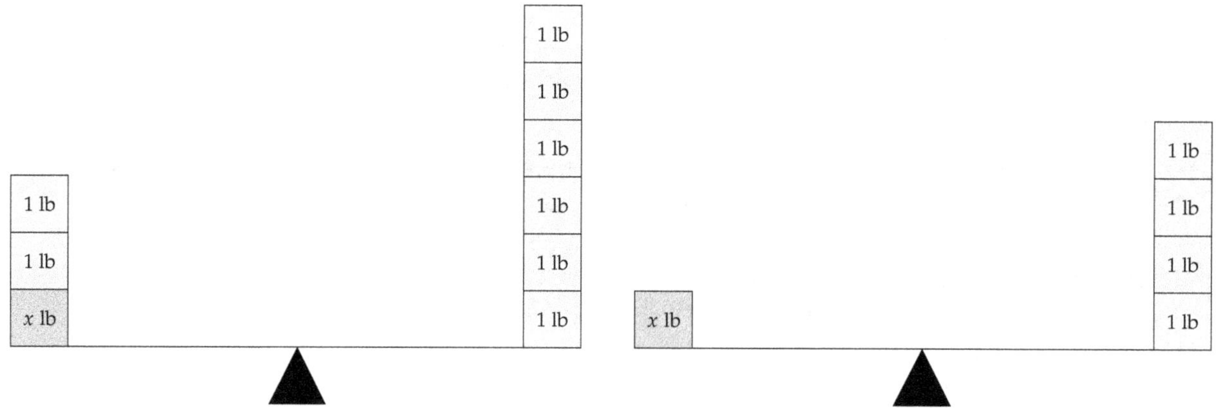

Figure 2.5.7: Scale to represent $x + 2 = 6$ | **Figure 2.5.8:** Scale to represent the solution to $x + 2 = 6$, after taking away 2 from each side

To find the weight of the unknown block, we can take away 2 one-pound blocks from *both* sides of the scale (to keep the scale balanced). Figure 2.5.8 shows the solution.

An equation is like a balanced scale, as the two sides of the equation are equal. In the same way that we can take away 2 lb from *both* sides of a balanced scale, we can subtract 2 on *both* sides of the equals sign. So instead of two pictures of balance scales, we can use algebra symbols and solve the equation $x + 2 = 6$ in the following manner:

$$\begin{aligned}
x + 2 &= 6 &&\text{a balanced scale}\\
x + 2 - 2 &= 6 - 2 &&\text{remove the same quantity from each side}\\
x &= 4 &&\text{still balanced; now it tells you the solution}
\end{aligned}$$

It's important to note that each line shows what is called an **equivalent equation**. In other words, each equation shown is algebraically equivalent to the one above it and the one below it and will have exactly the same solution(s). The final equivalent equation $x = 4$ tells us that the **solution** to the equation is 4. The **solution set** to this equation is the set that lists every solution to the equation. For this example, the solution set is $\{4\}$.

We have learned we can add or subtract the same number on both sides of the equals sign, just like we can add or remove the same amount of weight on a balanced scale. Can we multiply and divide the same number on both sides of the equals sign?

Let's look at Example 2.5.5 again: If a number times 2 is 6, what is the number? Another balance scale can help visualize this.

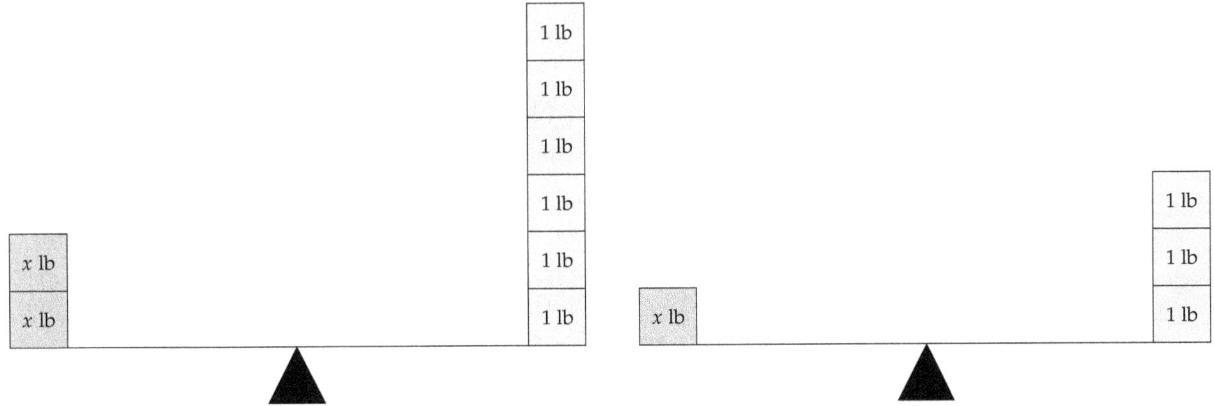

Figure 2.5.9: Scale to represent the equation $2x = 6$

Figure 2.5.10: Scale to represent the equation the solution to $2x = 6$, after cutting each side in half

Currently, the scale is balanced. If we cut the weight in half on both sides, the scale should still be balanced.

We can see from the scale that $x = 3$ is correct. Removing half of the weight on both sides of the scale is like dividing both sides of an equation by 2:

$$2x = 6$$
$$\frac{2x}{2} = \frac{6}{2}$$
$$x = 3$$

The equivalent equation in this example is $x = 3$, which tells us that the solution to the equation is 3 and the solution set is $\{3\}$.

Remark 2.5.11. Note that when we divide each side of an equation by a number, we use the fraction line in place of the division symbol. The fact that $\frac{6}{2} = 6 \div 2$ is a reminder that the fraction line and division symbol have the same purpose. The division symbol is rarely used in later math courses.

Similarly, we could multiply both sides of an equation by 2, just like we can keep a scale balanced if we double the weight on each side. We will summarize these properties.

Fact 2.5.12 Properties of Equivalent Equations. *If there is an equation $A = B$, we can do the following to obtain an equivalent equation:*

- *add the same number to each side: $A + c = B + c$*

- *subtract the same number from each side: $A - c = B - c$*

- *multiply each side of the equation by the same non-zero number: $A \cdot c = B \cdot c$*

- *divide each side of the equation by the same non-zero number: $\frac{A}{c} = \frac{B}{c}$*

With practice, you will learn when it is helpful to use each of these properties.

2.5.3 Solving One-Step Equations and Stating Solution Sets

Notice that when we solved equations in Subsection 2.5.2.2, the final equation looked like $x =$ number, where the variable x is separated from other numbers and stands alone on one side of the equals sign. The goal of solving any equation is to *isolate the variable* in this same manner.

Putting together both strategies (applying the opposite operation and balancing equations like a scale) that we just explored, we will summarize how to solve a one-step linear equation.

Algorithm 2.5.13 Steps to Solving Simple (One-Step) Linear Equations.

Apply *Apply the opposite operation to both sides of the equation. If a number was added to the variable, subtract that number, and vice versa. If a number was multiplied by the variable, divide by that number, and vice versa.*

Check *Check the solution. This means, verify that what you think is the solution actually solves the equation. For one reason, it's human to have made a simple arithmetic mistake, and by checking you will protect yourself from this. For another reason, there are situations where solving an equation will tell you that certain numbers are possible solutions, but they do not actually solve the original equation. Checking solutions will catch these situations.*

Summarize *State the solution set, or in the case of application problems, summarize the result using a complete sentence and appropriate units.*

Let's look at a few examples.

Example 2.5.14 Solve for y in the equation $7 + y = 3$.

Explanation.

To isolate y, we need to remove 7 from the left side. Since 7 is being *added* to y, we need to *subtract* 7 from each side of the equation.

$$7 + y = 3$$
$$7 + y - 7 = 3 - 7$$
$$y = -4$$

We should always check the solution when we solve equations. For this problem, we will substitute y in the original equation with -4:

$$7 + y = 3$$
$$7 + (-4) \stackrel{?}{=} 3$$
$$3 \stackrel{\checkmark}{=} 3$$

The solution -4 is checked, so the solution set is $\{-4\}$.

Checkpoint 2.5.15. Solve for z in the equation $-7.3 + z = 5.1$.

The solution is $\boxed{}$.

The solution set is $\boxed{}$.

Explanation. To remove the -7.3 from the left side, we need to *add* 7.3 to each side of the equation. (If we make the mistake of subtracting 7.3 from each side, we would have $-7.3 + z - 7.3 = 5.1 - 7.3$, which simplifies to $z - 14.3 = -2.2$, which would not isolate z.)

$$-7.3 + z = 5.1$$
$$-7.3 + z + 7.3 = 5.1 + 7.3$$
$$z = 12.4$$

We will check the solution by substituting z in the original equation with 12.4:

$$-7.3 + z = 5.1$$
$$-7.3 + (12.4) \overset{?}{=} 5.1$$
$$5.1 \overset{\checkmark}{=} 5.1$$

The solution 12.4 is checked and the solution set is $\{12.4\}$.

Checkpoint 2.5.16. Solve for a in the equation $10 = -2a$.

The solution is ⬚.

The solution set is ⬚.

Explanation. To isolate the variable a, we need to divide each side by -2 (because a is being *multiplied* by -2). One common mistake is to add 2 to each side. This would not isolate a, but would instead leave us with the expression $-2a + 2$ on the right-hand side.

$$10 = -2a$$
$$\frac{10}{-2} = \frac{-2a}{-2}$$
$$-5 = a$$

We will check the solution by substituting a in the original equation with -5:

$$10 = -2a$$
$$10 \overset{?}{=} -2(-5)$$
$$10 \overset{\checkmark}{=} 10$$

The solution -5 is checked and the solution set is $\{-5\}$.

Note that in solving the equation in Checkpoint 2.5.16 we found that $-5 = a$, which is equivalent to $a = -5$. We did not write $a = -5$ as an extra step though, as $-5 = a$ identified the solution.

Example 2.5.17 The formula for a circle's circumference is $C = \pi d$, where C stands for circumference, d stands for diameter, and π is a constant with the value of $3.1415926\ldots$. If a circle's circumference is 12π ft, find this circle's diameter and radius.

Explanation. The circumference is given as 12π feet. Approximating π with 3.14, this means the circumference is approximately 37.68 ft. It is nice to have a rough understanding of how long the circumference is, but if we use 3.14 instead of π, we are using a slightly smaller number than π, and the result of any calculations we do would not be as accurate. This is why we will use the symbol π throughout solving this equation and round only at the end in the conclusion summary (if necessary).

We will substitute C in the formula with 12π and solve for d:

$$C = \pi d$$
$$12\pi = \pi d$$
$$\frac{12\pi}{\pi} = \frac{\pi d}{\pi}$$
$$12 = d$$

So the circle's diameter is 12 ft. And since radius is half of diameter, the radius is 6 ft.

Example 2.5.18 Solve for b in $-b = 2$.

Explanation. Note that b is not yet isolated as there is a negative sign in front of it. One way to solve for b is to recognize that multiplying on both sides by -1 would clear away that negative sign:

$$-b = 2$$
$$-1 \cdot (-b) = -1 \cdot (2)$$
$$b = -2$$

We removed the negative sign from $-b$ using the fact that $-1 \cdot (-b) = b$. A second way to remove the negative sign -1 from $-b$ is to divide both sides by -1. If you view the original $-b$ as $-1 \cdot b$, then this approach resembles the solution from Checkpoint 2.5.16.

$$-b = 2$$
$$-1 \cdot b = 2$$
$$\frac{-1 \cdot b}{-1} = \frac{2}{-1}$$
$$b = -2$$

A third way to remove the original negative sign, is to recognize that the opposite operation of negation is negation. So negating both sides will work out too:

$$-b = 2$$
$$-(-b) = -2$$
$$b = -2$$

We will check the solution by substituting b in the original equation with -2:

$$-b = 2$$
$$-(-2) \stackrel{?}{=} 2$$

$$2 \stackrel{\checkmark}{=} 2$$

The solution -2 is checked and the solution set is $\{-2\}$.

2.5.4 Solving One-Step Equations Involving Fractions

When equations have fractions, solving them will make use of the same principles. You may need to use fraction arithmetic, and there may be special considerations that will make the calculations easier. So we have separated the following examples.

Example 2.5.19 Solve for g in $\frac{2}{3} + g = \frac{1}{2}$.

Explanation.

In Section 3.3, we will learn a skill to avoid fraction operations entirely in equations like this one. For now, let's solve the equation by using subtraction to isolate g:

$$\frac{2}{3} + g = \frac{1}{2}$$
$$\frac{2}{3} + g - \frac{2}{3} = \frac{1}{2} - \frac{2}{3}$$
$$g = \frac{3}{6} - \frac{4}{6}$$
$$g = -\frac{1}{6}$$

We will check the solution by substituting g in the original equation with $-\frac{1}{6}$:

$$\frac{2}{3} + g = \frac{1}{2}$$
$$\frac{2}{3} + \left(-\frac{1}{6}\right) \stackrel{?}{=} \frac{1}{2}$$
$$\frac{4}{6} + \left(-\frac{1}{6}\right) \stackrel{?}{=} \frac{1}{2}$$
$$\frac{3}{6} \stackrel{?}{=} \frac{1}{2}$$
$$\frac{1}{2} \stackrel{\checkmark}{=} \frac{1}{2}$$

The solution $-\frac{1}{6}$ is checked and the solution set is $\left\{-\frac{1}{6}\right\}$.

Checkpoint 2.5.20. Solve for q in the equation $q - \frac{3}{7} = \frac{3}{2}$.

The solution is [   ].

The solution set is [   ].

Explanation. To remove the $\frac{3}{7}$ from the left side, we need to *add* $\frac{3}{7}$ to each side of the equation.

$$q - \frac{3}{7} = \frac{3}{2}$$
$$q - \frac{3}{7} + \frac{3}{7} = \frac{3}{2} + \frac{3}{7}$$
$$q = \frac{27}{14}$$

We will check the solution by substituting q in the original equation with $\frac{27}{14}$:

$$q - \frac{3}{7} = \frac{3}{2}$$
$$\frac{27}{14} - \frac{3}{7} \overset{?}{=} \frac{3}{2}$$
$$\frac{27}{14} - \frac{6}{14} \overset{?}{=} \frac{3}{2}$$
$$\frac{21}{14} \overset{?}{=} \frac{3}{2}$$
$$\frac{3}{2} \overset{\checkmark}{=} \frac{3}{2}$$

The solution $\frac{27}{14}$ is checked and the solution set is $\left\{\frac{27}{14}\right\}$.

Example 2.5.21 Solve for c in $\frac{c}{5} = 4$.

Explanation.

Note that the fraction line here implies division, so our variable c is being divided by 5. The opposite operation is to *multiply* by 5:

$$\frac{c}{5} = 4$$
$$5 \cdot \frac{c}{5} = 5 \cdot 4$$
$$c = 20$$

We will check the solution by substituting c in the original equation with 20:

$$\frac{c}{5} = 4$$
$$\frac{20}{5} \overset{?}{=} 4$$
$$4 \overset{\checkmark}{=} 4$$

The solution 20 is checked and the solution set is $\{20\}$.

Example 2.5.22 Solve for d in $-\frac{1}{3}d = 6$.

Explanation. It's true that in this example, the variable d is *multiplied* by $-\frac{1}{3}$. This means that *dividing* each side by $-\frac{1}{3}$ would be a valid strategy for solving this equation. However, dividing by a fraction could lead to human error, so consider this alternative strategy.

Another way to be rid of the $-\frac{1}{3}$ is to multiply by -3. Indeed, $-\frac{1}{3}d$ is the same as $\frac{d}{-3}$, and when we view the expression this way, d is being *divided* by -3. So multiplying by -3 would be the opposite operation.

$$-\frac{1}{3}d = 6$$
$$(-3) \cdot \left(-\frac{1}{3}d\right) = (-3) \cdot 6$$
$$d = -18$$

If you choose to divide each side by $-\frac{1}{3}$, that will work out as well:

$$-\frac{1}{3}d = 6$$

$$\frac{-\frac{1}{3}d}{-\frac{1}{3}} = \frac{6}{-\frac{1}{3}}$$

$$d = \frac{6}{1} \cdot \frac{-3}{1}$$

$$d = -18$$

This gives the same solution.

We will check the solution by substituting d in the original equation with -18:

$$-\frac{1}{3}d = 6$$

$$-\frac{1}{3} \cdot (-18) \stackrel{?}{=} 6$$

$$6 \stackrel{\checkmark}{=} 6$$

The solution -18 is checked and the solution set is $\{-18\}$.

Example 2.5.23 Solve for x in $\frac{3x}{4} = 10$.

Explanation. The variable x appears to have *two* operations that apply to it: first multiplication by 3, and then division by 4. But note that

$$\frac{3x}{4} = \frac{3}{4} \cdot \frac{x}{1} = \frac{3}{4}x.$$

If we view the left side this way, we can get away with solving the equation in one step, by multiplying on each side by the reciprocal of $\frac{3}{4}$.

$$\frac{3x}{4} = 10$$

$$\frac{3}{4}x = 10$$

$$\frac{4}{3} \cdot \frac{3}{4}x = \frac{4}{3} \cdot 10$$

$$x = \frac{4}{3} \cdot \frac{10}{1}$$

$$x = \frac{40}{3}$$

We will check the solution by substituting x in the original equation with $\frac{40}{3}$:

$$\frac{3x}{4} = 10$$

$$\frac{3\left(\frac{40}{3}\right)}{4} \stackrel{?}{=} 10$$

$$\frac{40}{4} \stackrel{?}{=} 10$$

$$10 \stackrel{\checkmark}{=} 10$$

The solution $\frac{40}{3}$ is checked and the solution set is $\left\{\frac{40}{3}\right\}$.

Checkpoint 2.5.24. Solve for H in the equation $\frac{-7H}{12} = \frac{2}{3}$.

The solution is [ ].

The solution set is [ ].

Explanation. The left side is effectively the same things as $-\frac{7}{12}H$, so multiplying by $-\frac{12}{7}$ will isolate H.

$$\frac{-7H}{12} = \frac{2}{3}$$
$$-\frac{7}{12}H = \frac{2}{3}$$
$$\left(-\frac{12}{7}\right) \cdot \left(-\frac{7}{12}H\right) = \left(-\frac{12}{7}\right) \cdot \frac{2}{3}$$
$$H = -\frac{4}{7} \cdot \frac{2}{1}$$
$$H = -\frac{8}{7}$$

We will check the solution by substituting H in the original equation with $-\frac{8}{7}$:

$$\frac{-7H}{12} = \frac{2}{3}$$
$$\frac{-7\left(-\frac{8}{7}\right)}{12} \overset{?}{=} \frac{2}{3}$$
$$\frac{8}{12} \overset{?}{=} \frac{2}{3}$$
$$\frac{2}{3} \overset{\checkmark}{=} \frac{2}{3}$$

The solution $-\frac{8}{7}$ is checked and the solution set is $\left\{-\frac{8}{7}\right\}$.

Exercises

Review and Warmup

1. Add the following.

a. $-8 + (-1) = \boxed{}$

b. $-7 + (-7) = \boxed{}$

c. $-2 + (-9) = \boxed{}$

2. Add the following.

a. $-8 + (-2) = \boxed{}$

b. $-5 + (-3) = \boxed{}$

c. $-2 + (-7) = \boxed{}$

3. Add the following.

a. $5 + (-9) = \boxed{}$

b. $5 + (-2) = \boxed{}$

c. $7 + (-7) = \boxed{}$

4. Add the following.

a. $1 + (-6) = \boxed{}$

b. $7 + (-3) = \boxed{}$

c. $6 + (-6) = \boxed{}$

5. Add the following.

a. $-8 + 1 = \boxed{}$

b. $-4 + 10 = \boxed{}$

c. $-5 + 5 = \boxed{}$

6. Add the following.

a. $-10 + 2 = \boxed{}$

b. $-1 + 6 = \boxed{}$

c. $-5 + 5 = \boxed{}$

7. Evaluate the following.

a. $\dfrac{-63}{-7} = \boxed{}$

b. $\dfrac{30}{-5} = \boxed{}$

c. $\dfrac{-35}{7} = \boxed{}$

8. Evaluate the following.

a. $\dfrac{-36}{-6} = \boxed{}$

b. $\dfrac{28}{-7} = \boxed{}$

c. $\dfrac{-63}{7} = \boxed{}$

9. Do the following multiplications:

a. $15 \cdot \dfrac{4}{5} = \boxed{}$

b. $20 \cdot \dfrac{4}{5} = \boxed{}$

c. $25 \cdot \dfrac{4}{5} = \boxed{}$

10. Do the following multiplications:

a. $36 \cdot \dfrac{4}{9} = \boxed{}$

b. $45 \cdot \dfrac{4}{9} = \boxed{}$

c. $54 \cdot \dfrac{4}{9} = \boxed{}$

11. Evaluate the following.

a. $\dfrac{-3}{-1} = \boxed{}$

b. $\dfrac{7}{-1} = \boxed{}$

c. $\dfrac{130}{-130} = \boxed{}$

d. $\dfrac{-16}{-16} = \boxed{}$

e. $\dfrac{11}{0} = \boxed{}$

f. $\dfrac{0}{-7} = \boxed{}$

12. Evaluate the following.

a. $\dfrac{-2}{-1} = \boxed{}$

b. $\dfrac{4}{-1} = \boxed{}$

c. $\dfrac{180}{-180} = \boxed{}$

d. $\dfrac{-19}{-19} = \boxed{}$

e. $\dfrac{11}{0} = \boxed{}$

f. $\dfrac{0}{-3} = \boxed{}$

Solving One-Step Equations with Addition/Subtraction Solve the equation.

13. $x + 7 = 10$

14. $x + 5 = 12$

15. $y + 1 = -9$

16. $y + 8 = 1$

17. $5 = y + 8$

18. $-2 = r + 6$

19. $-9 = r - 7$

20. $-14 = t - 9$

21. $t + 78 = 0$

22. $x + 50 = 0$

23. $x - 3 = 2$

24. $y - 10 = -2$

25. $-16 = y - 7$

26. $-9 = y - 4$

27. $r - (-5) = 7$

28. $r - (-7) = 10$

29. $-1 = t - (-5)$

30. $-8 = t - (-1)$

31. $3 + x = -4$

32. $1 + x = -9$

33. $3 = -3 + y$

34. $4 = -6 + y$

35. $y + \dfrac{3}{4} = \dfrac{3}{4}$

36. $r + \dfrac{7}{10} = \dfrac{9}{10}$

37. $-\dfrac{2}{7} + r = -\dfrac{5}{14}$

38. $-\dfrac{8}{3} + t = -\dfrac{5}{6}$

39. $\dfrac{2}{7} + q = -\dfrac{1}{8}$

40. $\dfrac{6}{5} + x = -\dfrac{1}{4}$

Solving One-Step Equations with Multiplication/Division Solve the equation.

41. $2x = 22$

42. $8y = 56$

43. $40 = -8y$

44. $33 = -3y$

45. $0 = 13B$

46. $0 = -22C$

47. $\dfrac{1}{10}n = 5$

48. $\dfrac{1}{7}q = 8$

49. $\frac{4}{13}x = 8$

50. $\frac{7}{12}r = 28$

51. $\frac{5}{2}t = 4$

52. $\frac{5}{3}b = 2$

53. $\frac{9}{8} = -\frac{c}{10}$

54. $\frac{3}{2} = -\frac{B}{6}$

55. $3C = -8$

56. $9n = -2$

57. $-35 = -30p$

58. $-25 = -10x$

59. $-\frac{r}{18} = \frac{4}{9}$

60. $-\frac{t}{30} = \frac{8}{5}$

61. $-\frac{b}{8} = -\frac{3}{2}$

62. $-\frac{c}{24} = -\frac{7}{8}$

63. $-\frac{3}{5} = \frac{3B}{10}$

64. $-\frac{7}{2} = \frac{7C}{9}$

65. $\frac{x}{72} = \frac{5}{9}$

66. $\frac{x}{50} = \frac{3}{10}$

67. $\frac{9}{2} = \frac{x}{18}$

68. $\frac{2}{3} = \frac{x}{21}$

Comparisons Solve the equation.

69.
a. $4y = 32$
b. $4 + r = 32$

70.
a. $8y = 16$
b. $8 + t = 16$

71.
a. $28 = -7y$
b. $28 = -7 + r$

72.
a. $18 = -3r$
b. $18 = -3 + x$

73.
a. $-r = 5$
b. $-t = -5$

74.
a. $-t = 13$
b. $-r = -13$

75.
a. $-\frac{1}{2}p = 7$
b. $-\frac{1}{2}r = -7$

76.
a. $-\frac{1}{6}x = 5$
b. $-\frac{1}{6}n = -5$

77.
a. $36 = -\frac{9}{2}y$
b. $-36 = -\frac{9}{2}A$

78.
a. $16 = -\frac{4}{9}t$
b. $-16 = -\frac{4}{9}r$

79.
a. $8y = 32$
b. $12t = 68$

80.
a. $4y = 16$
b. $21x = 51$

81.
a. $30 = -6r$
b. $64 = -12x$

82.
a. $40 = -10r$
b. $80 = -35t$

Geometry Application Problems

83. A circle's circumference is 18π mm.

a. This circle's diameter is ______.

b. This circle's radius is ______.

84. A circle's circumference is 20π mm.

a. This circle's diameter is ______.

b. This circle's radius is ______.

85. A circle's circumference is 30 cm. Find the following values. Round your answer to at least 2 decimal places.

    a. This circle's diameter is __________.

    b. This circle's radius is __________.

86. A circle's circumference is 32 cm. Find the following values. Round your answer to at least 2 decimal places.

    a. This circle's diameter is __________.

    b. This circle's radius is __________.

87. A circle's circumference is 8π mm.

    a. This circle's diameter is __________.

    b. This circle's radius is __________.

88. A circle's circumference is 10π mm.

    a. This circle's diameter is __________.

    b. This circle's radius is __________.

89. A circle's circumference is 39 cm. Find the following values. Round your answer to at least 2 decimal places.

    a. This circle's diameter is __________.

    b. This circle's radius is __________.

90. A circle's circumference is 42 cm. Find the following values. Round your answer to at least 2 decimal places.

    a. This circle's diameter is __________.

    b. This circle's radius is __________.

91. A cylinder's base's radius is 7 m, and its volume is 392π m^3.

    This cylinder's height is __________.

92. A cylinder's base's radius is 4 m, and its volume is 144π m^3.

    This cylinder's height is __________.

93. A rectangle's area is 570 mm^2. Its height is 19 mm.

    Its base is __________.

94. A rectangle's area is 300 mm^2. Its height is 15 mm.

    Its base is __________.

95. A rectangular prism's volume is 3822 ft^3. The prism's base is a rectangle. The rectangle's length is 21 ft and the rectangle's width is 13 ft.

    This prism's height is __________.

96. A rectangular prism's volume is 10450 ft^3. The prism's base is a rectangle. The rectangle's length is 22 ft and the rectangle's width is 19 ft.

    This prism's height is __________.

97. A triangle's area is 187.5 m^2. Its base is 25 m.

    Its height is __________.

98. A triangle's area is 162.5 m^2. Its base is 25 m.

    Its height is __________.

Challenge

99. Write a linear equation whose solution is $x = 5$. You may not write an equation whose left side is just "x" or whose right side is just "x."

There are infinitely many correct answers to this problem. *Be creative.* After finding an equation that works, see if you can come up with a different one that also works.

100. Fill in the blanks with the numbers 38 and 77 (using each number only once) to create an equation where x has the greatest possible value.

a. $\boxed{} + x = \boxed{}$

b. $\boxed{} = \boxed{} \cdot x$

2.6 Solving One-Step Inequalities

We have learned how to check whether a specific number is a solution to an equation or inequality. In this section, we will begin learning how to *find* the solution(s) to basic inequalities ourselves.

With one small complication, we can use very similar properties to Fact 2.5.12 when we solve inequalities (as opposed to equations).

Here are some numerical examples.

Add to both sides If $2 < 4$, then $2 + 1 \overset{\checkmark}{<} 4 + 1$.

Subtract from both sides If $2 < 4$, then $2 - 1 \overset{\checkmark}{<} 4 - 1$.

Multiply on both sides by a *positive* number If $2 < 4$, then $3 \cdot 2 \overset{\checkmark}{<} 3 \cdot 4$.

Divide on both sides by a *positive* number If $2 < 4$, then $\frac{2}{2} \overset{\checkmark}{<} \frac{4}{2}$.

However, something interesting happens when we multiply or divide by the same *negative* number on both sides of an inequality: the direction reverses! To understand why, consider Figure 2.6.2, where the numbers 2 and 4 are multiplied by the negative number -1.

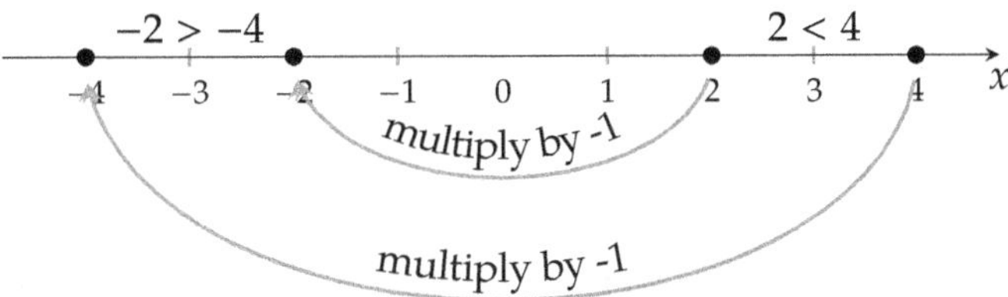

Figure 2.6.2: Multiplying two numbers by a negative number, and how their relationship changes

So even though $2 < 4$, if we multiply both sides by -1, we have $-2 \overset{no}{<} -4$. (The true inequality is $-2 > -4$.)

In general, we must apply the following property when solving an inequality.

Fact 2.6.3 Changing the Direction of the Inequality Sign. *When we multiply or divide each side of an inequality by the same* negative *number, the inequality sign must change direction. Do not* change the inequality sign when *multiplying/dividing by a positive number, or when adding/subtracting by any number.*

Example 2.6.4 Solve the inequality $-2x \geq 12$. State the solution set graphically, using interval notation, and using set-builder notation. (Interval notation and set-builder notation are discussed in Section 1.6.

Explanation. To solve this inequality, we will divide each side by -2:

$$-2x \geq 12$$
$$\frac{-2x}{-2} \leq \frac{12}{-2} \qquad\qquad \text{Note the change in direction.}$$
$$x \leq -6$$

Note that the inequality sign changed direction at the step where we divided each side of the inequality by a *negative* number.

When we solve a linear *equation*, there is usually exactly one solution. When we solve a linear *inequality*, there are usually infinitely many solutions. For this example, any number smaller than -6 or equal to -6 is a solution.

There are at least three ways to represent the solution set for the solution to an inequality: graphically, with set-builder notation, and with interval notation. Graphically, we represent the solution set as:

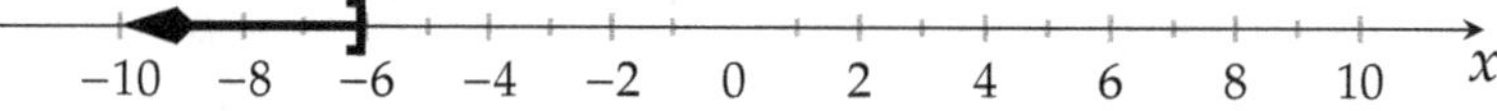

Using interval notation, we write the solution set as $(-\infty, -6]$. Using set-builder notation, we write the solution set as $\{x \mid x \le -6\}$.

As with equations, we should check solutions to catch both human mistakes as well as for possible extraneous solutions (numbers which were *possible* solutions according to algebra, but which actually do not solve the inequality).

Since there are infinitely many solutions, it's impossible to literally check them all. We found that all values of x for which $x \le -6$ are solutions. One approach is to check that -6 satisfies the inequality, and also that one number less than -6 (any number, your choice) is a solution.

$$-2x \ge 12 \qquad\qquad\qquad\qquad -2x \ge 12$$
$$-2(-6) \overset{?}{\ge} 12 \qquad\qquad\qquad -2(-7) \overset{?}{\ge} 12$$
$$12 \overset{\checkmark}{\ge} 12 \qquad\qquad\qquad\qquad 14 \overset{\checkmark}{\ge} 12$$

Thus both -6 and -7 are solutions. It's important to note this doesn't directly verify that *all* solutions to this inequality check. But it is evidence that our solution is correct, and it's valuable in that making these two checks would likely help us catch an error if we had made one. Consult your instructor to see if you're expected to check your answer in this manner.

Example 2.6.5 Solve the inequality $t + 7 < 5$. State the solution set graphically, using interval notation, and using set-builder notation.

Explanation. To solve this inequality, we will subtract 7 from each side. There is not much difference between this process and solving the *equation* $t + 7 = 5$, because we are not going to multiply or divide by negative numbers.

$$t + 7 < 5$$
$$t + 7 - 7 < 5 - 7$$
$$t < -2$$

Note again that the direction of the inequality did not change, since we did not multiply or divide each side of the inequality by a negative number at any point.

Graphically, we represent this solution set as:

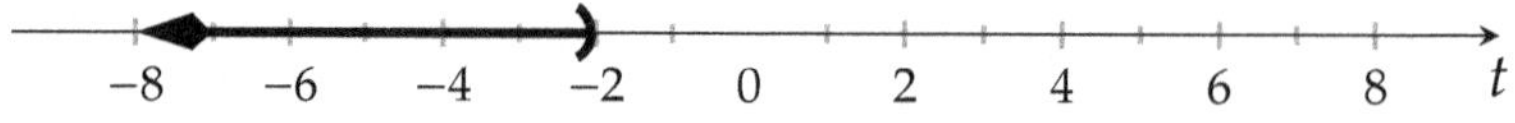

Using interval notation, we write the solution set as $(-\infty, -2)$. Using set-builder notation, we write the solution set as $\{t \mid t < -2\}$.

We should check that -2 is *not* a solution, but that some number less than -2 *is* a solution.

$$t + 7 < 5 \qquad\qquad t + 7 < 5$$
$$-2 + 7 \overset{?}{<} 5 \qquad\qquad -10 + 7 \overset{?}{<} 5$$
$$5 \overset{\text{no}}{<} 5 \qquad\qquad -3 \overset{\checkmark}{<} 5$$

So our solution is reasonably checked.

Checkpoint 2.6.6. Solve the inequality $x - 5 > -4$. State the solution set using interval notation and using set-builder notation.

In interval notation, the solution set is $\boxed{}$.

In set-builder notation, the solution set is $\boxed{}$.

Explanation. To solve this inequality, we will add 5 to each side.

$$x - 5 > -4$$
$$x - 5 + 5 > -4 + 5$$
$$x > 1$$

Note again that the direction of the inequality did not change, since we did not multiply or divide each side of the inequality by a negative number at any point.

Graphically, we represent this solution set as:

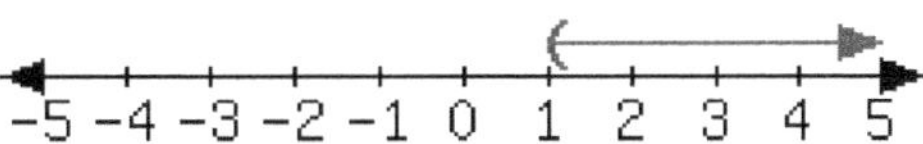

Using interval notation, we write the solution set as $(1, \infty)$. Using set-builder notation, we write the solution set as $\{x \mid x > 1\}$.

We should check that 1 is *not* a solution, but that some number greater than 1 *is* a solution.

$$x - 5 > -4 \qquad x - 5 > -4$$
$$1 - 5 \overset{?}{<} -4 \qquad 10 - 5 \overset{?}{<} -4$$
$$-4 \overset{\text{no}}{<} -4 \qquad 5 \overset{\checkmark}{<} -4$$

So our solution is reasonably checked.

Checkpoint 2.6.7. Solve the inequality $-\frac{1}{2}z \geq -1.74$. State the solution set using interval notation and using set-builder notation.

In interval notation, the solution set is $\boxed{}$.

In set-builder notation, the solution set is $\boxed{}$.

Explanation.   To solve this inequality, we will multiply by −2 to each side.

$$-\frac{1}{2}z \geq -1.74$$

$$(-2)\left(-\frac{1}{2}z\right) \leq (-2)(-1.74)$$

$$z \leq 3.48$$

In this exercise, we *did* multiply by a negative number and so the direction of the inequality sign changed. Graphically, we represent this solution set as:

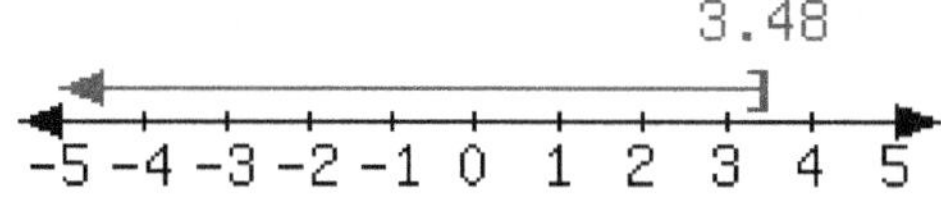

Using interval notation, we write the solution set as $(-\infty, 3.48]$. Using set-builder notation, we write the solution set as $\{z \mid z \leq 3.48\}$.

We should check that 3.48 *is* a solution, and also that some number less than 3.48 is a solution.

$$-\frac{1}{2}z \geq -1.74 \qquad -\frac{1}{2}z \geq -1.74$$

$$-\frac{1}{2}(3.48) \stackrel{?}{\geq} -1.74 \qquad -\frac{1}{2}(0) \stackrel{?}{\geq} -1.74$$

$$-1.74 \stackrel{\checkmark}{\geq} -1.74 \qquad 0 \stackrel{\checkmark}{\geq} -1.74$$

So our solution is reasonably checked.

Exercises

Review and Warmup

1. Add the following.

   a. $-10 + (-2) =$ ▢

   b. $-5 + (-3) =$ ▢

   c. $-3 + (-10) =$ ▢

2. Add the following.

   a. $-10 + (-3) =$ ▢

   b. $-7 + (-4) =$ ▢

   c. $-3 + (-8) =$ ▢

3. Add the following.

   a. $2 + (-7) =$ ▢

   b. $8 + (-3) =$ ▢

   c. $3 + (-3) =$ ▢

4. Add the following.

   a. $2 + (-8) =$ ▢

   b. $10 + (-4) =$ ▢

   c. $3 + (-3) =$ ▢

5. Add the following.

   a. $-10 + 3 =$ ▢

   b. $-1 + 7 =$ ▢

   c. $-8 + 8 =$ ▢

6. Add the following.

   a. $-7 + 3 =$ ▢

   b. $-2 + 9 =$ ▢

   c. $-8 + 8 =$ ▢

7. Evaluate the following.

a. $\dfrac{-20}{-4} = \boxed{}$

b. $\dfrac{28}{-7} = \boxed{}$

c. $\dfrac{-12}{6} = \boxed{}$

8. Evaluate the following.

a. $\dfrac{-6}{-3} = \boxed{}$

b. $\dfrac{50}{-5} = \boxed{}$

c. $\dfrac{-42}{6} = \boxed{}$

9. Do the following multiplications:

a. $16 \cdot \dfrac{5}{8} = \boxed{}$

b. $24 \cdot \dfrac{5}{8} = \boxed{}$

c. $32 \cdot \dfrac{5}{8} = \boxed{}$

10. Do the following multiplications:

a. $15 \cdot \dfrac{2}{5} = \boxed{}$

b. $20 \cdot \dfrac{2}{5} = \boxed{}$

c. $25 \cdot \dfrac{2}{5} = \boxed{}$

11. Evaluate the following.

a. $\dfrac{-9}{-1} = \boxed{}$

b. $\dfrac{10}{-1} = \boxed{}$

c. $\dfrac{190}{-190} = \boxed{}$

d. $\dfrac{-19}{-19} = \boxed{}$

e. $\dfrac{5}{0} = \boxed{}$

f. $\dfrac{0}{-10} = \boxed{}$

12. Evaluate the following.

a. $\dfrac{-8}{-1} = \boxed{}$

b. $\dfrac{8}{-1} = \boxed{}$

c. $\dfrac{130}{-130} = \boxed{}$

d. $\dfrac{-11}{-11} = \boxed{}$

e. $\dfrac{5}{0} = \boxed{}$

f. $\dfrac{0}{-5} = \boxed{}$

Solving One-Step Inequalities using Addition/Subtraction Solve this inequality.

13. $x + 2 > 7$

In set-builder notation, the solution set is $\boxed{}$.

In interval notation, the solution set is $\boxed{}$.

14. $x + 3 > 10$

In set-builder notation, the solution set is $\boxed{}$.

In interval notation, the solution set is $\boxed{}$.

15. $x - 3 \leq 8$

In set-builder notation, the solution set is $\boxed{}$.

In interval notation, the solution set is $\boxed{}$.

16. $x - 4 \leq 7$

In set-builder notation, the solution set is $\boxed{}$.

In interval notation, the solution set is $\boxed{}$.

17. $4 \leq x + 10$

In set-builder notation, the solution set is $\boxed{}$.

In interval notation, the solution set is $\boxed{}$.

18. $5 \leq x + 8$

In set-builder notation, the solution set is $\boxed{}$.

In interval notation, the solution set is $\boxed{}$.

19. $1 > x - 7$

In set-builder notation, the solution set is ☐.

In interval notation, the solution set is ☐.

20. $1 > x - 10$

In set-builder notation, the solution set is ☐.

In interval notation, the solution set is ☐.

Solving One-Step Inequalities using Multiplication/Division Solve this inequality.

21. $2x \leq 6$

In set-builder notation, the solution set is ☐.

In interval notation, the solution set is ☐.

22. $3x \leq 6$

In set-builder notation, the solution set is ☐.

In interval notation, the solution set is ☐.

23. $9x > 5$

In set-builder notation, the solution set is ☐.

In interval notation, the solution set is ☐.

24. $1x > 6$

In set-builder notation, the solution set is ☐.

In interval notation, the solution set is ☐.

25. $-4x \geq 8$

In set-builder notation, the solution set is ☐.

In interval notation, the solution set is ☐.

26. $-5x \geq 20$

In set-builder notation, the solution set is ☐.

In interval notation, the solution set is ☐.

27. $15 \geq -5x$

In set-builder notation, the solution set is ☐.

In interval notation, the solution set is ☐.

28. $4 \geq -2x$

In set-builder notation, the solution set is ☐.

In interval notation, the solution set is ☐.

29. $3 < -x$

In set-builder notation, the solution set is ☐.

In interval notation, the solution set is ☐.

30. $4 < -x$

In set-builder notation, the solution set is ☐.

In interval notation, the solution set is ☐.

31. $-x \leq 5$

In set-builder notation, the solution set is ☐.

In interval notation, the solution set is ☐.

32. $-x \leq 6$

In set-builder notation, the solution set is ☐.

In interval notation, the solution set is ☐.

33. $\dfrac{6}{5}x > 6$

    In set-builder notation, the solution set is ☐.

    In interval notation, the solution set is ☐.

34. $\dfrac{7}{10}x > 14$

    In set-builder notation, the solution set is ☐.

    In interval notation, the solution set is ☐.

35. $-\dfrac{8}{7}x \leq 32$

    In set-builder notation, the solution set is ☐.

    In interval notation, the solution set is ☐.

36. $-\dfrac{9}{4}x \leq 18$

    In set-builder notation, the solution set is ☐.

    In interval notation, the solution set is ☐.

37. $-3 < \dfrac{1}{10}x$

    In set-builder notation, the solution set is ☐.

    In interval notation, the solution set is ☐.

38. $-2 < \dfrac{2}{7}x$

    In set-builder notation, the solution set is ☐.

    In interval notation, the solution set is ☐.

39. $-9 < -\dfrac{3}{4}x$

    In set-builder notation, the solution set is ☐.

    In interval notation, the solution set is ☐.

40. $-8 < -\dfrac{4}{9}x$

    In set-builder notation, the solution set is ☐.

    In interval notation, the solution set is ☐.

41. $3x > -9$

    In set-builder notation, the solution set is ☐.

    In interval notation, the solution set is ☐.

42. $4x > -8$

    In set-builder notation, the solution set is ☐.

    In interval notation, the solution set is ☐.

43. $-16 < -4x$

    In set-builder notation, the solution set is ☐.

    In interval notation, the solution set is ☐.

44. $-15 < -5x$

    In set-builder notation, the solution set is ☐.

    In interval notation, the solution set is ☐.

45. $\dfrac{3}{10} \geq \dfrac{x}{20}$

    In set-builder notation, the solution set is ☐.

    In interval notation, the solution set is ☐.

46. $\dfrac{9}{2} \geq \dfrac{x}{8}$

    In set-builder notation, the solution set is ☐.

    In interval notation, the solution set is ☐.

47. $-\dfrac{z}{12} < -\dfrac{5}{2}$

In set-builder notation, the solution set is ☐.

In interval notation, the solution set is ☐.

48. $-\dfrac{z}{12} < -\dfrac{3}{4}$

In set-builder notation, the solution set is ☐.

In interval notation, the solution set is ☐.

Challenge

49. Choose the correct inequality or equal sign to make the relation true.

a. Let x and y be integers, such that $x < y$.

Then $x - y$ (☐ <   ☐ >   ☐ =) $y - x$.

b. Let x and y be integers, such that $1 < x < y$.

Then xy (☐ <   ☐ >   ☐ =) $x + y$.

c. Let x and y be rational numbers, such that $0 < x < y < 1$.

Then xy (☐ <   ☐ >   ☐ =) $x + y$.

d. Let x and y be integers, such that $x < y$.

Then $x + 2y$ (☐ <   ☐ >   ☐ =) $2x + y$.

2.7 Percentages

Percent-related problems arise in everyday life. This section reviews some basic caluclations that can be made with percentages.

In many situations when translating from English to math, the word "of" translates as multiplication. Also the word "is" (and many similar words related to "to be") translates to an equals sign. For example:

$$\text{One third of thirty is ten.}$$

$$\frac{1}{3} \cdot 30 = 10$$

Here is another example, this time involving a percentage. We know that "2 is 50% of 4," so we can say:

$$2 \text{ is } 50\% \text{ of } 4$$

$$2 = 0.5 \cdot 4$$

Example 2.7.2 Translate each statement involving percents below into an equation. Define any variables used. (Solving these equations is an exercise).

   a. How much is 30% of $24.00?

   b. $7.20 is what percent of $24.00?

   c. $7.20 is 30% of how much money?

Explanation.   Each question can be translated from English into a math equation by reading it slowly and looking for the right signals.

   a. The word "is" means about the same thing as the equals sign. "How much" is a question phrase, and we can let x be the unknown amount (in dollars). The word "of" translates to multiplication, as discussed earlier. So we have:

$$\begin{array}{ccccc} \text{how much} & \text{is} & 30\% & \text{of} & \$24 \\ | & | & | & | \; | \\ x & & = 0.30 & \cdot & 24 \end{array}$$

   b. Let P be the unknown value. We have:

$$\begin{array}{ccccc} \$7.20 & \text{is} & \text{what percent} & \text{of} & \$24 \\ | & | & | & | \; | \\ 7.2 & = & P & \cdot & 24 \end{array}$$

   With this setup, P is going to be a decimal value (0.30) that you would translate into a percentage (30%).

   c. Let x be the unknown amount (in dollars). We have:

$$\begin{array}{ccccc} \$7.20 & \text{is} & 30\% & \text{of} & \text{how much} \\ | & | & | & | & | \\ 7.2 & = & 0.30 & \cdot & x \end{array}$$

Checkpoint 2.7.3. Solve each equation from Example 2.7.2.

a. How much is 30% of \$24.00?

b. \$7.20 is what percent of \$24.00?

c. \$7.20 is 30% of how much money?

$$x = 0.3 \cdot 24$$

$$7.2 = P \cdot 24$$

$$7.2 = 0.3 \cdot x$$

x is [   ].

P is [   ].

x is [   ].

Explanation.

a. $x = 0.3 \cdot 24$
$$x = 8$$

b. $7.2 = P \cdot 24$
$$\frac{7.2}{24} = \frac{P \cdot 24}{24}$$
$$0.3 = P$$

c. $7.2 = 0.3 \cdot x$
$$\frac{7.2}{0.3} = \frac{0.3 \cdot x}{0.3}$$
$$24 = x$$

2.7.1 Setting up and Solving Percent Equations

An important skill for solving percent-related problems is to boil down a complicated word problem into a simple form like "2 is 50% of 4." Let's look at some further examples.

Example 2.7.4

In Fall 2016, Portland Community College had 89,900 enrolled students. According to Figure 2.7.5, how many black students were enrolled at PCC in Fall 2016?

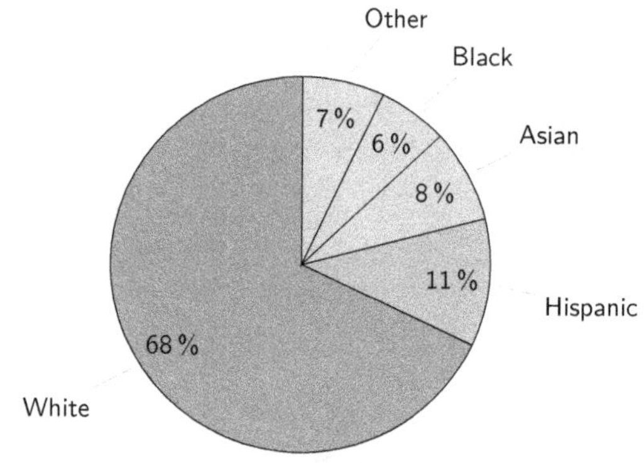

Figure 2.7.5: Racial breakdown of PCC students in Fall 2016

Explanation. After reading this word problem and the chart, we can translate the problem into "what is 6% of 89,900?" Let x be the number of black students enrolled at PCC in Fall 2016. We can set up and solve the equation:

$$\begin{array}{ccccc} \text{what} & \text{is} & 6\% & \text{of} & 89{,}900 \\ | & | & | & | & | \\ x & = & 0.06 & \cdot & 89900 \end{array}$$

$$x = 5394$$

There was not much "solving" to do, since the variable we wanted to isolate was already isolated.

As of Fall 2016, Portland Community College had 5394 black students. Note: this is not likely to be perfectly accurate, because the numbers we started with (89,900 enrolled students and 6%) appear to be rounded.

Example 2.7.6

The bar graph in Figure 2.7.7 displays how many students are in each class at a local high school. According to the bar graph, what percentage of the school's student population is freshman?

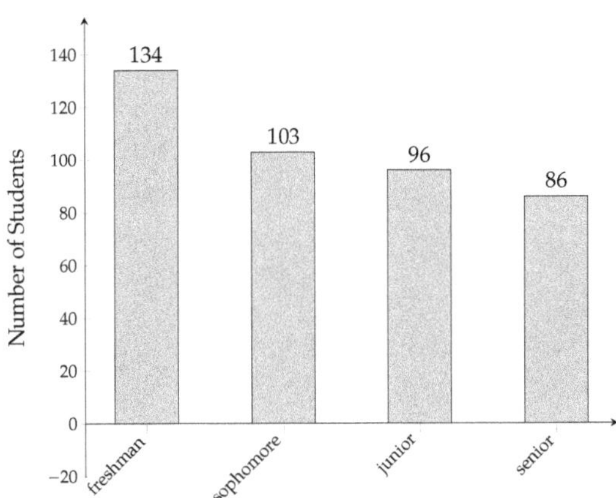

Figure 2.7.7: Number of students at a high school by class

Explanation. The school's total number of students is:

$$134 + 103 + 96 + 86 = 419$$

With that calculated, we can translate the main question:

"What percentage of the school's student population is freshman?"

into:

"What percent of 419 is 134?"

Using P to represent the unknown quantity, we write and solve the equation:

$$\underset{P}{\text{what percent}} \underset{}{\text{of}} \cdot \underset{419}{419} \underset{\text{is}}{=} \underset{134}{134}$$

$$\frac{P \cdot 419}{419} = \frac{134}{419}$$

$$P \approx 0.3198$$

$$P \approx 31.98\%$$

Approximately 31.98% of the school's student population is freshman.

Remark 2.7.8. When solving equations that do *not* have context we state the solution set. However, when solving an equation or inequality that arises in an application problem (such as the context of the high school in Example 2.7.6), it makes more sense to summarize our result with a sentence, using the context of the application. This allows us to communicate the full result, including appropriate units.

Example 2.7.9 Carlos just received his monthly paycheck. His gross pay (the amount before taxes and related things are deducted) was \$2,346.19, and his total tax and other deductions was \$350.21. The rest was deposited directly into his checking account. What percent of his gross pay went into his checking account?

Explanation. Train yourself to read the word problem and not try to pick out numbers to substitute into formulas. You may find it helps to read the problem over to yourself three or more times before you attempt to solve it. There are *three* dollar amounts to discuss in this problem, and many students fall into a trap of using the wrong values in the wrong places. There is the gross pay, the amount that was deducted, and the amount that was deposited. Only two of these have been explicitly written down. We need to use subtraction to find the dollar amount that was deposited:

$$2346.19 - 350.21 = 1995.98$$

Now, we can translate the main question:

"What percent of his gross pay went into his checking account?"

into:

"What percent of \$2346.19 is \$1995.98?"

Using P to represent the unknown quantity, we write and solve the equation:

$$\underset{P}{\text{what percent}} \underset{}{\text{of}} \underset{\cdot 2346.19}{\$2346.19} \underset{=}{\text{is}} \underset{1995.98}{\$1995.98}$$

$$\frac{P \cdot 2346.19}{2346.19} = \frac{1995.98}{2346.19}$$

$$P \approx 0.8507$$

$$P \approx 85.07\%$$

Approximately 85.07% of his gross pay went into his checking account.

Checkpoint 2.7.10. Alexis sells cars for a living, and earns 28% of the dealership's sales profit as commission. In a certain month, she plans to earn \$2200 in commissions. How much total sales profit does she need to bring in for the dealership?

Alexis needs to bring in [ ] in sales profit.

Explanation. Be careful that you do not calculate 28% of \$2200. That might be what a student would do who doesn't thoroughly read the question. If you have ever trained yourself to quickly find numbers in word problems and substitute them into formulas, you must *unlearn* this. The issue is that \$2200 is not the dealership's sales profit, and if you mistakenly multiply $0.28 \cdot 2200 = 616$, then \$616 makes no sense as an answer to this question. How could Alexis bring in only \$616 of sales profit, and be rewarded with \$2200 in commission?

We can translate the problem into "\$2200 is 28% of what?" Letting x be the sales profit for the dealership (in dollars), we can write and solve the equation:

$$\overset{\$2200}{\big|} \ \overset{\text{is}}{\big|} \ \overset{28\%}{\big|} \ \overset{\text{of}}{\big|} \ \overset{\text{what}}{\big|}$$

$$2200 = 0.28 \cdot x$$

$$\frac{2200}{0.28} = \frac{0.28x}{0.28}$$

$$7857.14 \approx x$$

$$x \approx 7857.14$$

To earn \$2200 in commission, Alexis needs to bring in approximately \$7857.14 of sales profit for the dealership.

Example 2.7.11

According to e-Literate, the average cost of a new college textbook has been increasing. Find the percentage of increase from 2009 to 2013.

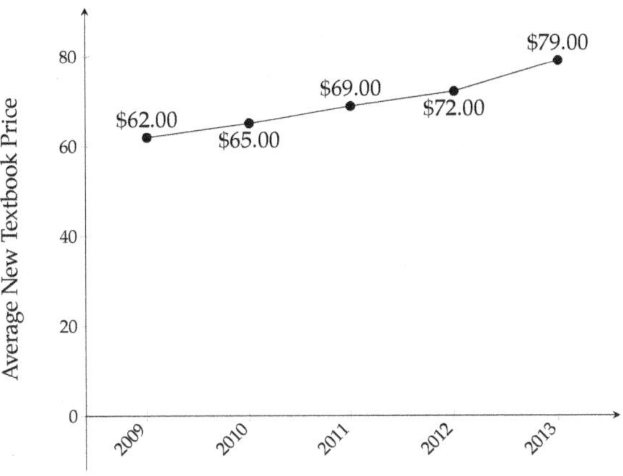

Figure 2.7.12: Average New Textbook Price from 2009 to 2013

Explanation. The actual amount of increase from 2009 to 2013 was $79 - 62 = 17$, dollars. We need to answer the question "\$17 is what percent of \$62?" Note that we are comparing the 17 to 62, not to 79. In these situations where one amount is the earlier amount, the earlier original amount is the one that represents 100%. Let P represent the percent of increase. We can set up and solve the equation:

$$\overset{\$17}{\big|} \ \overset{\text{is}}{\big|} \ \overset{\text{what percent}}{\big|} \ \overset{\text{of}}{\big|} \ \overset{\$62}{\big|}$$

$$17 = P \cdot 62$$

$$17 = 62P$$

$$\frac{17}{62} = \frac{62P}{62}$$

$$0.2742 \approx P$$

From 2009 to 2013, the average cost of a new textbook increased by approximately 27.42%.

Checkpoint 2.7.13. Last month, a full tank of gas for a car you drive cost you \$40.00. You hear on the news that gas prices have risen by 12%. By how much, in dollars, has the cost of a full tank gone up?

A full tank of gas now costs [   ] more than it did last month.

Explanation. Let x represent the amount of increase. We can set up and solve the equation:

$$
\begin{array}{ccccc}
12\% & \text{of} & \text{old cost} & \text{is} & \text{how much} \\
| & | & | & | & | \\
0.12 \cdot & & 40 & = & x
\end{array}
$$

$$4.8 = x$$

A full tank now costs \$4.80 more than it did last month.

Example 2.7.14 Enrollment at your neighborhood's elementary school two years ago was 417 children. After a 15% increase last year and a 15% decrease this year, what's the new enrollment?

Explanation. It is tempting to think that increasing by 15% and then decreasing by 15% would bring the enrollment right back to where it started. But the 15% decrease applies to the enrollment *after* it had already increased. So that 15% decrease is going to translate to *more* students lost than were gained.

Using 100% as corresponding to the enrollment from two years ago, the enrollment last year was 100% + 15% = 115% of that. But then using 100% as corresponding to the enrollment from last year, the enrollment this year was 100% − 15% = 85% of that. So we can set up and solve the equation

$$
\begin{array}{cccccccc}
\text{this year's enrollment} & \text{is} & 85\% & \text{of} & 115\% & \text{of} & \text{enrollment two years ago} \\
| & | & | & | & | & | & | \\
x & = & 0.85 \cdot & 1.15 \cdot & & & 417
\end{array}
$$

$$x = 0.85 \cdot 1.15 \cdot 417$$

$$x = 407.6175$$

We would round and report that enrollment is now 408 students. (The percentage rise and fall of 15% were probably rounded in the first place, which is why we did not end up with a whole number.)

Exercises

Review and Warmup

1. Change the following percentages into decimals:

17% = [   ]

54% = [   ]

2. Change the following percentages into decimals:

18% = [   ]

51% = [   ]

3. Convert the following decimals into percentages:

0.29 = [   ]

0.67 = [   ]

4. Convert the following decimals into percentages:

0.21 = [   ]

0.64 = [   ]

5. Change the following percentages into decimals:

3% = [____]

30% = [____]

100% = [____]

300% = [____]

6. Change the following percentages into decimals:

4% = [____]

40% = [____]

100% = [____]

400% = [____]

7. Convert the following decimals into percentages:

0.05 = [____]

0.5 = [____]

5 = [____]

1 = [____]

8. Convert the following decimals into percentages:

0.06 = [____]

0.6 = [____]

6 = [____]

1 = [____]

9. Convert the following decimals into percentages:

6.67 = [____]

0.667 = [____]

0.0667 = [____]

10. Convert the following decimals into percentages:

7.22 = [____]

0.722 = [____]

0.0722 = [____]

11. Change the following percentages into decimals:

895% = [____]

89.5% = [____]

8.95% = [____]

12. Change the following percentages into decimals:

959% = [____]

95.9% = [____]

9.59% = [____]

Basic Percentage Calculation

13. 3% of 200 is [____].

14. 8% of 300 is [____].

15. 60% of 400 is [____].

16. 30% of 490 is [____].

17. 780% of 590 is [____].

18. 530% of 690 is [____].

19. Answer with a percent.

176 is [____] of 220.

20. Answer with a percent.

720 is [____] of 800.

21. Answer with a percent.

142.1 is [____] of 49.

22. Answer with a percent.

21.6 is [____] of 18.

23. Answer with a percent.

12 is about [____] of 47.

24. Answer with a percent.

8 is about [____] of 44.

25. 14% of [____] is 68.6.

26. 72% of [____] is 424.8.

27. 4% of [____] is 27.6.

28. 9% of [____] is 71.1.

29. 420% of [____] is 3738.

30. 250% of [____] is 2450.

Applications

31. A town has 1400 registered residents. Among them, 39% were Democrats, 38% were Republicans. The rest were Independents. How many registered Independents live in this town?

There are [____] registered Independent residents in this town.

32. A town has 1800 registered residents. Among them, 36% were Democrats, 27% were Republicans. The rest were Independents. How many registered Independents live in this town?

There are [____] registered Independent residents in this town.

33. Martha is paying a dinner bill of $29.00. Martha plans to pay 12% in tips. How much tip will Martha pay?

Martha will pay [____] in tip.

34. Evan is paying a dinner bill of $33.00. Evan plans to pay 20% in tips. How much tip will Evan pay?

Evan will pay [____] in tip.

35. Ashley is paying a dinner bill of $36.00. Ashley plans to pay 16% in tips. How much in total (including bill and tip) will Ashley pay?

Ashley will pay [____] in total (including bill and tip).

36. Rita is paying a dinner bill of $40.00. Rita plans to pay 12% in tips. How much in total (including bill and tip) will Rita pay?

Rita will pay [____] in total (including bill and tip).

37. A watch's wholesale price was $440.00. The retailer marked up the price by 40%. What's the watch's new price (markup price)?

The watch's markup price is [____].

38. A watch's wholesale price was $460.00. The retailer marked up the price by 30%. What's the watch's new price (markup price)?

The watch's markup price is [____].

39. In the past few seasons' basketball games, Timothy attempted 170 free throws, and made 153 of them. What percent of free throws did Timothy make?

Timothy made [____] of free throws in the past few seasons.

40. In the past few seasons' basketball games, Nina attempted 430 free throws, and made 86 of them. What percent of free throws did Nina make?

Nina made [____] of free throws in the past few seasons.

41. A painting is on sale at $360.00. Its original price was $450.00. What percentage is this off its original price?

The painting was [____] off its original price.

42. A painting is on sale at $450.00. Its original price was $500.00. What percentage is this off its original price?

The painting was [____] off its original price.

43. The pie chart represents a collector's collection of signatures from various artists.

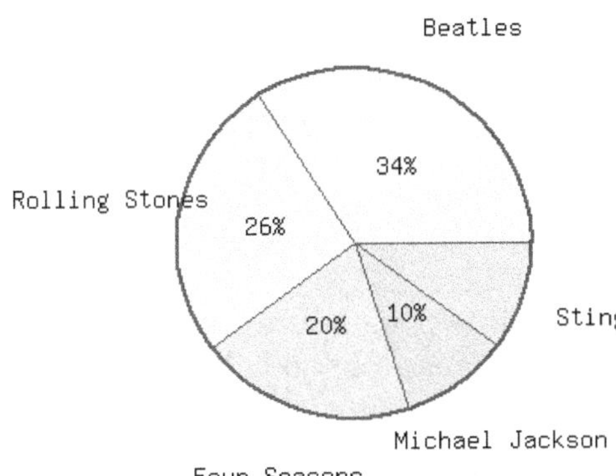

If the collector has a total of 1050 signatures, there are ____________ signatures by Sting.

44. The pie chart represents a collector's collection of signatures from various artists.

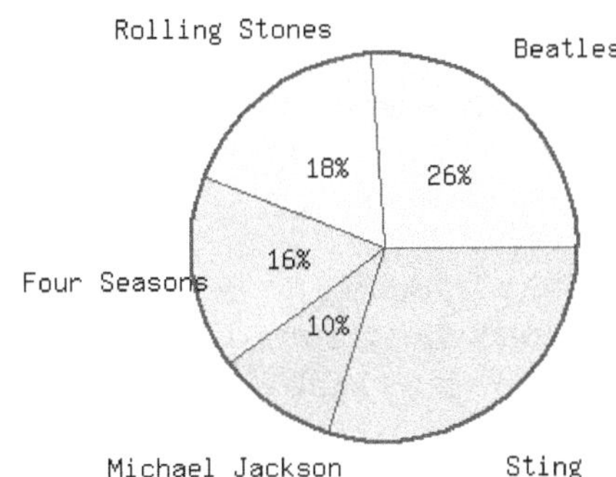

If the collector has a total of 1250 signatures, there are ____________ signatures by Sting.

45. In the last election, 34% of a county's residents, or 12240 people, turned out to vote. How many residents live in this county?

This county has ____________ residents.

46. In the last election, 59% of a county's residents, or 23836 people, turned out to vote. How many residents live in this county?

This county has ____________ residents.

47. 70.68 grams of pure alcohol was used to produce a bottle of 18.6% alcohol solution. What is the weight of the solution in grams?

The alcohol solution weighs ____________.

48. 43.6 grams of pure alcohol was used to produce a bottle of 10.9% alcohol solution. What is the weight of the solution in grams?

The alcohol solution weighs ____________.

49. Connor paid a dinner and left 17%, or $3.74, in tips. How much was the original bill (without counting the tip)?

The original bill (not including the tip) was ____.

50. Sydney paid a dinner and left 13%, or $3.38, in tips. How much was the original bill (without counting the tip)?

The original bill (not including the tip) was ____.

51. Rebecca sells cars for a living. Each month, she earns $2,000.00 of base pay, plus a certain percentage of commission from her sales.

One month, Rebecca made $52,500.00 in sales, and earned a total of $3,186.50 in that month (including base pay and commission). What percent commission did Rebecca earn?

Rebecca earned ______________ in commission.

52. Ryan sells cars for a living. Each month, he earns $2,000.00 of base pay, plus a certain percentage of commission from his sales.

One month, Ryan made $57,000.00 in sales, and earned a total of $4,644.80 in that month (including base pay and commission). What percent commission did Ryan earn?

Ryan earned ______________ in commission.

53. The following is a nutrition fact label from a certain macaroni and cheese box.

Nutrition Facts	
Serving Size 1 cup	
Servings Per Container 2	
Amount Per Serving	
Calories 300	Calories from Fat 110
	% Daily Value
Total Fat 7.8 g	12%
Saturated Fat 2 g	15%
Trans Fat 2 g	
Cholesterol 30 mg	10%
Sodium 400 mg	20%
Total Carbohydrate 30 g	11%
Dietary Fiber 0 g	0%
Sugars 5 g	
Protein 5 g	
Vitamin A 2 mg	3%
Vitamin C 2 mg	2.5%
Calcium 2 mg	20%
Iron 3 mg	4%

The highlighted row means each serving of macaroni and cheese in this box contains 7.8 g of fat, which is 12% of an average person's daily intake of fat. What's the recommended daily intake of fat for an average person?

The recommended daily intake of fat for an average person is ______________.

54. The following is a nutrition fact label from a certain macaroni and cheese box.

Nutrition Facts	
Serving Size 1 cup	
Servings Per Container 2	
Amount Per Serving	
Calories 300	Calories from Fat 110
	% Daily Value
Total Fat 14 g	20%
Saturated Fat 2 g	15%
Trans Fat 2 g	
Cholesterol 30 mg	10%
Sodium 400 mg	20%
Total Carbohydrate 30 g	11%
Dietary Fiber 0 g	0%
Sugars 5 g	
Protein 5 g	
Vitamin A 2 mg	3%
Vitamin C 2 mg	2.5%
Calcium 2 mg	20%
Iron 3 mg	4%

The highlighted row means each serving of macaroni and cheese in this box contains 14 g of fat, which is 20% of an average person's daily intake of fat. What's the recommended daily intake of fat for an average person?

The recommended daily intake of fat for an average person is ______________.

55. Sydney earned $278.07 of interest from a mutual fund, which was 0.69% of his total investment. How much money did Sydney invest into this mutual fund?

Sydney invested ______________ in this mutual fund.

56. Shane earned $102.81 of interest from a mutual fund, which was 0.23% of his total investment. How much money did Shane invest into this mutual fund?

Shane invested ______________ in this mutual fund.

57. A town has 5000 registered residents. Among them, there are 1950 Democrats and 1750 Republicans. The rest are Independents. What percentage of registered voters in this town are Independents?

In this town, [ ] of all registered voters are Independents.

58. A town has 1300 registered residents. Among them, there are 468 Democrats and 299 Republicans. The rest are Independents. What percentage of registered voters in this town are Independents?

In this town, [ ] of all registered voters are Independents.

59. A community college conducted a survey about the number of students riding each bus line available. The following bar graph is the result of the survey.

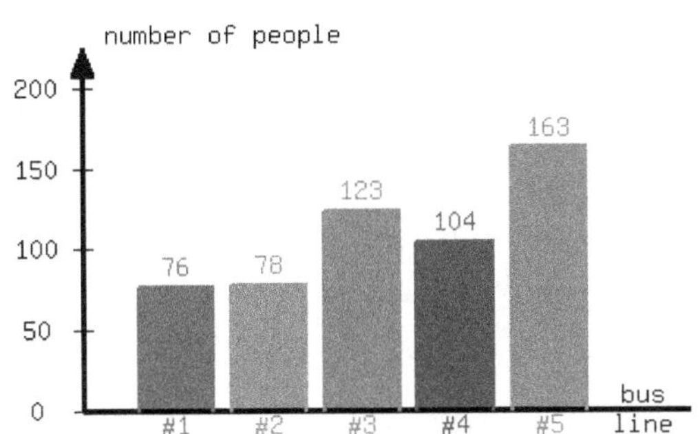

What percent of students ride Bus #1?

Approximately [ ] of students ride Bus #1.

60. A community college conducted a survey about the number of students riding each bus line available. The following bar graph is the result of the survey.

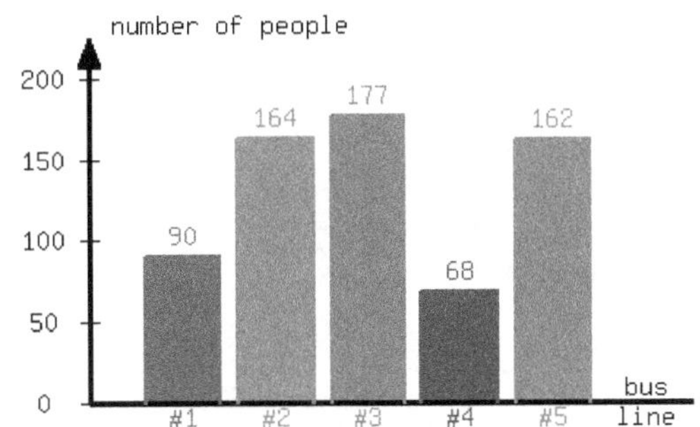

What percent of students ride Bus #1?

Approximately [ ] of students ride Bus #1.

Percent Increase/Decrease

61. The population of cats in a shelter decreased from 100 to 75. What is the percentage decrease of the shelter's cat population?

The percentage decrease is [ ].

62. The population of cats in a shelter decreased from 120 to 108. What is the percentage decrease of the shelter's cat population?

The percentage decrease is [ ].

63. The population of cats in a shelter increased from 55 to 73. What is the percentage increase of the shelter's cat population?

The percentage increase is approximately [ ].

64. The population of cats in a shelter increased from 63 to 75. What is the percentage increase of the shelter's cat population?

The percentage increase is approximately [ ].

65. Last year, a small town's population was 760. This year, the population decreased to 753. What is the percentage decrease?

The percentage decrease of the town's population was approximately ⬚.

66. Last year, a small town's population was 800. This year, the population decreased to 798. What is the percentage decrease?

The percentage decrease of the town's population was approximately ⬚.

67. Your salary used to be $39,000 per year.

You had to take a 2% pay cut. After the cut, your salary was ⬚ per year.

Then, you earned a 2% raise. After the raise, your salary was ⬚ per year.

68. Your salary used to be $31,000 per year.

You had to take a 2% pay cut. After the cut, your salary was ⬚ per year.

Then, you earned a 2% raise. After the raise, your salary was ⬚ per year.

69. This line graph shows a certain stock's price change over a few days.

From 11/1 to 11/5, what is the stock price's percentage change?

From 11/1 to 11/5, the stock price's percentage change was approximately ⬚.

70. This line graph shows a certain stock's price change over a few days.

From 11/1 to 11/5, what is the stock price's percentage change?

From 11/1 to 11/5, the stock price's percentage change was approximately ⬚.

71. A house was bought two years ago at the price of $110,000. Each year, the house's value decreased by 5%. What's the house's value this year?

The house's value this year is ⬚.

72. A house was bought two years ago at the price of $380,000. Each year, the house's value decreased by 6%. What's the house's value this year?

The house's value this year is ⬚.

2.8 Modeling with Equations and Inequalities

One purpose of learning math is to be able to model real-life situations and then use the model to ask and answer questions about the situation. In this lesson, we will examine the basics of modeling to set up an equation (or inequality).

2.8.1 Setting Up Equations for Rate Models

To set up an equation modeling a real world scenario, the first thing we need to do is identify what variable we will use. The variable we use will be determined by whatever is unknown in our problem statement. Once we've identified and defined our variable, we'll use the numerical information provided to set up our equation.

Example 2.8.2 A savings account starts with \$500. Each month, an automatic deposit of \$150 is made. Write an equation that represents the number of months it will take for the balance to reach \$1,700.

Explanation.

To determine this equation, we might start by making a table in order to identify a general pattern for the total amount in the account after m months.

Using this pattern, we can determine that an equation showing the unknown number of months, m, when the total savings equals \$1700 would look like this:

$$500 + 150m = 1700$$

Months Since Saving Started	Total Amount Saved (in Dollars)
0	500
1	$500 + 150 = 650$
2	$500 + 150(2) = 800$
3	$500 + 150(3) = 950$
4	$500 + 150(4) = 1100$
$\vdots$	$\vdots$
m	$500 + 150m$

Table 2.8.3: Amount in Savings Account

Remark 2.8.4. To determine the solution to the equation in Example 2.8.2, we could continue the pattern in Table 2.8.3:

We can see that the value of m that makes the equation true is 8 as $500 + 150(8) = 1700$. Thus it would take 8 months for an account starting with \$500 to reach \$1,700 if \$150 is saved each month.

Months Since Saving Started	Total Amount Saved (in Dollars)
5	$500 + 150(5) = 1250$
6	$500 + 150(6) = 1400$
7	$500 + 150(7) = 1550$
8	$500 + 150(8) = 1700$

Table 2.8.5: Amount in Savings Account

Here we are able to determine the solution by creating a table and using inputs that were whole numbers. Often the solution will not be something we can find this way. We will need to solve the equation using algebra, as we'll learn how to do in later sections. For this section, we'll only focus on setting up the equation.

Example 2.8.6 A bathtub contains 2.5 ft^3 of water. More water is being poured in at a rate of 1.75 ft^3 per minute. Write an equation representing when the amount of water in the bathtub will reach 6.25 ft^3.

Explanation.

Since this problem refers to *when* the amount of water will reach a certain amount, we immediately know that the unknown quantity is time. As the volume of water in the tub is measured in ft^3 per minute, we know that time needs to be measured in minutes. We'll define t to be the number of minutes that water is poured into the tub. To determine this equation, we'll start by making a table of values:

Minutes Water Has Been Poured	Total Amount of Water (in ft^3)
0	2.5
1	$2.5 + 1.75 = 4.25$
2	$2.5 + 1.75(2) = 6$
3	$2.5 + 1.75(3) = 7.75$
$\vdots$	$\vdots$
t	$2.5 + 1.75t$

Table 2.8.7: Amount of Water in the Bathtub

Using this pattern, we can determine that the equation representing when the amount will be 6.25 ft^3 is:

$$2.5 + 1.75t = 6.25$$

2.8.2 Setting Up Equations for Percent Problems

Section 2.7 reviewed some basics of working with percentages, and even solved some one-step equations that were set up using percentages. Here we look at some scenarios where there is an equation to set up based on percentages, but it is slightly more involved than a one-step equation.

Example 2.8.8 Jakobi's annual salary as a nurse in Portland, Oregon, is \$73,290. His salary increased by 4% from last year. Write a linear equation modeling this scenario, where the unknown value is Jakobi's salary last year.

Explanation. We need to know Jakobi's salary last year. So we'll introduce s, defined to be Jakobi's salary last year (in dollars). To set up the equation, we need to think about how he arrived at this year's salary. To get to this year's salary, his employer took last year's salary and added 4% to it. Conceptually, this means we have:

$$(\text{last year's salary}) + (4\% \text{ of last year's salary}) = (\text{this year's salary})$$

We'll represent 4% of last year's salary with $0.04s$ since 0.04 is the decimal representation of 4%. This means that the equation we set up is:

$$s + 0.04s = 73290$$

Checkpoint 2.8.9. Kirima offered to pay the bill and tip at a restaurant where she and her freinds had dinner. In total she paid \$150, which made the tip come out to a little more than 19%. We'd like to know what was the bill before tip. Set up an equation for this situation.

Explanation. A common mistake is to translate a question like this into "what is 19% of \$150?" as a way to calculate the tip amount, and then subtract that from \$150. But that is not how tipping works. The tip percentage is applied to the original bill, not the final total. If we let x represent the original bill, then:

$$\begin{array}{ccccccc} \text{bill} & \text{plus} & 19\% & \text{of} & \text{bill} & \text{is} & \$150 \\ x & + & 0.19 \cdot & & x & = & 150 \end{array}$$

Example 2.8.10 The price of a refrigerator after a 15% discount is $612. Write a linear equation modeling this scenario, where the original price of the refrigerator (before the discount was applied) is the unknown quantity.

Explanation. We'll let c be the original price of the refrigerator. To obtain the discounted price, we take the original price and subtract 15% of that amount. Conceptually, this looks like:

$$(\text{original price}) - (15\% \text{ of the original price}) = (\text{discounted price})$$

Since the amount of the discount is 15% of the original price, we'll represent this with $0.15c$. The equation we set up is then:

$$c - 0.15c = 612$$

Checkpoint 2.8.11. A shirt is on sale at 20% off. The current price is $51.00. Write an equation based on this scenario where the variable represents the shirt's original price.

Explanation. Let x represent the original price of the shirt. Since 20% is removed to bring the cost to $51, we can set up the equation:

$$\begin{array}{ccccccc} \text{original} & \text{minus} & 20\% & \text{of} & \text{original} & \text{is} & \$51 \\ x & - & 0.20 \cdot & & x & = & 51 \end{array}$$

2.8.3 Setting Up Equations for Geometry Problems

With geometry problems and algebra, there is often the possibility to draw some picture to help understand the scenario better. Additionally it is often necessary to rely on some formula from geometry, such as the formulas from Subsection 2.2.1.

Example 2.8.12 An Olympic-size swimming pool is rectangular and 50 m in length. We don't know its width, but we do know that it required 150 m of painter's tape to outline the edge of the pool during recent renovations. Use this information to set up an equation that models the width of the pool.

Explanation.

Since the pool's shape is a rectangle, it helps to sketch a rectangle representing the pool as in Figure 2.8.13. Since we know its length is 50 m, it is a good idea to label that in the sketch. The width is our unknown quantity, so we can use w as a variable to represent the pool's width in meters and label that too.

Figure 2.8.13: An Olympic-size pool

145

Since it required 150 m of painter's tape to outline the pool, we know the perimeter of the pool is 150 m. This suggests using the perimeter formula for a rectangle: $P = 2(\ell + w)$. (This formula was discussed in Subsection 2.2.1).

With this formula, we can substitute 150 in for P and 50 in for ℓ:

$$150 = 2(50 + w)$$

and this equation models the width of the pool.

Checkpoint 2.8.14. One sail on a sail boat is approximately shaped like a triangle. If the base length is 10 feet and the total sail area is 125 square feet, we can wonder how tall is the sail. Set up an equation to model the sail's height.

Explanation. Since the sail's shape is (approximately) a triangle, it helps to sketch a triangle representing the sail. Since we know its base width is 10 feet, it is a good idea to label that in the sketch. The heigth is our unknown quantity, so we can use h as a variable to represent the sail's height in feet and label that too.

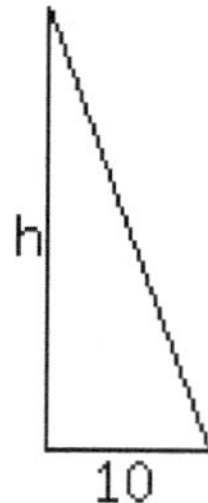

Since the total area is known to be 125 square feet, this suggests using the area formula for a triangle: $A = \frac{1}{2}bh$.

With this formula, we can substitute 125 in for A and 100 in for b:

$$125 = \frac{1}{2}(10)h$$

and this equation models the height of the pool.

2.8.4 Setting Up Inequalities for Models

In general, we'll model using inequalities when we want to determine a maximum or minimum value. To identify that an inequality is needed instead of an equality, we'll look for phrases like *at least*, *at most*, *at a minimum* or *at a maximum*.

Example 2.8.15 The car share company car2go has a one-time registration fee of $5 and charges $14.99 per hour for use of their vehicles. Hana wants to use car2go and has a maximum budget of $300. Write a linear inequality representing this scenario, where the unknown quantity is the number of hours she uses their vehicles.

Explanation. We'll let h be the number of hours that Hana uses car2go. We need the initial cost and the cost from the hourly charge to be less than or equal to $300, which we set up as:

$$5 + 14.99h \leq 300$$

Example 2.8.16 When an oil tank is decommissioned, it is drained of its remaining oil and then re-filled with an inert material, such as sand. A cylindrical oil tank has a volume of 275 gal and is being filled with sand at a rate of 700 gal per hour. Write a linear inequality representing this scenario, where the time it takes for the tank to overflow with sand is the unknown quantity.

Explanation. The unknown in this scenario is time, so we'll define t to be the number of hours that sand is poured into the tank. (Note that we chose hours based on the rate at which the sand is being poured.) We'll represent the amount of sand poured in as $700t$ as each hour an additional 700 gal are added. Given that we want to know when this amount exceeds 275 gal, we set this equation up as:

$$700t > 275$$

2.8.5 Translating Phrases into Mathematical Expressions and Equations/Inequalities

The following table shows how to translate common phrases into mathematical expressions:

English Phrases	Math Expressions
the sum of 2 and a number	$x + 2$ or $2 + x$
2 more than a number	$x + 2$ or $2 + x$
a number increased by 2	$x + 2$ or $2 + x$
a number and 2 together	$x + 2$ or $2 + x$
the difference between a number and 2	$x - 2$
the difference of 2 and a number	$2 - x$
2 less than a number	$x - 2$ (*not* $2 - x$)
a number decreased by 2	$x - 2$
2 decreased by a number	$2 - x$
2 subtracted from a number	$x - 2$
a number subtracted from 2	$2 - x$
the product of 2 and a number	$2x$
twice a number	$2x$
a number times 2	$x \cdot 2$ or $2x$
two thirds of a number	$\frac{2}{3}x$
25% of a number	$0.25x$
the quotient of a number and 2	$x/2$
the quotient of 2 and a number	$2/x$
the ratio of a number and 2	$x/2$
the ratio of 2 and a number	$2/x$

Table 2.8.17: Translating English Phrases into Math Expressions

We can extend this to setting up equations and inequalities. Let's look at some examples. The key is to break a complicated phrase or sentence into smaller parts, identifying key vocabulary such as "is," "of," "greater than," "at most," etc.

English Sentences	Math Equations and Inequalities
The sum of 2 and a number is 6.	$x + 2 = 6$
2 less than a number is at least 6.	$x - 2 \geq 6$
Twice a number is at most 6.	$2x \leq 6$
6 is the quotient of a number and 2.	$6 = \frac{x}{2}$
4 less than twice a number is greater than 10.	$2x - 4 > 10$
Twice the difference between 4 and a number is 10.	$2(4 - x) = 10$
The product of 2 and the sum of 3 and a number is less than 10.	$2(x + 3) < 10$
The product of 2 and a number, subtracted from 5, yields 8.	$5 - 2x = 8$
Two thirds of a number subtracted from 10 is 2.	$10 - \frac{2}{3}x = 2$
25% of the sum of 7 and a number is 2.	$0.25(x + 7) = 2$

Table 2.8.18: Translating English Sentences into Math Equations

Exercises

Review and Warmup

1. Identify a variable you might use to represent each quantity. And identify what units would be most appropriate.

    a. Let ___ be the area of a house, measured in ________.

    b. Let ___ be the age of a dog, measured in ________.

    c. Let ___ be the amount of time passed since a driver left Seattle, Washington, bound for Portland, Oregon, measured in ________.

2. Identify a variable you might use to represent each quantity. And identify what units would be most appropriate.

    a. Let ___ be the age of a person, measured in ________.

    b. Let ___ be the distance traveled by a driver that left Portland, Oregon, bound for Boise, Idaho, measured in ________.

    c. Let ___ be the surface area of the walls of a room, measured in ________.

Modeling with Linear Equations

3. Chris's annual salary as a radiography technician is $39,858.00. His salary increased by 2.2% from last year. What was his salary last year?

Assume his salary last year was s dollars. Write an equation to model this scenario. There is no need to solve it.

4. Sherial's annual salary as a radiography technician is $42,630.00. Her salary increased by 1.5% from last year. What was her salary last year?

Assume her salary last year was s dollars. Write an equation to model this scenario. There is no need to solve it.

5. A bicycle for sale costs $194.04, which includes 7.8% sales tax. What was the cost before sales tax?

Assume the bicycle's price before sales tax is p dollars. Write an equation to model this scenario. There is no need to solve it.

6. A bicycle for sale costs $224.07, which includes 6.7% sales tax. What was the cost before sales tax?

Assume the bicycle's price before sales tax is p dollars. Write an equation to model this scenario. There is no need to solve it.

7. The price of a washing machine after 10% discount is $216.00. What was the original price of the washing machine (before the discount was applied)?

Assume the washing machine's price before the discount is p dollars. Write an equation to model this scenario. There is no need to solve it.

8. The price of a washing machine after 30% discount is $189.00. What was the original price of the washing machine (before the discount was applied)?

Assume the washing machine's price before the discount is p dollars. Write an equation to model this scenario. There is no need to solve it.

9. The price of a restaurant bill, including an 15% gratuity charge, was $115.00. What was the price of the bill before gratuity was added?

Assume the bill without gratuity is b dollars. Write an equation to model this scenario. There is no need to solve it.

10. The price of a restaurant bill, including an 11% gratuity charge, was $11.10. What was the price of the bill before gratuity was added?

Assume the bill without gratuity is b dollars. Write an equation to model this scenario. There is no need to solve it.

11. In May 2016, the median rent price for a one-bedroom apartment in a city was reported to be $908.10 per month. This was reported to be an increase of 0.9% over the previous month. Based on this reporting, what was the median price of a one-bedroom apartment in April 2016?

Assume the median price of a one-bedroom apartment in April 2016 was p dollars. Write an equation to model this scenario. There is no need to solve it.

12. In May 2016, the median rent price for a one-bedroom apartment in a city was reported to be $1,006.00 per month. This was reported to be an increase of 0.6% over the previous month. Based on this reporting, what was the median price of a one-bedroom apartment in April 2016?

Assume the median price of a one-bedroom apartment in April 2016 was p dollars. Write an equation to model this scenario. There is no need to solve it.

13. Izabelle is driving an average of 42 miles per hour, and she is 58.8 miles away from home. After how many hours will she reach his home?

    Assume Izabelle will reach home after h hours. Write an equation to model this scenario. There is no need to solve it.

14. Blake is driving an average of 46 miles per hour, and he is 156.4 miles away from home. After how many hours will he reach his home?

    Assume Blake will reach home after h hours. Write an equation to model this scenario. There is no need to solve it.

15. Uhaul charges an initial fee of \$32.65 and then \$0.68 per mile to rent a 15-foot truck for a day. If the total bill is \$116.29, how many miles were driven?

    Assume m miles were driven. Write an equation to model this scenario. There is no need to solve it.

16. Uhaul charges an initial fee of \$34.85 and then \$0.53 per mile to rent a 15-foot truck for a day. If the total bill is \$132.90, how many miles were driven?

    Assume m miles were driven. Write an equation to model this scenario. There is no need to solve it.

17. A cat litter box has a rectangular base that is 24 inches by 24 inches. What will the height of the cat litter be if 4 cubic feet of cat litter is poured? (Hint: $1 \text{ ft}^3 = 1728 \text{ in}^3$)

    Assume h inches will be the height of the cat litter if 4 cubic feet of cat litter is poured. Write an equation to model this scenario. There is no need to solve it.

18. A cat litter box has a rectangular base that is 24 inches by 18 inches. What will the height of the cat litter be if 6 cubic feet of cat litter is poured? (Hint: $1 \text{ ft}^3 = 1728 \text{ in}^3$)

    Assume h inches will be the height of the cat litter if 6 cubic feet of cat litter is poured. Write an equation to model this scenario. There is no need to solve it.

Modeling with Linear Inequalities

19. A truck that hauls water is capable of carrying a maximum of 1500 lb. Water weighs $8.3454\frac{\text{lb}}{\text{gal}}$, and the plastic tank on the truck that holds water weighs 53 lb. Assume the truck can carry a maximum of g gallons of water. Write an *inequality* to model this scenario. There is no need to solve it.

20. A truck that hauls water is capable of carrying a maximum of 2600 lb. Water weighs $8.3454\frac{\text{lb}}{\text{gal}}$, and the plastic tank on the truck that holds water weighs 59 lb. Assume the truck can carry a maximum of g gallons of water. Write an *inequality* to model this scenario. There is no need to solve it.

21. Grant's maximum lung capacity is 5.2 liters. If his lungs are full and he exhales at a rate of 0.8 liters per second, write an *inequality* that models when he will still have at least 0.4 liters of air left in his lungs. There is no need to solve it.

22. Izabelle's maximum lung capacity is 5.6 liters. If her lungs are full and she exhales at a rate of 0.8 liters per second, write an *inequality* that models when she will still have at least 2.8 liters of air left in his lungs. There is no need to solve it.

23. A swimming pool is being filled with water from a garden hose at a rate of 8 gallons per minute. If the pool already contains 60 gallons of water and can hold up to 276 gallons, set up an *inequality* modeling how much time can pass without the pool overflowing. There is no need to solve it.

24. A swimming pool is being filled with water from a garden hose at a rate of 6 gallons per minute. If the pool already contains 70 gallons of water and can hold up to 154 gallons, set up an *inequality* modeling how much time can pass without the pool overflowing. There is no need to solve it.

25. An engineer is designing a cylindrical spring-form pan (the kind of pan a cheesecake is baked in). The pan needs to be able to hold a volume at least 398 cubic inches and have a diameter of 13 inches. Write an *inequality* modeling possible height of the pan. There is no need to solve it.

26. An engineer is designing a cylindrical spring-form pan (the kind of pan a cheesecake is baked in). The pan needs to be able to hold a volume at least 338 cubic inches and have a diameter of 14 inches. Write an *inequality* modeling possible height of the pan. There is no need to solve it.

Translating English Phrases into Math Expressions and Equations Translate the following phrase or sentence into a math expression or equation (whichever is appropriate).

27. three more than a number

28. ten less than a number

29. the sum of a number and six

30. the difference between a number and three

31. the difference between nine and a number

32. the difference between six and a number

33. two subtracted from a number

34. nine added to a number

35. five decreased by a number

36. two increased by a number

37. a number decreased by eight

38. a number increased by five

39. two times a number, increased by five

40. eight times a number, decreased by ten

41. five less than four times a number

42. one less than eight times a number

43. eight more than the quotient of three and a number

44. four less than the quotient of seven and a number

45. Two times a number is sixteen.

46. Seven times a number is twenty-eight.

47. The sum of fifty-six and a number is seventy-three.

48. The difference between thirty-three and a number is twenty-eight.

49. The quotient of a number and seventeen is fourteen seventeenths.

50. The quotient of a number and twenty-five is one twenty-fifth.

51. The quotient of twenty-six and a number is thirteen twenty-fifths.

52. The quotient of eighteen and a number is nine twenty-thirds.

53. The sum of three times a number and eleven is fifty-six.

54. The sum of eight times a number and twenty-three is thirty-one.

55. One less than five times a number yields fifty-nine.

56. Two less than three times a number yields ninety-one.

57. The product of seven and a number, increased by eight, yields ninety-nine.

58. The product of five and a number, added to four, yields 234.

59. The product of three and a number increased by seven, yields 123.

60. The product of seven and a number added to three, yields 168.

61. one sixth of a number

62. one half of a number

63. twenty-three thirty-eighths of a number

64. thirteen forty-firsts of a number

65. a number decreased by one eleventh of itself

66. a number decreased by seven thirtieths of itself

67. A number increased by two ninths is one ninth of that number.

68. A number decreased by one sixth is one eighth of that number.

69. One more than the product of three elevenths and a number yields three tenths of that number.

70. Five more than the product of one fifth and a number gives two sevenths of that number.

Challenge

71. Last year, Joan received a 2.5% raise. This year, she received a 4% raise. Her current wage is $11.46 an hour. What was her wage before the two raises?

2.9 Introduction to Exponent Rules

In this section, we will look at some rules or properties we use when simplifying expressions that involve multiplication and exponents.

2.9.1 Exponent Basics

Before we discuss any exponent rules, we need to quickly remind ourselves of some important concepts and vocabulary.

When working with expressions with exponents, we have the following vocabulary:

$$\text{base}^{\text{exponent}} = \text{power}$$

For example, when we calculate $8^2 = 64$, the **base** is 8, the **exponent** is 2, and the expression 8^2 is called the 2nd **power** of 8.

The other foundational concept is that if an exponent is a positive integer, the power can be rewritten as repeated multiplication of the base. For example, the 4th power of 3 can be written as 4 factors of 3 like so:

$$3^4 = 3 \cdot 3 \cdot 3 \cdot 3$$

2.9.2 Exponent Rules

Product Rule   If we write out $3^5 \cdot 3^2$ without using exponents, we'd have:

$$3^5 \cdot 3^2 = (3 \cdot 3 \cdot 3 \cdot 3 \cdot 3) \cdot (3 \cdot 3)$$

If we then count how many 3s are being multiplied together, we find we have $5 + 2 = 7$, a total of seven 3s.

$$3^5 \cdot 3^2 = 3^{5+2}$$
$$= 3^7$$

Example 2.9.2 Simplify $x^2 \cdot x^3$.

To simplify $x^2 \cdot x^3$, we write this out in its expanded form, as a product of x's, we have

$$x^2 \cdot x^3 = (x \cdot x)(x \cdot x \cdot x)$$
$$= x \cdot x \cdot x \cdot x \cdot x$$
$$= x^5$$

Note that we obtained the exponent of 5 by adding 2 and 3.

This is our first rule, the **Product Rule**: when multiplying two expressions that have the same base, we can simplify the product by adding the exponents.

$$x^m \cdot x^n = x^{m+n} \tag{2.9.1}$$

Checkpoint 2.9.3. Use the properties of exponents to simplify the expression.

$x^{19} \cdot x^{13}$

Explanation. We *add* the exponents as follows:

$$x^{19} \cdot x^{13} = x^{19+13}$$
$$= x^{32}$$

Power to a Power Rule   The second rule is an extension of the first rule. If we write out $\left(3^5\right)^2$ without using exponents, we'd have 3^5 multiplied by itself:

$$\left(3^5\right)^2 = \left(3^5\right) \cdot \left(3^5\right)$$
$$= (3 \cdot 3 \cdot 3 \cdot 3 \cdot 3) \cdot (3 \cdot 3 \cdot 3 \cdot 3 \cdot 3)$$

If we again count how many 3s are being multiplied, we have a total of two groups each with five 3s. So we'd have $2 \cdot 5 = 10$ instances of a 3.

$$\left(3^5\right)^2 = 3^{2 \cdot 5}$$
$$= 3^{10}$$

Example 2.9.4 Simplify $\left(x^2\right)^3$.

To simplify $\left(x^2\right)^3$, we write this out in its expanded form, as a product of x's, we have

$$\left(x^2\right)^3 = \left(x^2\right) \cdot \left(x^2\right) \cdot \left(x^2\right)$$
$$= (x \cdot x) \cdot (x \cdot x) \cdot (x \cdot x)$$
$$= x^6$$

Note that we obtained the exponent of 6 by multiplying 2 and 3.

We have our second rule, the **Power to a Power Rule**: when a base is raised to an exponent and that expression is raised to another exponent, we multiply the exponents.

$$\left(x^m\right)^n = x^{m \cdot n}$$

Checkpoint 2.9.5. Use the properties of exponents to simplify the expression.

$\left(r^2\right)^5$

Explanation. We *multiply* the exponents as follows:

$$\left(r^2\right)^5 = r^{2 \cdot 5}$$
$$= r^{10}$$

Product to a Power Rule   The third exponent rule deals with having multiplication inside a set of parentheses and an exponent outside the parentheses. If we write out $(3t)^5$ without using an exponent, we'd have $3t$ multiplied by itself five times:

$$(3t)^5 = (3t)(3t)(3t)(3t)(3t)$$

Keeping in mind that there is multiplication between every 3 and t and multiplication between all of the parentheses, we can reorder and regroup the factors:

$$\begin{aligned}
(3t)^5 &= (3 \cdot t) \cdot (3 \cdot t) \cdot (3 \cdot t) \cdot (3 \cdot t) \cdot (3 \cdot t) \\
&= (3 \cdot 3 \cdot 3 \cdot 3 \cdot 3) \cdot (t \cdot t \cdot t \cdot t \cdot t) \\
&= 3^5 t^5
\end{aligned}$$

We essentially applied the outer exponent to each factor inside the parentheses.

Example 2.9.6 Simplify $(xy)^5$.

To simplify $(xy)^5$, we write this out in its expanded form, as a product of x's and y's, we have

$$\begin{aligned}
(xy)^5 &= (x \cdot y) \cdot (x \cdot y) \cdot (x \cdot y) \cdot (x \cdot y) \cdot (x \cdot y) \\
&= (x \cdot x \cdot x \cdot x \cdot x) \cdot (y \cdot y \cdot y \cdot y \cdot y) \\
&= x^5 y^5
\end{aligned}$$

Note that the exponent on xy can simply be applied to both x and y.

This is our third rule, the **Product to a Power Rule**: when a product is raised to an exponent, we can apply the exponent to each factor in the product.

$$(x \cdot y)^n = x^n \cdot y^n$$

Checkpoint 2.9.7. Use the properties of exponents to simplify the expression.

$(5x)^2$

Explanation. We *multiply* the exponents and apply the rule $(ab)^m = a^m \cdot b^m$ as follows:

$$\begin{aligned}
(5x)^2 &= (5)^2 x^2 \\
&= 25x^2
\end{aligned}$$

If a and b are real numbers, and n and m are positive integers, then we have the following rules:

Product Rule $a^n \cdot a^m = a^{n+m}$

Power to a Power Rule $(a^n)^m = a^{n \cdot m}$

Product to a Power Rule $(ab)^n = a^n \cdot b^n$

List 2.9.8: Summary of the Rules of Exponents for Multiplication

Many examples we'll come across will make use of more than one exponent rule. In deciding which exponent rule to work with first, it's important to remember that the order of operations still applies.

Example 2.9.9 Simplify the following expressions.

a. $\left(3^7 r^5\right)^4$

b. $\left(t^3\right)^2 \cdot \left(t^4\right)^5$

Explanation.

a. Since we cannot simplify anything inside the parentheses, we'll begin simplifying this expression using the Product to a Power Rule. We'll apply the outer exponent of 4 to each factor inside the parentheses. Then we'll use the Power to a Power Rule to finish out simplification process:

$$\left(3^7 r^5\right)^4 = \left(3^7\right)^4 \cdot \left(r^5\right)^4$$
$$= 3^{7\cdot 4} \cdot r^{5\cdot 4}$$
$$= 3^{28} r^{20}$$

b. According to the order of operations, we should first simplify any exponents before carrying out any multiplication. Therefore, we'll begin simplifying this by applying the Power to a Power Rule and then finish using the Product Rule:

$$\left(t^3\right)^2 \cdot \left(t^4\right)^5 = t^{3\cdot 2} \cdot t^{4\cdot 5}$$
$$= t^6 \cdot t^{20}$$
$$= t^{6+20}$$
$$= t^{26}$$

Remark 2.9.10. We cannot simplify an expression like $x^2 y^3$ using the Product Rule, as the factors x^2 and y^3 do not have the same base.

Exercises

Review and Warmup   Evaluate the following.

1. a. $3^2 =$
    b. $4^3 =$
    c. $(-4)^2 =$
    d. $(-5)^3 =$

2. a. $3^2 =$
    b. $2^3 =$
    c. $(-2)^2 =$
    d. $(-3)^3 =$

3. a. $1^7 =$
    b. $(-1)^{15} =$
    c. $(-1)^{14} =$
    d. $0^{19} =$

4. a. $1^8 =$
    b. $(-1)^{13} =$
    c. $(-1)^{16} =$
    d. $0^{17} =$

5. a. $(-3)^2 =$
    b. $-10^2 =$

6. a. $(-1)^2 =$
    b. $-2^2 =$

7. a. $(-1)^3 =$
    b. $-2^3 =$

8. a. $(-4)^3 =$
    b. $-1^3 =$

Exponent Rules   Use the properties of exponents to simplify the expression.

9. $3 \cdot 3^9$

10. $4 \cdot 4^6$

11. $5^2 \cdot 5^7$

12. $5^8 \cdot 5^2$

13. $r^{13} \cdot r^9$

14. $t^{15} \cdot t^3$

15. $y^{17} \cdot y^{15} \cdot y^8$

16. $x^{19} \cdot x^8 \cdot x^{16}$

17. $\left(2^2\right)^3$

18. $\left(4^8\right)^7$

19. $\left(x^4\right)^5$

20. $\left(y^6\right)^{12}$

21. $(4t)^3$

22. $(3y)^3$

23. $(5rx)^4$

24. $(4xt)^4$

25. $\left(4r^{12}\right)^2$

26. $\left(5y^2\right)^4$

27. $(5r^4) \cdot (10r^{12})$

28. $(9t^7) \cdot (-4t^5)$

29. $\left(-\dfrac{x^9}{3}\right) \cdot \left(\dfrac{x^{18}}{5}\right)$

30. $\left(-\dfrac{x^{11}}{6}\right) \cdot \left(-\dfrac{x^{11}}{4}\right)$

31. $\left(-10t^8\right)^2$

32. $\left(-7r^9\right)^3$

Use the properties of exponents to simplify the expression.

33. $(-3y^{17}) \cdot (-9y^{10}) \cdot (-y^9)$

34. $(-2r^{19}) \cdot (-5r^4) \cdot (4r^4)$

35.   a. $\left(-2x^2\right)^2 = $ ⬚

     b. $-\left(2x^2\right)^2 = $ ⬚

36.   a. $\left(-6r^3\right)^2 = $ ⬚

     b. $-\left(6r^3\right)^2 = $ ⬚

Challenge

37.   a. Let $x^7 \cdot x^a = x^{17}$. Let's say that a is a natural number. How many possibilities are there for a?

     b. Let $x^b \cdot x^c = x^{75}$. Let's say that b and c are natural numbers. How many possibilities are there for b?

     c. Let $x^d \cdot x^e = x^{800}$. Let's say that d and e are natural numbers. How many possibilities are there for d?

38. Choose the correct inequality or equal sign to make the relation true.

    2^{500} ($\square <$ $\square >$ $\square =$) 5^{200}

2.10 Simplifying Expressions

We know that if we have two apples and add three more, then our result is the same as if we'd had three apples and added two more. In this section, we'll formally define and extend these basic properties we know about numbers to variable expressions.

2.10.1 Identities and Inverses

We will start with some definitions. The number 0 is called the **additive identity**. If the sum of two numbers is the additive identity, 0, these two numbers are called **additive inverses**. For example, 2 is the additive inverse of -2, and the additive inverse of -2 is 2.

Similarly, the number 1 is called the **multiplicative identity**. If the product of two numbers is the multiplicative identity, 1, these two numbers are called **multiplicative inverses**. For example, 2 is the multiplicative inverse of $\frac{1}{2}$, and the multiplicative inverse of $-\frac{2}{3}$ is $-\frac{3}{2}$. The multiplicative inverse is also called **reciprocal**.

2.10.2 Introduction to Algebraic Properties

Commutative Property   When we compute the area of a rectangle, we generally multiply the length by the width. Does the result change if we multiply the width by the length?

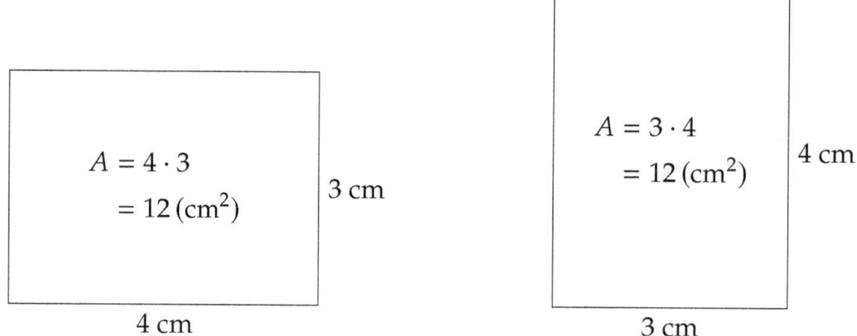

Figure 2.10.2: Horizontal and Vertical Rectangles

We can see $3 \cdot 2 = 2 \cdot 3$. If we denote the length of a rectangle with ℓ and the width with w, this implies $\ell w = w\ell$. This is referred to as the **commutative property of multiplication**. The commutative property also applies to addition, as in $1+2 = 2+1$, where it is called the **commutative property of addition**. However, there is no commutative property of subtraction or division, as $2 - 1 \neq 1 - 2$, and $\frac{4}{2} \neq \frac{2}{4}$.

Associative Property   Let's extend that example to a rectangular prism with width $w = 4\,\text{cm}$, depth $d = 3\,\text{cm}$, and height $h = 2\,\text{cm}$. To compute the volume of this solid, we multiply the width, depth and height, which we write as wdh.

In the following figure, on the left side, we multiply the length and width first, and then multiply the height; on the right side, we multiply the width and height first, and then multiply the length. Let's compare the products.

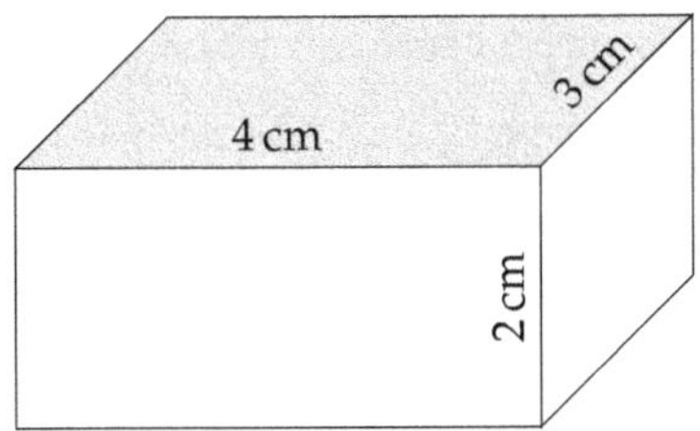

Figure 2.10.3: $(4 \cdot 3) \cdot 2 = 24$

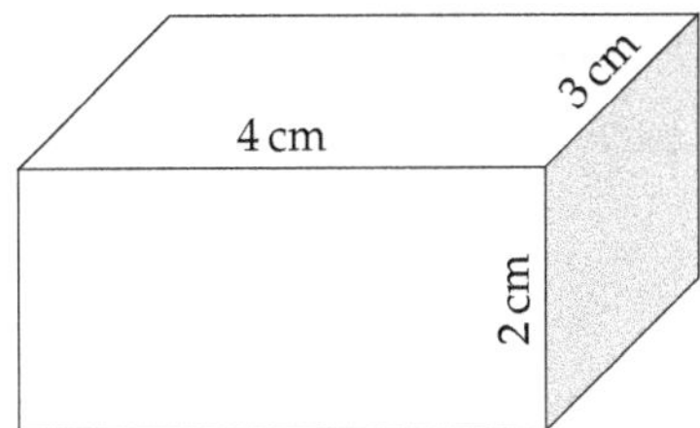

Figure 2.10.4: $4 \cdot (3 \cdot 2) = 24$

We can see $(wd)h = w(dh)$. This is known as the **associative property of multiplication**. The associative property also applies to addition, as in $(1 + 2) + 3 = 1 + (2 + 3)$, which is called the **associative property of addition**. However, there is no associative property of subtraction, as $(3 - 2) - 1 \neq 3 - (2 - 1)$.

Distributive Property   The final property we'll explore is called the **distributive property**, which involves both multiplication and addition. To conceptualize this property, let's consider what happens if we buy 3 boxes that each contain one apple and one pear. This will have the same total cost as if we'd bought 3 apples and 3 pears. We write this algebraically:

$$3(a + p) = 3a + 3p.$$

Visually, we can see that it's just a means of re-grouping: $3(\text{🍎} + \text{🍐}) = 3(\text{🍎}) + 3(\text{🍐})$.

2.10.3 Summary of Algebraic Properties

Let a, b, and c represent real numbers, variables, or algebraic expressions. Then the following properties hold:

Commutative Property of Multiplication $a \cdot b = b \cdot a$

Associative Property of Multiplication $a \cdot (b \cdot c) = (a \cdot b) \cdot c$

Commutative Property of Addition $a + b = b + a$

Associative Property of Addition $a + (b + c) = (a + b) + c$

Distributive Property $a(b + c) = ab + ac$

List 2.10.5: Algebraic Properties

Let's practice these properties in the following exercises.

Checkpoint 2.10.6.

a. Use the commutative property of multiplication to write an equivalent expression to $53m$.

b. Use the associative property of multiplication to write an equivalent expression to $3(5n)$.

c. Use the commutative property of addition to write an equivalent expression to $q + 84$.

   d. Use the associative property of addition to write an equivalent expression to $x + (20 + c)$.

   e. Use the distributive property to write an equivalent expression to $3(r + 7)$ that has no grouping symbols.

Explanation.

   a. To use the commutative property of multiplication, we change the order in which two factors are multiplied:

$$53m$$
$$= m \cdot 53.$$

   b. To use the associative property of multiplication, we leave factors written in their original order, but change the grouping symbols so that a different multiplication has higher priority:

$$3(5n)$$
$$= (3 \cdot 5)n.$$

You may further simplify by carrying out the multiplication between the two numbers:

$$3(5n)$$
$$= (3 \cdot 5)n$$
$$= 15n.$$

   c. To use the commutative property of addition, we change the order in which two terms are added:

$$q + 84$$
$$= 84 + q.$$

   d. To use the associative property of addition, we leave terms written in their original order, but change the grouping symbols so that a different addition has higher priority:

$$x + (20 + c)$$
$$= (x + 20) + c.$$

   e. To use the distributive property, we multiply the number outside the parentheses, 3, with each term inside the parentheses:

$$3(r + 7)$$
$$= 3 \cdot r + 3 \cdot 7$$
$$= 3r + 21.$$

2.10.4 Applying the Commutative, Associative, and Distributive Properties

Like Terms One of the main ways that we will use the commutative, associative, and distributive properties is to simplify expressions. In order to do this, we need to recognize **like terms**, as discussed in Section 2.3. We combine like terms when we take an expression lilke $2a + 3a$ and write the result as $5a$. The formal process actually involves using the distributive property:

$$2a + 3a = (2 + 3)a$$
$$= 5a$$

In practice, however, it's more helpful to think of this as having 2 of an object and then an additional 3 of that same object. In total, we then have 5 of that object.

Example 2.10.7 Where possible, simplify the following expressions by combining like terms.

a. $6c + 12c - 5c$ b. $-5q^2 - 3q^2$ c. $x - 5y + 4x$ d. $2x - 3y + 4z$

Explanation.

a. All three terms are like terms, so they may combined. We combine them two at a time:

$$6c + 12c - 5c = 18c - 5c$$
$$= 13c$$

b. The two terms $-5q^2$ and $-3q^2$ are like terms, so we may combine them:

$$-5q^2 - 3q^2 = -8q^2$$

c. The two terms x and $4x$ are like terms, while the other term is different. Using the associative and commutative properties of addition in the first step allows us to place the two like terms next to each other, and then combine them:

$$x - 5y + 4x = x + 4x + (-5y)$$
$$= 1x + 4x + (-5y)$$
$$= 5x - 5y$$

Note the expression x is the same as $1x$. Usually we don't write the "1" as it is implied. However, it's helpful when combining like terms to remember that $x = 1x$. (Similarly, $-x$ is equal to $-1x$, which can be helpful when combining $-x$ with like terms.)

d. The expression $2x - 3y + 4z$ cannot be simplified as there are no like terms.

Adding Expressions When we add an expression like $4x - 5$ to an expression like $3x - 7$, we write them as follows:

$$(4x - 5) + (3x - 7)$$

In order to remove the given sets of parentheses and apply the commutative property of addition, we will rewrite the subtraction operation as "adding the opposite":

$$4x + (-5) + 3x + (-7)$$

At this point we can apply the commutative property of addition and then combine like terms. Here's how the entire problem will look:

$$(4x - 5) + (3x - 7) = 4x + (-5) + 3x + (-7)$$
$$= 4x + 3x + (-5) + (-7)$$

$$= 7x + (-12)$$
$$= 7x - 12$$

Remark 2.10.8. Once we become more comfortable simplifying such expressions, we will simply write this kind of simplification in one step:

$$(4x - 5) + (3x - 7) = 7x - 12$$

Example 2.10.9 Use the associative, commutative, and distributive properties to simplify the following expressions as much as possible.

a. $(2x + 3) + (4x + 5)$
b. $(-5x + 3) + (4x - 7)$

Explanation.

a. We will remove parentheses, and then combine like terms:

$$(2x + 3) + (4x + 5) = 2x + 3 + 4x + 5$$
$$= 2x + 4x + 3 + 5$$
$$= 6x + 8$$

b. We will remove parentheses, and then combine like terms:

$$(-5x + 3) + (4x - 7) = -5x + 3 + 4x + (-7)$$
$$= -x + (-4)$$
$$= -x - 4$$

Applying the Distributive Property with Negative Coefficients Applying the distributive property in an expression such as $2(3x + 4)$ is fairly straightforward, in that this becomes $2(3x) + 2(4)$ which we then simplify to $6x + 8$. Applying the distributive property is a little trickier when subtraction or a negative constant is involved, for example, with the expression $2(3x - 4)$. Recalling that subtraction is defined as "adding the opposite," we can change the subtraction of positive 4 to the addition of negative 4:

$$2\big(3x + (-4)\big)$$

Now when we distribute, we obtain:

$$2(3x) + 2(-4)$$

As a final step, we see that this simplifies to:

$$6x - 8$$

Remark 2.10.10. We can also extend the distributive property to use subtraction, and state that $a(b - c) = ab - ac$. With this property, we would simplify $2(3x - 4)$ more efficiently:

$$2(3x - 4) = 2(3x) - 2(4)$$
$$= 6x - 8$$

In general, we will use this approach.

Example 2.10.11 Apply the distributive property to each expression and simplify it as much as possible.

a. $-3(5x + 7)$

b. $2(-4x - 1)$

Explanation.

a. We will distribute -3 to the $5x$ and 7:

$$-3(5x + 7) = -3(5x) + (-3)(7)$$
$$= -15x - 21$$

b. We will distribute 2 to the $-4x$ and -1:

$$2(-4x - 1) = 2(-4x) - 2(1)$$
$$= -8x - 2$$

Checkpoint 2.10.12. Use the distributive property to write an equivalent expression to $-4(x - 2)$ that has no grouping symbols.

Explanation. To use the distributive property, we multiply the number outside the parentheses, -4, with each term inside the parentheses:

$$-4(x - 2) = -4 \cdot x - 4(-2)$$
$$= -4x + 8$$

Subtracting Expressions To subtract one expression from another expression, such as $(5x + 9) - (3x + 2)$, we will again rely on the fact that subtraction is defined as "adding the opposite." To add the *opposite* of an expression, we will technically distribute a constant factor of -1 and simplify from there:

$$(5x + 9) - (3x + 2) = (5x + 9) + (-1)(3x + 2)$$
$$= 5x + 9 + (-1)(3x) + (-1)(2)$$
$$= 5x + 9 + (-3x) + (-2)$$
$$= 2x + 7$$

Remark 2.10.13. The above example demonstrates *how* we apply the distributive property in order to subtract two expressions. But in practice, it can be pretty cumbersome. A shorter (and often clearer) approach is to instead subtract every term in the expression we are subtracting, which is shown like this:

$$(5x + 9) - (3x + 2) = 5x + 9 - 3x - 2$$
$$= 2x + 7$$

In general, we'll use this approach.

Example 2.10.14 Use the associative, commutative, and distributive properties to simplify the following expressions as much as possible.

a. $(-6x + 4) - (3x - 7)$

b. $(-2x - 5) - (-4x - 6)$

Explanation.

a. We will remove parentheses using the distributive property, and then combine like terms:

$$(-6x + 4) - (3x - 7) = -6x + 4 - 3x - (-7)$$

$$= -6x + 4 - 3x + 7$$
$$= -9x + 11$$

b. We will remove parentheses using the distributive property, and then combine like terms:

$$(-2x - 5) - (-4x - 6) = -2x - 5 - (-4x) - (-6)$$
$$= -2x - 5 + 4x + 6$$
$$= 2x + 1$$

2.10.5 The Role of the Order of Operations in Applying the Commutative, Associative, and Distributive Properties

When simplifying an expression such as $3 + 4(5x + 7)$, we need to respect the order of operations. Since the terms inside the parentheses are not like terms, there is nothing to simplify there. The next highest priority operation is multiplying the 4 by $(5x + 7)$. This must be done *before* anything happens with the adding of that 3. We cannot say $3 + 4(5x + 7) = 7(5x + 7)$, because that would mean we treated the addition as having higher priority than the multiplication.

So to simplify $3 + 4(5x + 7)$, we will first examine the multiplication of 4 with $(5x + 7)$, and here we may apply the distributive property. After that, we will use the commutative and associative properties:

$$3 + 4(5x + 7) = 3 + 4(5x) + 4(7)$$
$$= 3 + 20x + 28$$
$$= 20x + 3 + 28$$
$$= 20x + 31$$

Example 2.10.15 Simplify the following expressions using the commutative, associative, and distributive properties.

a. $4 - (3x - 9)$
b. $5x + 9(-2x + 3)$
c. $5(x - 9) + 4(x + 4)$

Explanation.

a. We will remove parentheses using the distributive property, and then combine like terms:

$$4 - (3x - 9) = 4 - 3x - (-9)$$
$$= 4 - 3x + 9$$
$$= -3x + 13$$

b. We will remove parentheses using the distributive property, and then combine like terms:

$$5x + 9(-2x + 3) = 5x + 9(-2x) + 9(3)$$
$$= 5x - 18x + 27$$
$$= -13x + 27$$

c. We will remove parentheses using the distributive property, and then combine like terms:

$$5(x - 9) + 4(x + 4) = 5x - 45 + 4x + 16$$
$$= 9x - 29$$

Checkpoint 2.10.16. Use the distributive property to simplify $3 - 9(9 - 8r)$ completely.

Explanation. We first use distributive property to get rid of parentheses, and then combine like terms:

$$3 - 9(9 - 8r) = 3 + (-9)(9 - 8r)$$
$$= 3 + (-9)(9) + (-9)(-8r)$$
$$= 3 - 81 + 72r$$
$$= -78 + 72r$$
$$= 72r - 78$$

Note that either of the last two expressions are acceptable final answers.

2.10.6 Rules of Exponents and Simplifying

In Section 2.9, we introduced three exponent rules. We continue to use these rules when simplifying expressions. Sometimes though, students incorrectly apply "rules" of exponents where they have misremembered the actual rule. Let's summarize what we can and cannot do.

When we add/subtract two expressions, we can only combine *like* terms. For example:

- $3x - x = 2x$
- $t^2 + t^2 = 2t^2$
- $q^2 + q$ cannot be combined.

However, we can multiply two expressions regardless of whether or not they are like terms. For example:

- $x \cdot x = x^2$
- $t^2 \cdot t^3 = t^5$
- $(q^2)(q) = q^3$

Consider:

- When we combine like terms that have a variable, the exponent doesn't change, as in $x^2 + x^2 = 2x^2$.
- When we multiply powers of a variable that use the same variable, the exponent *will* change, as in $(x^2)(x^2) = x^4$.
- We *cannot* combine "unlike terms," as something like $x^2 + x$ is as simplified as it can be.
- We *can* multiply powers with different exponents, as in $(x^2)(x) = x^3$.

The next few examples test your understanding of these concepts.

Example 2.10.17 Simplify the following expressions using the rules of exponents and the distributive property.

    a. $3x^2 + 2x + x^2$ b. $(3x^2)(2x)(x^2)$ c. $2x(3x + 4)$ d. $x^3 - 3x^2(5x - 2)$

Explanation.

a. We will combine like terms $3x^2$ and x^2:

$$3x^2 + 2x + x^2 = 4x^2 + 2x$$

b. We will apply the Product Rule:

$$(3x^2)(2x)(x^2) = 6x^5$$

c. To simplify $2x(3x + 4)$, we want to first distribute $2x$, and then we can apply the Product Rule:

$$2x(3x + 4) = 2x(3x) + 2x(4)$$
$$= 6x^2 + 8x$$

d. We will use the distributive property first, apply the Product Rule, and combine like terms:

$$x^3 - 3x^2(5x - 2) = x^3 - 3x^2(5x) - (-3x^2)(2)$$
$$= x^3 - 15x^3 + 6x^2$$
$$= -14x^3 + 6x^2$$

Exercises

Review and Warmup

1. Count the number of terms in each expression.

   a. $-6x - 7 + 3t - 8z^2$

   b. $-6t + 1$

   c. $4s^2 - 6y - 6s^2$

   d. $2s - 2z - 5y^2 + 7s$

2. Count the number of terms in each expression.

   a. $-4x^2 + z + y - 9t$

   b. $8x^2 - 6t + 2s^2 + 5t$

   c. $-7x^2$

   d. $-8z^2 + 5y - 5$

3. List the terms in each expression.

   a. $-2x$

   b. $-8.5s^2 - 1.2s + 8t^2$

   c. $3.8t + 6.4 - 0.3x$

   d. $-2.3y - 7.8$

4. List the terms in each expression.

   a. $-0.4x^2 - 0.3$

   b. $-3.2t + 6.8t$

   c. $-0.2y + 0.2s - 2.6x + 8.3s$

   d. $-3.5y^2$

5. List the terms in each expression.

   a. $1.2x + 7.9t$

   b. $-5.2y + 5.6 - 4.3s^2 - 6$

   c. $-4.8y + 0.9z$

   d. $8.4y - 0.1z - 8.5t$

6. List the terms in each expression.

   a. $2.8x^2 - 2x^2$

   b. $-7.2s + 4.3t^2 - 8.3z$

   c. $5.8s - 7.1 - 6.5x + 7.4z$

   d. $8.4x^2 - z$

7. Simplify each expression, if possible, by combining like terms.

  a. $5x + 6y^2 + 9y$

  b. $3s + 6s$

  c. $-z + 5z^2$

  d. $x^2 + 4x^2$

8. Simplify each expression, if possible, by combining like terms.

  a. $6x^2 - 4s^2 + 7s^2$

  b. $2s^2 + 2y^2 - 7 + 7y$

  c. $-3y^2 + 6z + 8z$

  d. $-4s - 2x - 8s^2 + 2s^2$

These exercises involve the concepts of like terms and the commutative, associative, and distributive properties.

9. The additive inverse of 4 is

10. The additive inverse of 7 is

11. The multiplicative inverse of 9 is

12. The multiplicative inverse of -10 is

13. Use the associative property of addition to write an equivalent expression to $r + (38 + q)$.

14. Use the associative property of addition to write an equivalent expression to $t + (2 + n)$.

15. Use the associative property of addition to write an equivalent expression to $10 + (9 + b)$.

16. Use the associative property of addition to write an equivalent expression to $3 + (17 + c)$.

17. Use the associative property of multiplication to write an equivalent expression to $4(5m)$.

18. Use the associative property of multiplication to write an equivalent expression to $8(3b)$.

19. Use the commutative property of addition to write an equivalent expression to $n + 8$.

20. Use the commutative property of addition to write an equivalent expression to $p + 73$.

21. Use the commutative property of addition to write an equivalent expression to $10x + 38$.

22. Use the commutative property of addition to write an equivalent expression to $5y + 3$.

23. Use the commutative property of addition to write an equivalent expression to $8(t + 69)$.

24. Use the commutative property of addition to write an equivalent expression to $3(b + 34)$.

25. Use the commutative property of multiplication to write an equivalent expression to $99c$.

26. Use the commutative property of multiplication to write an equivalent expression to $92x$.

27. Use the commutative property of multiplication to write an equivalent expression to $43 + 7a$.

28. Use the commutative property of multiplication to write an equivalent expression to $95 + 9n$.

29. Use the commutative property of multiplication to write an equivalent expression to $4(p + 60)$.

30. Use the commutative property of multiplication to write an equivalent expression to $8(x + 25)$.

31. Use the distributive property to write an equivalent expression to $10(y + 2)$ that has no grouping symbols.

32. Use the distributive property to write an equivalent expression to $6(t + 6)$ that has no grouping symbols.

33. Use the distributive property to write an equivalent expression to $-9(b + 9)$ that has no grouping symbols.

34. Use the distributive property to write an equivalent expression to $-3(c - 3)$ that has no grouping symbols.

35. Use the distributive property to write an equivalent expression to $-(m - 2)$ that has no grouping symbols.

36. Use the distributive property to write an equivalent expression to $-(r - 7)$ that has no grouping symbols.

37. Use the distributive property to simplify $9 + 7(9 + 8n)$ completely.

38. Use the distributive property to simplify $6 + 2(7 + 7p)$ completely.

39. Use the distributive property to simplify $3 - 7(-4 + 3x)$ completely.

40. Use the distributive property to simplify $8 - 3(-10 + 3y)$ completely.

41. Use the distributive property to simplify $5 - (-5 + 5t)$ completely.

42. Use the distributive property to simplify $2 - (4 + 2a)$ completely.

43. Use the distributive property to simplify $8 - (-6c - 8)$ completely.

44. Use the distributive property to simplify $6 - (2b + 9)$ completely.

45. Use the distributive property to simplify $\frac{10}{7}(2 - 5x)$ completely.

46. Use the distributive property to simplify $\frac{8}{5}(-3 + n)$ completely.

47. Use the distributive property to simplify $\frac{5}{8}\left(-9 + \frac{3}{2}p\right)$ completely.

48. Use the distributive property to simplify $\frac{10}{3}\left(7 + \frac{3}{4}x\right)$ completely.

49. The expression $y + t + c$ would be ambiguous if we did not have a left-to-right reading convention. Use grouping symbols to emphasize the order that these additions should be carried out.

Use the associative property of addition to write an equivalent (but different) algebraic expression.

50. The expression $t + a + x$ would be ambiguous if we did not have a left-to-right reading convention. Use grouping symbols to emphasize the order that these additions should be carried out.

Use the associative property of addition to write an equivalent (but different) algebraic expression.

51. A student has (correctly) simplified an algebraic expression in the following steps. Between each pair of steps, identify the algebraic property that justifies moving from one step to the next.

$$9(a + 5) + 9a$$

($\square$ commutative property of addition $\square$ commutative property of multiplication $\square$ associative property of addition $\square$ associative property of multiplication $\square$ distributive property)

$$= (9a + 45) + 9a$$

($\square$ commutative property of addition $\square$ commutative property of multiplication $\square$ associative property of addition $\square$ associative property of multiplication $\square$ distributive property)

$$= (45 + 9a) + 9a$$

($\square$ commutative property of addition $\square$ commutative property of multiplication $\square$ associative property of addition $\square$ associative property of multiplication $\square$ distributive property)

$$= 45 + (9a + 9a)$$

($\square$ commutative property of addition $\square$ commutative property of multiplication $\square$ associative property of addition $\square$ associative property of multiplication $\square$ distributive property)

$$= 45 + (9 + 9)a$$

$$= 45 + 18a$$

($\square$ commutative property of addition $\square$ commutative property of multiplication $\square$ associative property of addition $\square$ associative property of multiplication $\square$ distributive property)

$$= 18a + 45$$

52. A student has (correctly) simplified an algebraic expression in the following steps. Between each pair of steps, identify the algebraic property that justifies moving from one step to the next.

$$6(c + 8) + 7c$$

($\square$ commutative property of addition $\square$ commutative property of multiplication $\square$ associative property of addition $\square$ associative property of multiplication $\square$ distributive property)

$$= (6c + 48) + 7c$$

($\square$ commutative property of addition $\square$ commutative property of multiplication $\square$ associative property of addition $\square$ associative property of multiplication $\square$ distributive property)

$$= (48 + 6c) + 7c$$

(□ commutative property of addition   □ commutative property of multiplication   □ associative property of addition   □ associative property of multiplication   □ distributive property)

$$= 48 + (6c + 7c)$$

(□ commutative property of addition   □ commutative property of multiplication   □ associative property of addition   □ associative property of multiplication   □ distributive property)

$$= 48 + (6 + 7) c$$

$$= 48 + 13c$$

(□ commutative property of addition   □ commutative property of multiplication   □ associative property of addition   □ associative property of multiplication   □ distributive property)

$$= 13c + 48$$

53. The number of students enrolled in math courses at Portland Community College has grown over the years. The formulas

$$M = 0.47x + 3.7 \quad W = 0.38x + 4.3 \quad N = 0.02x + 0.1$$

describe the numbers (of thousands) of men, women, and gender-non-binary students enrolled in math courses at PCC x years after 2005. (Note this is an exercise using randomized numbers, not actual data.) Give a simplified formula for the total number T of thousands of students at PCC taking math classes x years after 2005. Be sure to give the entire formula, starting with T=.

54. The number of students enrolled in math courses at Portland Community College has grown over the years. The formulas

$$M = 0.51x + 5.7 \quad W = 0.51x + 3.4 \quad N = 0.02x + 0.2$$

describe the numbers (of thousands) of men, women, and gender-non-binary students enrolled in math courses at PCC x years after 2005. (Note this is an exercise using randomized numbers, not actual data.) Give a simplified formula for the total number T of thousands of students at PCC taking math classes x years after 2005. Be sure to give the entire formula, starting with T=.

Find the product of the *mo*nomial and the *bi*nomial.

55. $6x (x - 6) =$

56. $8x (x + 4) =$

57. $2x (-10x - 5) =$

58. $3x (-5x + 5) =$

59. $-6x^2 (x + 4) =$

60. $-3x^2 (x - 8) =$

61. $6y^2 (-3y^2 + 4y) =$

62. $-3r^2 (-10r^2 + 8r) =$

Simplify the following expressions if possible.

63. a. $m^2 + m^2 =$ ☐

b. $(m^2)(m^2) =$ ☐

c. $m^2 + m^3 =$ ☐

d. $(m^2)(m^3) =$ ☐

64. a. $n^2 + n^2 =$ ☐

b. $(n^2)(n^2) =$ ☐

c. $n^2 + n^4 =$ ☐

d. $(n^2)(n^4) =$ ☐

65. a. $p + p =$ ☐

b. $(p)(p) =$ ☐

c. $p + p^2 =$ ☐

d. $(p)(p^2) =$ ☐

66. a. $x^3 + x^3 =$ ☐

b. $(x^3)(x^3) =$ ☐

c. $x^3 + x^4 =$ ☐

d. $(x^3)(x^4) =$ ☐

67. a. $-y - 4y =$ ☐

b. $(-y)(-4y) =$ ☐

c. $-y - 2y^4 =$ ☐

d. $(-y)(-2y^4) =$ ☐

68. a. $-4t^3 - 2t^3 =$ ☐

b. $(-4t^3)(-2t^3) =$ ☐

c. $-4t^3 - 2t =$ ☐

d. $(-4t^3)(-2t) =$ ☐

69. a. $2a^2 + 4a^2 =$ ☐

b. $(2a^2)(4a^2) =$ ☐

c. $2a^2 + 4a =$ ☐

d. $(2a^2)(4a) =$ ☐

70. a. $-c - c =$ ☐

b. $(-c)(-c) =$ ☐

c. $-c + 2c^3 =$ ☐

d. $(-c)(2c^3) =$ ☐

71. a. $2q - q^2 - 2q =$ ☐

b. $(2q)(-q^2)(-2q) =$ ☐

72. a. $-3r^3 - 2r^4 + 2r^3 =$ ☐

b. $(-3r^3)(-2r^4)(2r^3) =$ ☐

Simplify the following expression.

73. $2n^2(5n^4)^2 =$ ☐

74. $5p^5(-4p^2)^2 =$ ☐

75. $2x^2c^2(4xc^2)^2 =$ ☐

76. $-3y^5t^4(-2yt^5)^4 =$ ☐

77. $(-5t^3)(-4t^4) + (t^4)(-5t^3) =$ ☐

78. $(3a^5)(-3a^5) - (4a^4)(2a^6) =$ ☐

79. $(-2c^5)(-2c^4)^3 - (c)(4c^8) =$ ☐

80. $(3p^4)(p^5)^4 - (2p^2)(p^7) =$ ☐

81. $(4q^4)(-2q^4)^2 + (q^2)^5(5q^2) =$ ☐

82. $(3m^5)(2m^4)^4 + (m^2)^2(3m^{17}) =$ ☐

83. $(3p^3)^2 n^9 + 3(p^2n^3)^3 =$ ☐

84. $(3x^4)^3 c^{12} + 4(x^3c^3)^4 =$ ☐

85. Use the distributive property to write an equivalent expression to $-9y(7y + 2)$ that has no grouping symbols.

86. Use the distributive property to write an equivalent expression to $-4t(4t - 10)$ that has no grouping symbols.

87. Use the distributive property to write an equivalent expression to $-7a^2(a - 1)$ that has no grouping symbols.

88. Use the distributive property to write an equivalent expression to $-10c^4(c + 8)$ that has no grouping symbols.

89. Use the distributive property to simplify $4 + 6m(2 + 3m)$ completely.

90. Use the distributive property to simplify $8 + 3b(10 + 7b)$ completely.

91. Use the distributive property to simplify $2m - 9m(8 + 10m^2)$ completely.

92. Use the distributive property to simplify $8p - 6p(3 + 10p^3)$ completely.

93. Use the distributive property to simplify $5x^3 - 2x^3(-3 + 9x^2)$ completely.

94. Use the distributive property to simplify $10y^3 - 7y^3(-9 + 9y^3)$ completely.

95. Fully simplify $3(7x - 8) + 6(3x - 7)$.

96. Fully simplify $4(3x + 2) - 6(7x - 1)$.

97. Fully simplify $-4(9x - 6) + 7(-x - 4)$.

98. Fully simplify $5(6x + 1) - 8(4x - 6)$.

Challenge

99. Fill in the blanks with algebraic expressions that make the equation true. You may not use 0 or 1 in any of the blank spaces. Here is an example: $? + ? = 8x$.

One possible answer is: $3x + 5x = 8x$.

There are infinitely many correct answers to this problem. I encourage you to be creative. After finding a correct answer, see if you can come up with a different answer that is also correct.

a. $\boxed{} + \boxed{} = -13x$

b. $\boxed{} + \boxed{} = -13x^{30}$

c. $\boxed{} \cdot \boxed{} \cdot \boxed{} = 2x^{80}$

2.11 Variables, Expressions, and Equations Chapter Review

2.11.1 Variables and Evaluating Expressions

In Section 2.1 we covered the definitions of variables and expressions, and how to evaluate an expression with a particular number. We learned the formulas for perimeter and area of rectangles, triangles, and circles.

Evaluating Expressions When we evaluate an expression's value, we substitute each variable with its given value.

Example 2.11.1 Evaluate the value of $\frac{5}{9}(F - 32)$ if $F = 212$.

$$\frac{5}{9}(F - 32) = \frac{5}{9}(212 - 32)$$
$$= \frac{5}{9}(180)$$
$$= 100$$

Substituting a Negative Number When we substitute a variable with a negative number, it's important to use parentheses around the number.

Example 2.11.2 Evaluate the following expressions if $x = -3$.

a. $x^2 = (-3)^2$
$\quad = 9$

b. $x^3 = (-3)^3$
$\quad = (-3)(-3)(-3)$
$\quad = -27$

c. $-x^2 = -(-3)^2$
$\quad = -9$

d. $-x^3 = -(-3)^3$
$\quad = -(-27)$
$\quad = 27$

2.11.2 Geometry Formulas

In Section 2.2 we established the following formulas.

Perimeter of a Rectangle $P = 2(\ell + w)$

Area of a Rectangle $A = \ell w$

Area of a Triangle $A = \frac{1}{2}bh$

Circumference of a Circle $c = 2\pi r$

Area of a Circle $A = \pi r^2$

Volume of a Rectangular Prism $V = wdh$

Volume of a Cylinder $V = \pi r^2 h$

Volume of a Rectangular Prism or Cylinder $V = Bh$

2.11.3 Combining Like Terms

In Section 2.3 we covered the definitions of a term and how to combine like terms.

Example 2.11.3 List the terms in the expression $5x - 3y + \frac{2w}{3}$.

Explanation. The expression has three terms that are being added, $5x$, $-3y$ and $\frac{2w}{3}$.

Example 2.11.4 Simplify the expression $5x - 3x^2 + 2x + 5x^2$, if possible, by combining like terms.

Explanation. This expression has four terms: $5x$, $-3x^2$, $2x$, and $5x^2$. Both $5x$ and $2x$ are like terms; also $-3x^2$ and $5x^2$ are like terms. When we combine like terms, we get:

$$5x - 3x^2 + 2x + 5x^2 = 7x + 2x^2$$

Note that we cannot combine $7x$ and $2x^2$ because x and x^2 represent different quantities.

2.11.4 Equations and Inequalities as True/False Statements

In Section 2.4 we covered the definitions of an equation and an inequality, as well as how to verify if a particular number is a solution to them.

Checking Possible Solutions Given an equation or an inequality (with one variable), checking if some particular number is a solution is just a matter of replacing the value of the variable with the specified number and determining if the resulting equation/inequality is true or false. This may involve some amount of arithmetic simplification.

Example 2.11.5 Is -5 a solution to $2(x + 3) - 2 = 4 - x$?

Explanation. To find out, substitute in -5 for x and see what happens.

$$2(x + 3) - 2 = 4 - x$$
$$2((-5) + 3) - 2 \overset{?}{=} 4 - (-5)$$
$$2(-2) - 2 \overset{?}{=} 9$$
$$-4 - 2 \overset{?}{=} 9$$
$$-6 \overset{\text{no}}{=} 9$$

So no, -5 is not a solution to $2(x + 3) - 2 = 4 - x$.

2.11.5 Solving One-Step Equations

In Section 2.5 we covered to to add, subtract, multiply, or divide on both sides of an equation to isolate the variable, summarized in Fact 2.5.12. We also learned how to state our answer, either as a solution or a solution set. Last we discussed how to solve equations with fractions.

Solving One-Step Equations When we solve linear equations, we use Properties of Equivalent Equations and follow an algorithm to solve a linear equation.

Example 2.11.6 Solve for g in $\frac{1}{2} = \frac{2}{3} + g$.

Explanation. We will subtract $\frac{2}{3}$ on both sides of the equation:

$$\frac{1}{2} = \frac{2}{3} + g$$
$$\frac{1}{2} - \frac{2}{3} = \frac{2}{3} + g - \frac{2}{3}$$
$$\frac{3}{6} - \frac{4}{6} = g$$
$$-\frac{1}{6} = g$$

We will check the solution by substituting g in the original equation with $-\frac{1}{6}$:

$$\frac{1}{2} = \frac{2}{3} + g$$
$$\frac{1}{2} \overset{?}{=} \frac{2}{3} + \left(-\frac{1}{6}\right)$$
$$\frac{1}{2} \overset{?}{=} \frac{4}{6} + \left(-\frac{1}{6}\right)$$
$$\frac{1}{2} \overset{?}{=} \frac{3}{6}$$
$$\frac{1}{2} \overset{\checkmark}{=} \frac{1}{2}$$

The solution $-\frac{1}{6}$ is checked and the solution set is $\left\{-\frac{1}{6}\right\}$.

2.11.6 Solving One-Step Inequalities

In Section 2.6 we covered how solving inequalities is very much like how we solve equations, except that if we multiply or divide by a negative we switch the inequality sign.

Solving One-Step Inequalities When we solve linear inequalities, we also use Properties of Equivalent Equations with one small complication: When we multiply or divide by the same *negative* number on both sides of an inequality, the direction reverses!

Example 2.11.7 Solve the inequality $-2x \geq 12$. State the solution set with both interval notation and set-builder notation.

Explanation. To solve this inequality, we will divide each side by -2:

$$-2x \geq 12$$
$$\frac{-2x}{-2} \leq \frac{12}{-2} \qquad \text{Note the change in direction.}$$
$$x \leq -6$$

- The inequality's solution set in interval notation is $(-\infty, -6]$.

- The inequality's solution set in set-builder notation is $\{x \mid x \le -6\}$.

2.11.7 Percentages

In Section 2.7 we covered how to translate sentences with percentages into equations that we can solve.

Solving One-Step Equations Involving Percentages   An important skill for solving percent-related problems is to boil down a complicated word problem into a simple form like "2 is 50% of 4."

Example 2.11.8 What percent of 2346.19 is 1995.98?

Using P to represent the unknown quantity, we write and solve the equation:

$$\underbrace{\text{what percent}}_{P} \quad \underbrace{\text{of}}_{\cdot} \quad \underbrace{\$2346.19}_{2346.19} \quad \underbrace{\text{is}}_{=} \quad \underbrace{\$1995.98}_{1995.98}$$

$$P \cdot 2346.19 = 1995.98$$
$$\frac{P \cdot 2346.19}{2346.19} = \frac{1995.98}{2346.19}$$
$$P = 0.85073\ldots$$
$$P \approx 85.07\%$$

In summary, 1995.98 is approximately 85.07% of 2346.19.

2.11.8 Modeling with Equations and Inequalities

In Section 2.8 we covered how to translate phrases into mathematics, and how to set up equations and inequalities for application models.

Modeling with Equations and Inequalities   To set up an equation modeling a real world scenario, the first thing we need to do is identifying what variable we will use. The variable we use will be determined by whatever is unknown in our problem statement. Once we've identified and defined our variable, we'll use the numerical information provided in the equation to set up our equation.

Example 2.11.9 A bathtub contains $2.5\,\text{ft}^3$ of water. More water is being poured in at a rate of $1.75\,\text{ft}^3$ per minute. When will the amount of water in the bathtub reach $6.25\,\text{ft}^3$?

Explanation.   Since the question being asked in this problem starts with "when," we immediately know that the unknown is time. As the volume of water in the tub is measured in ft^3 per minute, we know that time needs to be measured in minutes. We'll defined t to be the number of minutes that water is poured into the tub. Since each minute there are $1.75\,\text{ft}^3$ of water added, we will add the expression $1.75t$ to 2.5 to obtain the total amount of water. Thus the equation we set up is:

$$2.5 + 1.75t = 6.25$$

2.11.9 Introduction to Exponent Rules

In Section 2.9 we covered the rules of exponents for multiplication.

Example 2.11.10 Simplify the following expressions using the rules of exponents:

a. $-2t^3 \cdot 4t^5$
b. $5\left(v^4\right)^2$
c. $-(3u)^2$
d. $(-3z)^2$

Explanation.

a. $-2t^3 \cdot 4t^5 = -8t^8$
b. $5\left(v^4\right)^2 = 5v^8$
c. $-(3u)^2 = -9u^4$
d. $(-3z)^2 = 9z^4$

2.11.10 Simplifying Expressions

In Section 2.10 we covered the definitions of the identities and inverses, and the various algebraic properties. We then learned about the order of operations.

Example 2.11.11 Use the associative, commutative, and distributive properties to simplify the expression $5x + 9(-2x + 3)$ as much as possible.

Explanation. We will remove parentheses by the distributive property, and then combine like terms:

$$5x + 9(-2x + 3) = 5x + 9(-2x + 3)$$
$$= 5x + 9(-2x) + 9(3)$$
$$= 5x - 18x + 27$$
$$= -13x + 27$$

Exercises

A trapezoid's area can be calculated by the formula $A = \frac{1}{2}(b_1 + b_2)h$, where A stands for area, b_1 for the first base's length, b_2 for the second base's length, and h for height.

1. Find the area of the trapeoid below.

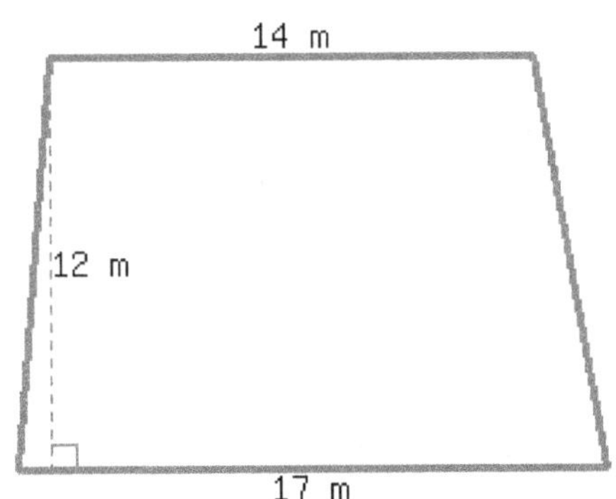

2. Find the area of the trapeoid below.

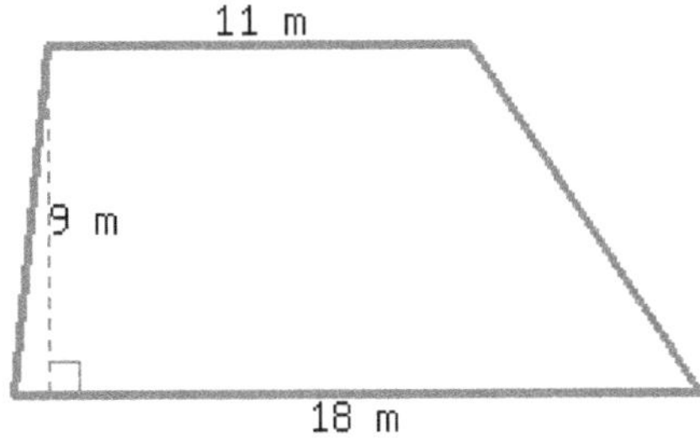

To convert a temperature measured in degrees Fahrenheit to degrees Celsius, there is a formula:

$$C = \frac{5}{9}(F - 32)$$

where C represents the temperature in degrees Celsius and F represents the temperature in degrees Fahrenheit.

3. If a temperature is 122°F, what is that temperature measured in Celsius?

4. If a temperature is 14°F, what is that temperature measured in Celsius?

5. Evaluate the expression x^2:

a. When $x = 6$, $x^2 = $ ____________

b. When $x = -6$, $x^2 = $ ____________

6. Evaluate the expression y^2:

a. When $y = 3$, $y^2 = $ ____________

b. When $y = -9$, $y^2 = $ ____________

7. Evaluate the expression y^3:

a. When $y = 5$, $y^3 = $ ____________

b. When $y = -3$, $y^3 = $ ____________

8. Evaluate the expression r^3:

a. When $r = 3$, $r^3 = $ ____________

b. When $r = -5$, $r^3 = $ ____________

9. List the terms in each expression.

a. $4t + 2z + 6$

b. $7z^2$

c. $9t + y$

d. $2t + 7t$

10. List the terms in each expression.

a. $8t^2 + 6 + 2x^2 - t^2$

b. $5x^2 - 6y^2 + 7y$

c. $-5z^2 + 3t^2 - 7y^2$

d. $-2y^2 + 4 - 3s + 2t$

11. Simplify each expression, if possible, by combining like terms.

a. $8t - t + 3t + 9t$

b. $-8z^2 + 5z^2 + 6z^2$

c. $3z - 3z$

d. $-3x^2 - 2 - 7x$

12. Simplify each expression, if possible, by combining like terms.

a. $5t^2 - 8y^2 + 4 + 2t$

b. $7t - 2t^2$

c. $-7z^2 - 2z^2 + 8x^2$

d. $-6s + x$

13. Simplify each expression, if possible, by combining like terms.

   a. $\frac{4}{3}t + \frac{4}{9}$

   b. $-\frac{1}{7}y^2 + y^2 - \frac{4}{3}y^2 + \frac{2}{7}s^2$

   c. $-\frac{3}{8}y + \frac{3}{2}z^2 - 3z^2 + 2x$

   d. $-\frac{2}{3}t - s$

14. Simplify each expression, if possible, by combining like terms.

   a. $-\frac{1}{6}t - \frac{9}{4}t$

   b. $-\frac{3}{8}s - z - \frac{2}{5}s$

   c. $\frac{1}{3}y^2 - \frac{9}{5}y^2$

   d. $-\frac{2}{9}y - \frac{1}{3}y^2 + \frac{1}{4}y^2 + 9y$

15. Is -2 a solution for x in the equation $2x + 2 = 2 - (5 + x)$? Evaluating the left and right sides gives:

$$2x + 2 \quad \overset{?}{=} \quad 2 - (5 + x)$$

$$\underline{\hspace{2cm}} \quad \overset{?}{=} \quad \underline{\hspace{2cm}}$$

So -2 ($\square$ is $\quad \square$ is not) a solution to $2x + 2 = 2 - (5 + x)$.

16. Is -1 a solution for x in the equation $4x - 4 = -3 - (4 + x)$? Evaluating the left and right sides gives:

$$4x - 4 \quad \overset{?}{=} \quad -3 - (4 + x)$$

$$\underline{\hspace{2cm}} \quad \overset{?}{=} \quad \underline{\hspace{2cm}}$$

So -1 ($\square$ is $\quad \square$ is not) a solution to $4x - 4 = -3 - (4 + x)$.

17. Is 1 a solution for x in the inequality $-4x^2 + 5x \le 2x - 7$? Evaluating the left and right sides gives:

$$-4x^2 + 5x \quad \overset{?}{\underset{\le}{\le}} \quad 2x - 7$$

$$\underline{\hspace{2cm}} \quad \overset{?}{\underset{\le}{\le}} \quad \underline{\hspace{2cm}}$$

So 1 ($\square$ is $\quad \square$ is not) a solution to $-4x^2 + 5x \le 2x - 7$.

18. Is 2 a solution for x in the inequality $-2x^2 + 5x \le 2x - 12$? Evaluating the left and right sides gives:

$$-2x^2 + 5x \quad \overset{?}{\underset{\le}{\le}} \quad 2x - 12$$

$$\underline{\hspace{2cm}} \quad \overset{?}{\underset{\le}{\le}} \quad \underline{\hspace{2cm}}$$

So 2 ($\square$ is $\quad \square$ is not) a solution to $-2x^2 + 5x \le 2x - 12$.

Solve the equation.

19. $t + 7 = 2$

20. $t + 4 = 1$

21. $-10 = t - 6$

22. $-9 = x - 7$

23. $96 = -8x$

24. $24 = -3y$

25. $\dfrac{5}{13}c = 25$

26. $\dfrac{4}{7}A = 12$

27. The pie chart represents a collector's collection of signatures from various artists.

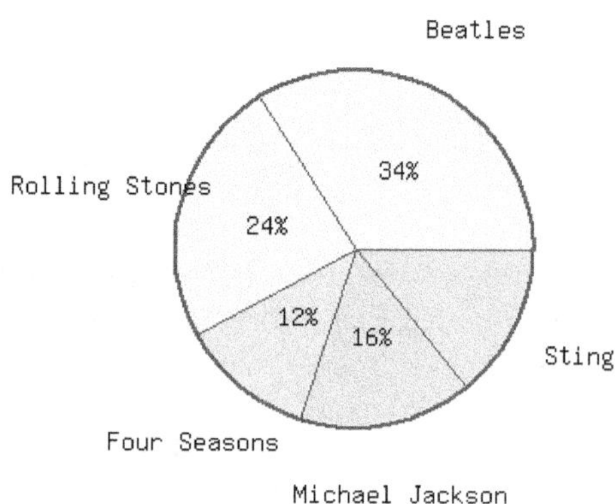

If the collector has a total of 1450 signatures, there are [   ] signatures by Sting.

28. The pie chart represents a collector's collection of signatures from various artists.

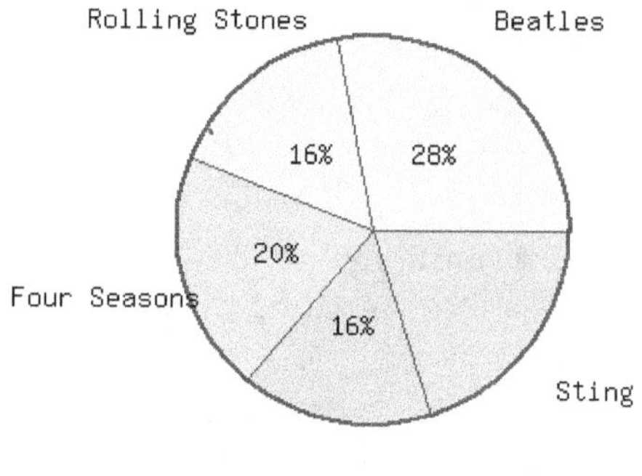

If the collector has a total of 1650 signatures, there are [   ] signatures by Sting.

29. A community college conducted a survey about the number of students riding each bus line available. The following bar graph is the result of the survey.

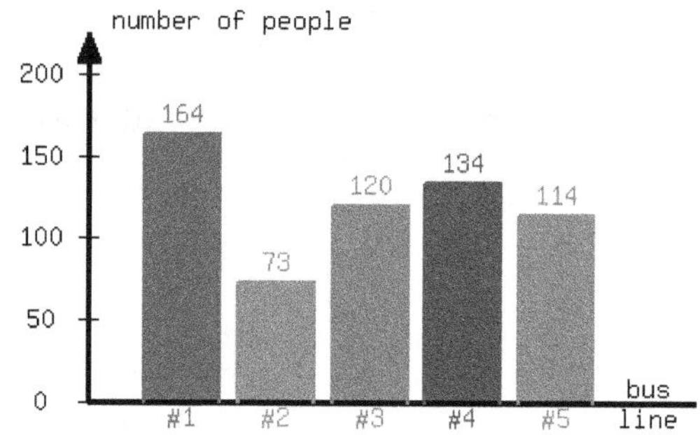

What percent of students ride Bus #1?

Approximately [   ] of students ride Bus #1.

30. A community college conducted a survey about the number of students riding each bus line available. The following bar graph is the result of the survey.

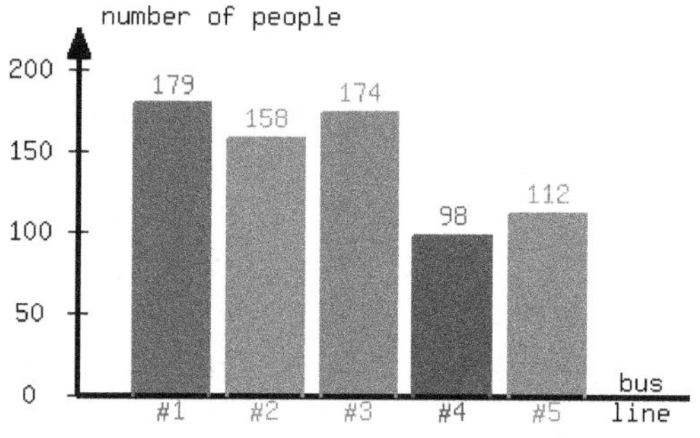

What percent of students ride Bus #1?

Approximately [   ] of students ride Bus #1.

31. The following is a nutrition fact label from a certain macaroni and cheese box.

Nutrition Facts	
Serving Size 1 cup	
Servings Per Container 2	
Amount Per Serving	
Calories 300	Calories from Fat 110
	% Daily Value
Total Fat 7 g	14%
Saturated Fat 2 g	15%
Trans Fat 2 g	
Cholesterol 30 mg	10%
Sodium 400 mg	20%
Total Carbohydrate 30 g	11%
Dietary Fiber 0 g	0%
Sugars 5 g	
Protein 5 g	
Vitamin A 2 mg	3%
Vitamin C 2 mg	2.5%
Calcium 2 mg	20%
Iron 3 mg	4%

The highlighted row means each serving of macaroni and cheese in this box contains 7 g of fat, which is 14% of an average person's daily intake of fat. What's the recommended daily intake of fat for an average person?

The recommended daily intake of fat for an average person is ______.

32. The following is a nutrition fact label from a certain macaroni and cheese box.

Nutrition Facts	
Serving Size 1 cup	
Servings Per Container 2	
Amount Per Serving	
Calories 300	Calories from Fat 110
	% Daily Value
Total Fat 5.5 g	10%
Saturated Fat 2 g	15%
Trans Fat 2 g	
Cholesterol 30 mg	10%
Sodium 400 mg	20%
Total Carbohydrate 30 g	11%
Dietary Fiber 0 g	0%
Sugars 5 g	
Protein 5 g	
Vitamin A 2 mg	3%
Vitamin C 2 mg	2.5%
Calcium 2 mg	20%
Iron 3 mg	4%

The highlighted row means each serving of macaroni and cheese in this box contains 5.5 g of fat, which is 10% of an average person's daily intake of fat. What's the recommended daily intake of fat for an average person?

The recommended daily intake of fat for an average person is ______.

33. Jerry used to make 13 dollars per hour. After he earned his Bachelor's degree, his pay rate increased to 48 dollars per hour. What is the percentage increase in Jerry's salary?

The percentage increase is ______.

34. Eileen used to make 14 dollars per hour. After she earned her Bachelor's degree, her pay rate increased to 49 dollars per hour. What is the percentage increase in Eileen's salary?

The percentage increase is ______.

35. After a 10% increase, a town has 550 people. What was the population before the increase?

Before the increase, the town's population was ______.

36. After a 70% increase, a town has 1020 people. What was the population before the increase?

Before the increase, the town's population was ______.

37. A bicycle for sale costs $254.88, which includes 6.2% sales tax. What was the cost before sales tax?

Assume the bicycle's price before sales tax is p dollars. Write an equation to model this scenario. There is no need to solve it.

38. A bicycle for sale costs $283.77, which includes 5.1% sales tax. What was the cost before sales tax?

Assume the bicycle's price before sales tax is p dollars. Write an equation to model this scenario. There is no need to solve it.

39. The property taxes on a 2100-square-foot house are $4,179.00 per year. Assuming these taxes are proportional, what are the property taxes on a 1700-square-foot house?

Assume property taxes on a 1700-square-foot house is t dollars. Write an equation to model this scenario. There is no need to solve it.

40. The property taxes on a 1600-square-foot house are $1,600.00 per year. Assuming these taxes are proportional, what are the property taxes on a 2000-square-foot house?

Assume property taxes on a 2000-square-foot house is t dollars. Write an equation to model this scenario. There is no need to solve it.

41. A swimming pool is being filled with water from a garden hose at a rate of 5 gallons per minute. If the pool already contains 30 gallons of water and can hold 135 gallons, after how long will the pool overflow?

Assume m minutes later, the pool would overflow. Write an equation to model this scenario. There is no need to solve it.

42. A swimming pool is being filled with water from a garden hose at a rate of 8 gallons per minute. If the pool already contains 40 gallons of water and can hold 280 gallons, after how long will the pool overflow?

Assume m minutes later, the pool would overflow. Write an equation to model this scenario. There is no need to solve it.

43. Use the commutative property of addition to write an equivalent expression to $5b + 31$.

44. Use the commutative property of addition to write an equivalent expression to $6q + 79$.

45. Use the associative property of multiplication to write an equivalent expression to $3(4r)$.

46. Use the associative property of multiplication to write an equivalent expression to $4(7m)$.

47. Use the distributive property to write an equivalent expression to $10(p + 2)$ that has no grouping symbols.

48. Use the distributive property to write an equivalent expression to $7(q + 6)$ that has no grouping symbols.

49. Use the distributive property to simplify $4 + 9(2 + 4y)$ completely.

50. Use the distributive property to simplify $9 + 4(9 + 3r)$ completely.

51. Use the distributive property to simplify $6 - 4(1 - 6a)$ completely.

52. Use the distributive property to simplify $3 - 9(-5 - 6b)$ completely.

53. Use the properties of exponents to simplify the expression.

$$r^{12} \cdot r^{17}$$

54. Use the properties of exponents to simplify the expression.

$$t^{14} \cdot t^{11}$$

55. Use the properties of exponents to simplify the expression.

$$\left(y^{10}\right)^3$$

56. Use the properties of exponents to simplify the expression.

$$\left(t^{11}\right)^{10}$$

57. Use the properties of exponents to simplify the expression.

$$(3x)^4$$

58. Use the properties of exponents to simplify the expression.

$$(2r)^2$$

59. Use the properties of exponents to simplify the expression.

$$(-2t^5)\cdot(4t^{16})$$

60. Use the properties of exponents to simplify the expression.

$$(-5y^7)\cdot(3y^9)$$

61. Use the properties of exponents to simplify the expression.

a. $\left(-2b^3\right)^6 =$

b. $-\left(2b^3\right)^6 =$

62. Use the properties of exponents to simplify the expression.

a. $\left(-2a^3\right)^2 =$

b. $-\left(2a^3\right)^2 =$

63. Simplify the following expression.

$$\left(3r^3\right)^4\left(r^2\right)^2 =$$

64. Simplify the following expression.

$$\left(5t^2\right)^2\left(t^5\right)^2 =$$

65. Simplify the following expressions if possible.

a. $p^2 + 2p^2 =$

b. $(p^2)(2p^2) =$

c. $p^2 - 4p^3 =$

d. $(p^2)(-4p^3) =$

66. Simplify the following expressions if possible.

a. $-2q + 4q =$

b. $(-2q)(4q) =$

c. $-2q - q^4 =$

d. $(-2q)(-q^4) =$

Find the product of the *mono*mial and the *bi*nomial.

67. $-10x^2\left(3x^2 + 5x\right) =$

68. $7x^2\left(9x^2 + 10x\right) =$

CHAPTER 3

Linear Equations and Inequalities

3.1 Solving Multistep Linear Equations

We have learned how to solve one-step equations in Section. In this section, we will learn how to solve multistep equations.

3.1.1 Solving Two-Step Equations

Example 3.1.2 A water tank can hold 140 gallons of water, but it has only 5 gallons of water. A tap was turned on, pouring 15 gallons of water into the tank every minute. After how many minutes will the tank be full? Let's find a pattern first.

Minutes since Tap Was Turned on	Amount of Water in the Tank (in Gallons)
0	5
1	$15 \cdot 1 + 5 = 20$
2	$15 \cdot 2 + 5 = 35$
3	$15 \cdot 3 + 5 = 50$
4	$15 \cdot 4 + 5 = 65$
$\vdots$	$\vdots$
m	$15m + 5$

Table 3.1.3: Amount of Water in the Tank

We can see that after m minutes, the tank has $15m + 5$ gallons of water. This makes sense since the tap pours $15m$ gallons of water into the tank in m minutes and it had 5 gallons to start with. To find when the tank will be full (with 140 gallons of water), we can write the equation

$$15m + 5 = 140$$

First, we need to isolate the variable term, $15m$, in the equation. In other words, we need to remove 5 from the left side of the equals sign. We can do this by subtracting 5 from both sides of the equation. Once the variable term is isolated, we can eliminate the coefficient and solve for m. The full process is:

$$15m + 5 = 140$$
$$15m + 5 - 5 = 140 - 5$$
$$15m = 135$$
$$\frac{15m}{15} = \frac{135}{15}$$
$$m = 9$$

Next, we need to substitute m with 9 in the equation $15m + 5 = 140$ to check the solution:

$$15m + 5 = 140$$

$$15(9) + 5 \overset{?}{=} 140$$

$$135 + 5 \overset{\checkmark}{=} 140$$

The solution 9 is checked. In summary, the tank will be full after 9 minutes.

In solving the two-step equation in Example 3.1.2, we first isolated the variable expression $15m$ and then eliminated the coefficient of 15 by dividing each side of the equation by 15. These two steps will be at the heart of our approach to solving linear equations. For more complicated equations, we may need to simplify some of the expressions first. Below is a general approach to solving linear equations that we will use as we solve more and more complicated equations.

Simplify Simplify the expressions on each side of the equation by distributing and combining like terms.

Isolate Use addition or subtraction to separate the variable terms and constant terms (numbers) so that they are on different sides of the equation.

Eliminate Use multiplication or division to eliminate the variable term's coefficient.

Check Check the solution. Substitute values into the original equation and use the order of operations to simplify both sides. It's important to use the order of operations alone rather than properties like the distributive law. Otherwise you might repeat the same arithmetic errors made while solving and fail to catch an incorrect solution.

Summarize State the solution set or (in the case of an application problem) summarize the result in a complete sentence using appropriate units.

List 3.1.4: Steps to Solve Linear Equations

Let's look at some more examples.

Example 3.1.5 Solve for y in the equation $7 - 3y = -8$.

Explanation. To solve, we will first separate the variable terms and constant terms into different sides of the equation. Then we will eliminate the variable term's coefficient.

$$7 - 3y = -8$$
$$7 - 3y - 7 = -8 - 7$$
$$-3y = -15$$
$$\frac{-3y}{-3} = \frac{-15}{-3}$$
$$y = 5$$

Checking the solution $y = 5$:

$$7 - 3y = -8$$
$$7 - 3(5) \overset{?}{=} -8$$
$$7 - 15 \overset{\checkmark}{=} -8$$

Therefore the solution to the equation $7 - 3y = -8$ is 5 and the solution set is $\{5\}$.

3.1.2 Solving Multistep Linear Equations

Example 3.1.6 Ahmed has saved $2,500.00 in his savings account and is going to start saving $550.00 per month. Julia has saved $4,600.00 in her savings account and is going to start saving $250.00 per month. If this situation continues, how many months later would Ahmed catch up with Julia in savings?

Ahmed saves $550.00 per month, so he can save $550m$ dollars in m months. With the $2,500.00 he started with, after m months he has $550m + 2500$ dollars. Similarly, after m months, Julia has $250m + 4600$ dollars. To find when those two accounts will have the same amount of money, we write the equation

$$550m + 2500 = 250m + 4600.$$

Checking the solution 7:

$$550m + 2500 = 250m + 4600$$
$$550m + 2500 - 2500 = 250m + 4600 - 2500$$
$$550m = 250m + 2100$$
$$550m - 250m = 250m + 2100 - 250m$$
$$300m = 2100$$
$$\frac{300m}{300} = \frac{2100}{300}$$
$$m = 7$$

$$550m + 2500 = 250m + 4600$$
$$550(7) + 2500 \overset{?}{=} 250(7) + 4600$$
$$3850 + 2500 \overset{?}{=} 1750 + 4600$$
$$6350 \overset{\checkmark}{=} 6350$$

In summary, Ahmed will catch up with Julia's savings in 7 months.

Example 3.1.7 Solve for x in $5 - 2x = 5x - 9$.

Explanation.

$$5 - 2x = 5x - 9$$
$$5 - 2x - 5 = 5m - 9 - 5$$
$$-2x = 5x - 14$$
$$-2x - 5x = 5x - 14 - 5x$$
$$-7x = -14$$
$$\frac{-7x}{-7} = \frac{-14}{-7}$$
$$x = 2$$

Checking the solution 2:

$$5 - 2x = 5x - 9$$

$$5 - 2(2) \stackrel{?}{=} 5(2) - 9$$

$$5 - 4 \stackrel{?}{=} 10 - 9$$

$$1 \stackrel{\checkmark}{=} 1$$

Therefore the solution is 2 and the solution set is $\{2\}$.

In Example 3.1.7, we could have moved variable terms to the right side of the equals sign, and number terms to the left side. We chose not to. There's no reason we *couldn't* have moved variable terms to the right side though. Let's compare:

$$5 - 2x = 5x - 9$$
$$5 - 2x + 9 = 5x - 9 + 9$$
$$14 - 2x = 5x$$
$$14 - 2x + 2x = 5x + 2x$$
$$14 = 7x$$
$$\frac{14}{7} = \frac{7x}{7}$$
$$2 = x$$

Lastly, we could save a step by moving variable terms and number terms in one step:

$$5 - 2x = 5x - 9$$
$$5 - 2x + 2x + 9 \quad 5x - 9 + 2x + 9$$
$$14 = 7x$$
$$\frac{14}{7} = \frac{7x}{7}$$
$$2 = x$$

Remark 3.1.8. This textbook will move variable terms and number terms separately throughout this chapter. Check with your instructor for their expectations.

Checkpoint 3.1.9. Solve the equation.

$7a + 3 = a + 45$

Explanation. The first step is to subtract terms in order to separate the variable and non-variable terms.

$$7a + 3 = a + 45$$
$$7a + 3 - \mathbf{a} - \mathbf{3} = a + 45 - \mathbf{a} - \mathbf{3}$$
$$6a = 42$$
$$\frac{6a}{6} = \frac{42}{6}$$
$$a = 7$$

The solution to this equation is 7. To stress that this is a value assigned to a, some report $a = 7$. We can also say that the solution set is $\{7\}$, or that $a \in \{7\}$. If we substitute 7 in for a in the original equation $7a + 3 = a + 45$, the equation will be true. Please check this on your own; it is an important habit.

The next example requires combining like terms.

Example 3.1.10 Solve for n in $n - 9 + 3n = n - 3n$.

Explanation.

To start solving this equation, we'll need to combine like terms. After this, we can put all terms containing n on one side of the equation and finish solving for n.

$$n - 9 + 3n = n - 3n$$
$$4n - 9 = -2n$$
$$4n - 9 - 4n = -2n - 4n$$
$$-9 = -6n$$
$$\frac{-9}{-6} = \frac{-6n}{-6}$$
$$n = \frac{3}{2}$$

Checking the solution $\frac{3}{2}$:

$$n - 9 + 3n = n - 3n$$
$$\frac{3}{2} - 9 + 3\left(\frac{3}{2}\right) \stackrel{?}{=} \frac{3}{2} - 3\left(\frac{3}{2}\right)$$
$$\frac{3}{2} - 9 + \frac{9}{2} \stackrel{?}{=} \frac{3}{2} - \frac{9}{2}$$
$$\frac{12}{2} - 9 \stackrel{?}{=} -\frac{6}{2}$$
$$6 - 9 \stackrel{?}{=} -3$$
$$-3 \stackrel{?}{=} -3$$

The solution to the equation $n - 9 + 3n = n - 3n$ is $\frac{3}{2}$ and the solution set is $\left\{\frac{3}{2}\right\}$.

Checkpoint 3.1.11. Solve the equation.

$$-1 + 7 = -8b - b - 3$$

Explanation. The first step is simply to combine like terms.

$$-1 + 7 = -8b - b - 3$$
$$6 = -9b - 3$$
$$6 + 3 = -9b - 3 + 3$$
$$9 = -9b$$
$$\frac{9}{-9} = \frac{-9b}{-9}$$
$$-1 = b$$
$$b = -1$$

The solution to this equation is -1. To stress that this is a value assigned to b, some report $b = -1$. We can also say that the solution set is $\{-1\}$, or that $b \in \{-1\}$. If we substitute -1 in for b in the original equation $-1 + 7 = -8b - b - 3$, the equation will be true. Please check this on your own; it is an important habit.

Example 3.1.12 Azul is designing a rectangular garden and they have 40 meters of wood for the border. Their garden's length must be 4 meters less than three times the width, and its perimeter must be 40 meters. Find the garden's length and width.

Explanation. Reminder: A rectangle's perimeter formula is $P = 2(L+W)$, where P stands for perimeter,

L stands for length and *W* stands for width.

Let Azul's garden width be *W* meters. We can then represent the length as $3W - 4$ meters since we are told that it is 4 meters less than three times the width. It's given that the perimeter is 40 meters. Substituting those values into the formula, we have:

$$P = 2(L + W)$$
$$40 = 2(3W - 4 + W)$$
$$40 = 2(4W - 4) \qquad \text{Like terms were combined.}$$

The next step to solve this equation is to remove the parentheses by distribution.

$$40 = 2(4W - 4)$$
$$40 = 8W - 8$$
$$40 + 8 = 8W - 8 + 8$$
$$48 = 8W$$
$$\frac{48}{8} = \frac{8W}{8}$$
$$6 = W.$$

Checking the solution $W = 6$:

$$40 = 2(4W - 4)$$
$$40 \overset{?}{=} 2(4(6) - 4)$$
$$40 \overset{\checkmark}{=} 2(20).$$

To determine the length, recall that this was represented by $3W - 4$, which is:

$$3W - 4 = 3(6) - 4$$
$$= 14.$$

Thus, the width of Azul's garden is 6 meters and the length is 14 meters.

Checkpoint 3.1.13. A rectangle's perimeter is 52 m. Its width is 10 m. Use an equation to solve for the rectangle's length.

Its length is [   ].

Explanation. When we deal with a geometric figure, it's always a good idea to sketch it to help us think. Let the length be *x* meters.

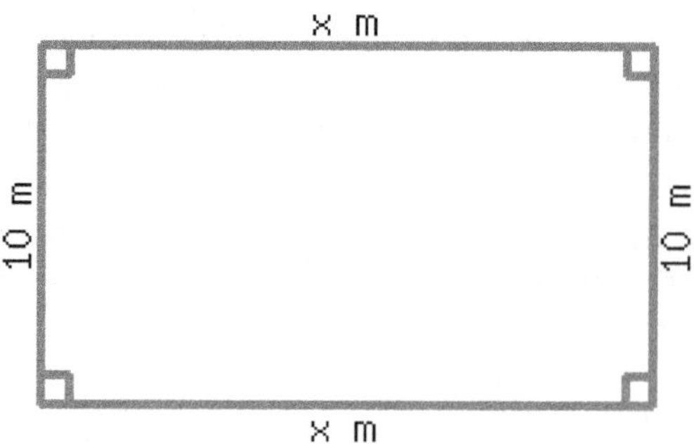

The perimeter is given as 52 m. Adding up the rectangle's 4 sides gives the perimeter. The equation is:

$$\begin{aligned}
x + x + 10 + 10 &= 52 \\
2x + 20 &= 52 \\
2x + 20 - 20 &= 52 - 20 \\
2x &= 32 \\
\frac{2x}{2} &= \frac{32}{2} \\
x &= 16
\end{aligned}$$

So the rectangle's length is 16 m. Don't forget the unit m.

We should be careful when we distribute a negative sign into the parentheses, like in the next example.

Example 3.1.14 Solve for a in $4 - (3 - a) = -2 - 2(2a + 1)$.

Explanation. To solve this equation, we will simplify each side of the equation, manipulate it so that all variable terms are on one side and all constant terms are on the other, and then solve for a:

Checking the solution -1:

$$\begin{aligned}
4 - (3 - a) &= -2 - 2(2a + 1) \\
4 - 3 + a &= -2 - 4a - 2 \\
1 + a &= -4 - 4a \\
1 + a + 4a &= -4 - 4a + 4a \\
1 + 5a &= -4 \\
1 + 5a - 1 &= -4 - 1 \\
5a &= -5 \\
\frac{5a}{5} &= \frac{-5}{5} \\
a &= -1
\end{aligned}$$

$$\begin{aligned}
4 - (3 - a) &= -2 - 2(2a + 1) \\
4 - (3 - (-1)) &\stackrel{?}{=} -2 - 2(2(-1) + 1) \\
4 - (4) &\stackrel{?}{=} -2 - 2(-1) \\
0 &\stackrel{\checkmark}{=} 0
\end{aligned}$$

Therefore the solution to the equation is -1 and the solution set is $\{-1\}$.

3.1.3 Differentiating between Simplifying Expressions, Evaluating Expressions and Solving Equations

Let's look at the following similar, yet different examples.

Example 3.1.15 Simplify the expression $10 - 3(x + 2)$.

Explanation.

$$\begin{aligned}
10 - 3(x + 2) &= 10 - 3x - 6 \\
&= -3x + 4
\end{aligned}$$

An equivalent result is $4 - 3x$. Note that our final result is an *expression*.

Example 3.1.16 Evaluate the expression $10 - 3(x + 2)$ when $x = 2$ and when $x = 3$.

Explanation.   We will substitute $x = 2$ into the expression:

$$
\begin{aligned}
10 - 3(x + 2) &= 10 - 3(2 + 2) \\
&= 10 - 3(4) \\
&= 10 - 12 \\
&= -2
\end{aligned}
$$

When $x = 2$, $10 - 3(x + 2) = -2$.

Similarly, we will substitute $x = 3$ into the expression:

$$
\begin{aligned}
10 - 3(x + 2) &= 10 - 3(3 + 2) \\
&= 10 - 3(5) \\
&= 10 - 15 = -5
\end{aligned}
$$

When $x = 3$, $10 - 3(x + 2) = -5$.

Note that the final results here are *values of the original expression.*

Example 3.1.17 Solve the equation $10 - 3(x + 2) = x - 16$.

Explanation.

$$
\begin{aligned}
10 - 3(x + 2) &= x - 16 \\
10 - 3x - 6 &= x - 16 \\
-3x + 4 &= x - 16 \\
-3x + 4 - 4 &= x - 16 - 4 \\
-3x &= x - 20 \\
-3x - x &= x - 20 - x \\
-4x &= -20 \\
\frac{-4x}{-4} &= \frac{-20}{-4} \\
x &= 5
\end{aligned}
$$

Checking the solution $x = 5$:

$$
\begin{aligned}
10 - 3(x + 2) &= x - 16 \\
10 - 3(5 + 2) &\overset{?}{=} 5 - 16 \\
10 - 3(7) &\overset{?}{=} -11 \\
10 - 21 &\overset{\checkmark}{=} -11
\end{aligned}
$$

We have checked that $x = 5$ is a solution of the equation $10 - 3(x + 2) = x - 16$.

Note that the final results here are *solutions* to the equations.

> - An expression like $10 - 3(x + 2)$ can be simplified to $-3x + 4$ (as in Example 3.1.15), but we cannot solve for x in an expression.
>
> - As x takes different values, an expression has different values. In Example 3.1.16, when $x = 2$, $10 - 3(x + 2) = -2$; but when $x = 3$, $10 - 3(x + 2) = -5$.
>
> - An equation connects two expressions with an equals sign. In Example 3.1.17, $10 - 3(x + 2) = x - 16$ has the expression $10 - 3(x + 2)$ on the left side of equals sign, and the expression $x - 16$ on the right side.
>
> - When we solve the equation $10 - 3(x + 2) = x - 16$, we are looking for a number which makes those two expressions have the same value. In Example 3.1.17, we found the solution to be $x = 5$, which makes both $10 - 3(x + 2) = -11$ and $x - 16 = -11$, as shown in the checking part.

List 3.1.18: A summary the differences among simplifying expressions, evaluating expressions and solving equations:

Exercises

Warmup and Review   Solve the equation.

1. $r + 3 = -3$ **2.** $r + 9 = 6$ **3.** $t - 6 = -2$ **4.** $t - 2 = 6$

5. $44 = -4x$ **6.** $42 = -7x$ **7.** $\dfrac{10}{3}a = 2$ **8.** $\dfrac{5}{9}b = 9$

Solving Two-Step Equations   Solve the equation.

9. $6A + 4 = 58$ **10.** $2B + 2 = 8$ **11.** $8m - 6 = 18$ **12.** $5n - 5 = -50$

13. $-9 = 2q + 3$ **14.** $-23 = 8y + 1$ **15.** $-26 = 5r - 6$ **16.** $6 = 2a - 4$

17. $-5b + 3 = 48$ **18.** $-8A + 1 = 49$ **19.** $-2B - 8 = -28$ **20.** $-5m - 5 = 5$

21. $17 = -n + 9$ **22.** $5 = -q + 3$ **23.** $7y + 35 = 0$ **24.** $4r + 40 = 0$

Application Problems for Solving Two-Step Equations

25. A gym charges members $25 for a registration fee, and then $38 per month. You became a member some time ago, and now you have paid a total of $557 to the gym. How many months have passed since you joined the gym?

    [ ] months have passed since you joined the gym.

26. Your cell phone company charges a $18 monthly fee, plus $0.15 per minute of talk time. One month your cell phone bill was $90. How many minutes did you spend talking on the phone that month?

    You spent [ ] talking on the phone that month.

27. A school purchased a batch of T-shirts from a company. The company charged $7 per T-shirt, and gave the school a $60 rebate. If the school had a net expense of $2,460 from the purchase, how many T-shirts did the school buy?

    The school purchased [ ] T-shirts.

28. Joshua hired a face-painter for a birthday party. The painter charged a flat fee of $80, and then charged $5.50 per person. In the end, Joshua paid a total of $217.50. How many people used the face-painter's service?

    [ ] people used the face-painter's service.

29. A certain country has 676.8 million acres of forest. Every year, the country loses 7.52 million acres of forest mainly due to deforestation for farming purposes. If this situation continues at this pace, how many years later will the country have only 368.48 million acres of forest left? (Use an equation to solve this problem.)

    After [ ] years, this country would have 368.48 million acres of forest left.

30. Heather has $87 in her piggy bank. She plans to purchase some Pokemon cards, which costs $1.55 each. She plans to save $62.20 to purchase another toy. At most how many Pokemon cards can he purchase?

    Write an equation to solve this problem.

    Heather can purchase at most [ ] Pokemon cards.

Solving Equations with Variable Terms on Both Sides   Solve the equation.

31. $9q + 10 = q + 50$ 32. $8x + 5 = x + 26$ 33. $-6r + 9 = -r - 1$

34. $-8a + 3 = -a - 39$ **35.** $5 - 7b = 6b + 96$ **36.** $2 - 2A = 6A + 82$

37. $5B + 7 = 9B + 10$ **38.** $4m + 4 = 2m + 3$ **39.** a. $7n + 3 = 3n + 39$

  b. $3x + 3 = 7x - 29$

40. a. $9q + 10 = 3q + 34$

  b. $3C + 10 = 9C - 44$

Application Problems for Solving Equations with Variable Terms on Both Sides   Use a linear equation to solve the word problem.

41. Two trees are 6 feet and 11.5 feet tall. The shorter tree grows 2.5 feet per year; the taller tree grows 2 feet per year. How many years later would the shorter tree catch up with the taller tree?

It would take the shorter tree ________ years to catch up with the taller tree.

42. Massage Heaven and Massage You are competitors. Massage Heaven has 3400 registered customers, and it gets approximately 900 newly registered customers every month. Massage You has 10600 registered customers, and it gets approximately 450 newly registered customers every month. How many months would it take Massage Heaven to catch up with Massage You in the number of registered customers?

These two companies would have approximately the same number of registered customers ________ months later.

43. Two truck rental companies have different rates. V-Haul has a base charge of $60.00, plus $0.60 per mile. W-Haul has a base charge of $52.60, plus $0.65 per mile. For how many miles would these two companies charge the same amount?

If a driver drives ________ miles, those two companies would charge the same amount of money.

44. Massage Heaven and Massage You are competitors. Massage Heaven has 9200 registered customers, but it is losing approximately 400 registered customers every month. Massage You has 1200 registered customers, and it gets approximately 400 newly registered customers every month. How many months would it take Massage Heaven to catch up with Massage You in the number of registered customers?

These two companies would have approximately the same number of registered customers ________ months later.

45. Tammy has $85.00 in her piggy bank, and she spends $4.00 every day.

Laurie has $8.00 in her piggy bank, and she saves $1.50 every day.

If they continue to spend and save money this way, how many days later would they have the same amount of money in their piggy banks?

________ days later, Tammy and Laurie will have the same amount of money in their piggy banks.

46. Lindsay has $95.00 in her piggy bank, and she spends $4.00 every day.

Derick has $18.00 in his piggy bank, and he saves $3.00 every day.

If they continue to spend and save money this way, how many days later would they have the same amount of money in their piggy banks?

________ days later, Lindsay and Derick will have the same amount of money in their piggy banks.

Solving Linear Equations with Like Terms Solve the equation.

47. $4m + 7m + 2 = 112$

48. $9n + 2n + 2 = 90$

49. $6q + 6 + 4 = 40$

50. $3x + 10 + 3 = 22$

51. $-2 + 4 = -3r - r - 30$

52. $-6 + 8 = -5t - t - 46$

53. $2y + 3 - 7y = 38$

54. $5r + 7 - 7r = 27$

55. $-6r + 10 + r = -20$

56. $-3r + 5 + r = 5$

57. $45 = -8n - 9 - n$

58. $-58 = -5q - 4 - q$

59. $2 - x - x = -7 + 3$

60. $8 - r - r = -2 + 14$

61. $3 - 2t - 8 = -5$

62. $2 - 9b - 10 = -8$

63. $A - 8 - 5A = -6 - 8A + 26$

64. $B - 4 - 8B = -4 - 2B + 25$

65. $-10m + 2m = 10 - 2m - 40$

66. $-9n + 4n = 8 - 2n - 11$

67. $4q + 10 = -5q + 10 - 2q$

68. $10x + 5 = -3x + 5 - 2x$

69. $-8 + 9 = 7r - 6 - 10r + 4 + 2r$

70. $-2 + (-1) = 4t - 6 - 7t + 2 + 2t$

Application Problems for Solving Linear Equations with Like Terms

71. A 138-meter rope is cut into two segments. The longer segment is 28 meters longer than the shorter segment. Write and solve a linear equation to find the length of each segment. Include units.

The segments are [____________] and [____________] long.

72. In a doctor's office, the receptionist's annual salary is $142,000 less than that of the doctor. Together, the doctor and the receptionist make $208,000 per year. Find each person's annual income.

The receptionist's annual income is [____________]. The doctor's annual income is [____________].

73. Phil and Penelope went picking strawberries. Phil picked 116 fewer strawberries than Penelope did. Together, they picked 214 strawberries. How many strawberries did Penelope pick?

Penelope picked [____________] strawberries.

74. Virginia and Ross collect stamps. Ross collected 27 fewer than five times the number of Virginia's stamps. Altogether, they collected 1005 stamps. How many stamps did Virginia and Ross collect?

Virginia collected [____________] stamps. Ross collected [____________] stamps.

75. Diane and Tracei sold girl scout cookies. Diane's sales were $37 more than three times of Tracei's. Altogether, their sales were $437. How much did each girl sell?

Diane's sales were [____________]. Tracei's sales were [____________].

76. A hockey team played a total of 191 games last season. The number of games they won was 11 more than five times of the number of games they lost.

Write and solve an equation to answer the following questions.

The team lost [____________] games. The team won [____________] games.

77. After a 55% increase, a town has 155 people. What was the population before the increase?

Before the increase, the town's population was [____________].

78. After a 35% increase, a town has 270 people. What was the population before the increase?

Before the increase, the town's population was [____________].

Solving Linear Equations Involving Distribution   Solve the equation.

79. $2(t + 2) = 24$

80. $8(b + 9) = 112$

81. $5(c - 6) = 5$

82. $2(B - 3) = -18$

83. $24 = -8(m + 7)$

84. $-30 = -5(n + 1)$

85. $12 = -2(q - 5)$

86. $128 = -8(x - 9)$

87. $-(r - 4) = 8$

88. $-(t - 8) = -2$

89. $-14 = -(7 - b)$

90. $-2 = -(3 - c)$

91. $10(10B - 8) = 520$

92. $7(5C - 8) = -56$

93. $2 = -2(9 - 2n)$

94. $110 = -5(3 - 5q)$

95. $3 + 9(x + 8) = 111$

96. $1 + 6(r + 7) = 31$

97. $5 - 8(t + 7) = 13$

98. $3 - 10(b + 7) = -137$

99. $22 = 2 - 4(c - 7)$

100. $97 = 9 - 8(B - 7)$

101. $3 - 6(C - 7) = 105$

102. $1 - 8(n - 7) = 17$

103. $3 = 9 - (5 - q)$

104. $-1 = 8 - (3 - x)$

105. $1 - (r + 10) = -18$

106. $4 - (t + 7) = -6$

107.   a. $5 + (b + 4) = 12$
   b. $5 - (b + 4) = 12$

108.   a. $2 + (c + 1) = -6$
   b. $2 - (c + 1) = -6$

109. $4(B + 5) - 10(B - 7) = 90$

110. $3(C + 10) - 8(C - 2) = 46$

111. $5 + 8(n - 5) = -28 - (7 - 2n)$

112. $4 + 9(p - 10) = -84 - (2 - 2p)$

113. $7(x - 2) - x = 67 - 3(7 + 3x)$

114. $10(r - 6) - r = -90 - 3(2 + 3r)$

115. $7(-10t + 10) = 14(-2 - 6t)$

116. $3(-10b + 6) = 6(-9 - 6b)$

117. $23 + 6(6 - 4c) = -4(c - 13) + 7$

118. $12 + 4(3 - 3B) = -4(B - 4) + 8$

Application Problems for Solving Linear Equations Involving Distribution

119. A rectangle's perimeter is 78 cm. Its base is 27 cm.

Its height is [ ].

120. A rectangle's perimeter is 58 m. Its width is 12 m. Use an equation to solve for the rectangle's length.

Its length is [ ].

121. A rectangle's perimeter is 116 in. Its length is 8 in longer than its width. Use an equation to find the rectangle's length and width.

Its width is [ ].

Its length is [ ].

122. A rectangle's perimeter is 120 cm. Its length is 2 times as long as its width. Use an equation to find the rectangle's length and width.

It's width is [ ].

Its length is [ ].

123. A rectangle's perimeter is 106 ft. Its length is 2 ft shorter than four times its width. Use an equation to find the rectangle's length and width.

Its width is [ ].

Its length is [ ].

124. A rectangle's perimeter is 184 ft. Its length is 4 ft longer than three times its width. Use an equation to find the rectangle's length and width.

Its width is [ ].

Its length is [ ].

Comparisons

125. Solve the equation.

   a. $-b + 7 = 7$

   b. $-y + 7 = -7$

   c. $-r - 7 = 7$

   d. $-a - 7 = -7$

126. Solve the equation.

   a. $-c + 4 = 4$

   b. $-m + 4 = -4$

   c. $-B - 4 = 4$

   d. $-y - 4 = -4$

127. a. Solve the following linear equation:

   $r - 2 = 8$

   b. Evaluate the following expression when $r = 10$:

   $r - 2 = $ [ ]

128. a. Solve the following linear equation:

   $r - 8 = -3$

   b. Evaluate the following expression when $r = 5$:

   $r - 8 = $ [ ]

129. a. Solve the following linear equation:

$$4(t + 6) - 4 = 36$$

   b. Evaluate the following expression when $t = 4$:

   $4(t + 6) - 4 = $ [______________]

   c. Simplify the following expression:

   $4(t + 6) - 4 = $ [______________]

130. a. Solve the following linear equation:

$$3(t - 5) + 9 = 6$$

   b. Evaluate the following expression when $t = 4$:

   $3(t - 5) + 9 = $ [______________]

   c. Simplify the following expression:

   $3(t - 5) + 9 = $ [______________]

131. Choose True or False for the following questions about the difference between expressions and equations.

   a. We can evaluate $-10x - 10$ when $x = 1$
   (□ True   □ False)

   b. $-10x - 10 = -10x - 10$ is an equation.
   (□ True   □ False)

   c. We can evaluate $-10x - 10 = -10x - 10$ when $x = 1$ (□ True   □ False)

   d. We can check whether $x = 1$ is a solution of $-10x - 10$.   (□ True   □ False)

   e. $-10x - 10$ is an expression.   (□ True   □ False)

   f. We can check whether $x = 1$ is a solution of $-10x - 10 = -10x - 10$.   (□ True   □ False)

   g. $-10x - 10 = -10x - 10$ is an expression.
   (□ True   □ False)

   h. $-10x - 10$ is an equation.   (□ True   □ False)

132. Choose True or False for the following questions about the difference between expressions and equations.

   a. $-7x + 4$ is an expression.   (□ True   □ False)

   b. $4x - 7$ is an equation.   (□ True   □ False)

   c. $-7x + 4 = 4x - 7$ is an equation.
   (□ True   □ False)

   d. We can check whether $x = 1$ is a solution of $-7x + 4$.   (□ True   □ False)

   e. $-7x + 4 = 4x - 7$ is an expression.
   (□ True   □ False)

   f. We can evaluate $-7x + 4 = 4x - 7$ when $x = 1$ (□ True   □ False)

   g. We can evaluate $-7x + 4$ when $x = 1$
   (□ True   □ False)

   h. We can check whether $x = 1$ is a solution of $-7x + 4 = 4x - 7$.   (□ True   □ False)

Challenge

133. Think of a number. Add four to your number. Now double that. Then add six. Then halve it. Finally, subtract 7. What is the result? Do you always get the same result, regardless of what number you start with? How does this work? Explain using algebra.

134. Write a linear equation whose solution is $x = -9$.

Note that you may not write an equation whose left side is just "x" or whose right side is just "x."

There are infinitely many correct answers to this problem. Be creative. After finding an equation that works, see if you can come up with a different one that also works.

3.2 Solving Multistep Linear Inequalities

We have learned how to solve one-step inequalities in Section. In this section, we will learn how to solve multistep inequalities.

3.2.1 Solving Multistep Inequalities

When solving a linear inequality, we follow the same steps in List 3.1.4. The only difference in our steps to solving is that when we multiply or divide by a negative number on both sides of an inequality, the direction of the inequality symbol must switch. We will look at some examples.

Simplify Simplify the expressions on each side of the inequality by distributing and combining like terms.

Isolate Use addition or subtraction to isolate the variable terms and constant terms (numbers) so that they are on different sides of the inequality symbol.

Eliminate Use multiplication or division to eliminate the variable term's coefficient. If each side of the inequality is multiplied or divided by a negative number, switch the direction of the inequality symbol.

Check When specified, verify the infinite solution set by checking multiple solutions.

Summarize State the solution set or (in the case of an application problem) summarize the result in a complete sentence using appropriate units.

List 3.2.2: Steps to Solve Linear Inequalities

Example 3.2.3 Solve for t in the inequality $-3t + 5 \geq 11$. Write the solution set in both set-builder notation and interval notation.

Explanation.

$$-3t + 5 \geq 11$$
$$-3t + 5 - 5 \geq 11 - 5$$
$$-3t \geq 6$$
$$\frac{-3t}{-3} \leq \frac{6}{-3}$$
$$t \leq -2$$

Note that when we divided both sides of the inequality by -3, we had to switch the direction of the inequality symbol.

The solution set in set-builder notation is $\{t \mid t \leq -2\}$.

The solution set in interval notation is $(-\infty, -2]$.

Remark 3.2.4. Since the inequality solved in Example 3.2.3 has infinitely many solutions, it's difficult to check. We found that all values of t for which $t \le -2$ are solutions, so one approach is to check if -2 is a solution and additionally if one other number less than -2 is a solution.

Here, we'll check that -2 satisfies this inequality:

$$-3t + 5 \ge 11$$
$$-3(-2) + 5 \overset{?}{\ge} 11$$
$$6 + 5 \overset{?}{\ge} 11$$
$$11 \overset{\checkmark}{\ge} 11$$

Next, we can check another number smaller than -2, such as -5:

$$-3t + 5 \ge 11$$
$$-3(-5) + 5 \overset{?}{\ge} 11$$
$$15 + 5 \overset{?}{\ge} 11$$
$$20 \overset{\checkmark}{\ge} 11$$

Thus both -2 and -5 are solutions. It's important to note that this doesn't directly verify that *all* solutions to this inequality check. It's valuable though in that it would likely help us catch an error if we had made one. Consult your instructor to see if you're expected to check your answer in this manner.

Example 3.2.5 Solve for z in the inequality $(6z + 5) - (2z - 3) < -12$. Write the solution set in both set-builder notation and interval notation.

Explanation.

$$(6z + 5) - (2z - 3) < -12$$
$$6z + 5 - 2z + 3 < -12$$
$$4z + 8 < -12$$
$$4z + 8 - 8 < -12 - 8$$
$$4z < -20$$
$$\frac{4z}{4} < \frac{-20}{4}$$
$$z < -5$$

Note that we divided both sides of the inequality by 4 and since this is a positive number we *did not* need to switch the direction of the inequality symbol.

The solution set in set-builder notation is $\{z \mid z < -5\}$.

The solution set in interval notation is $(-\infty, -5)$.

Example 3.2.6 Solve for x in $-2 - 2(2x + 1) > 4 - (3 - x)$. Write the solution set in both set-builder notation and interval notation.

Explanation.

$$-2 - 2(2x + 1) > 4 - (3 - x)$$
$$-2 - 4x - 2 > 4 - 3 + x$$
$$-4x - 4 > x + 1$$
$$-4x - 4 - x > x + 1 - x$$

$$-5x - 4 > 1$$
$$-5x - 4 + 4 > 1 + 4$$
$$-5x > 5$$
$$\frac{-5x}{-5} < \frac{5}{-5}$$
$$x < -1$$

Note that when we divided both sides of the inequality by -5, we had to switch the direction of the inequality symbol.

The solution set in set-builder notation is $\{x \mid x < -1\}$.

The solution set in interval notation is $(-\infty, -1)$.

Example 3.2.7 When a stopwatch started, the pressure inside a gas container was 4.2 atm (standard atmospheric pressure). As the container was heated, the pressure increased by 0.7 atm per minute. The maximum pressure the container can handle was 21.7 atm. Heating must be stopped once the pressure reaches 21.7 atm. In what time interval was the container safe?

Explanation. The pressure increases by 0.7 atm per minute, so it increases by $0.7m$ after m minutes. Counting in the original pressure of 4.2 atm, pressure in the container can be modeled by $0.7m + 4.2$, where m is the number of minutes since the stop watch started.

The container is safe when the pressure is 21.7 atm or lower. We can write and solve this inequality:

$$0.7m + 4.2 \le 21.7$$
$$0.7m + 4.2 - 4.2 \le 21.7 - 4.2$$
$$0.7m \le 17.5$$
$$\frac{0.7m}{0.7} \le \frac{17.5}{0.7}$$
$$m \le 25$$

In summary, the container was safe as long as $m \le 25$. Assuming that m also must be greater than or equal to zero, this means $0 \le m \le 25$. We can write this as the time interval as $[0, 25]$. Thus the container was safe between 0 minutes and 25 minutes.

Exercises

Review and Warmup

1. Solve this inequality.

$x + 3 > 9$

In set-builder notation, the solution set is ____________.

In interval notation, the solution set is ____________.

2. Solve this inequality.

$x + 3 > 7$

In set-builder notation, the solution set is ____________.

In interval notation, the solution set is ____________.

3. Solve this inequality.

$4 > x - 10$

In set-builder notation, the solution set is ____________.

In interval notation, the solution set is ____________.

4. Solve this inequality.

$$5 > x - 8$$

In set-builder notation, the solution set is ______.

In interval notation, the solution set is ______.

5. Solve this inequality.

$$5x \leq 10$$

In set-builder notation, the solution set is ______.

In interval notation, the solution set is ______.

6. Solve this inequality.

$$2x \leq 8$$

In set-builder notation, the solution set is ______.

In interval notation, the solution set is ______.

7. Solve this inequality.

$$6 \geq -2x$$

In set-builder notation, the solution set is ______.

In interval notation, the solution set is ______.

8. Solve this inequality.

$$6 \geq -3x$$

In set-builder notation, the solution set is ______.

In interval notation, the solution set is ______.

9. Solve this inequality.

$$\frac{4}{9}x > 12$$

In set-builder notation, the solution set is ______.

In interval notation, the solution set is ______.

10. Solve this inequality.

$$\frac{5}{6}x > 5$$

In set-builder notation, the solution set is ______.

In interval notation, the solution set is ______.

11. A swimming pool is being filled with water from a garden hose at a rate of 5 gallons per minute. If the pool already contains 60 gallons of water and can hold 160 gallons, after how long will the pool overflow?

Assume m minutes later, the pool would overflow. Write an equation to model this scenario. There is no need to solve it.

12. An engineer is designing a cylindrical springform pan. The pan needs to be able to hold a volume of 305 cubic inches and have a diameter of 12 inches. What's the minimum height it can have? (Hint: The formula for the volume of a cylinder is $V = \pi r^2 h$).

Assume the pan's minimum height is h inches. Write an equation to model this scenario. There is no need to solve it.

Solving Multistep Linear Inequalities Solve this inequality.

13. $9x + 6 > 42$

In set-builder notation, the solution set is ______.

In interval notation, the solution set is ______.

14. $10x + 3 > 83$

In set-builder notation, the solution set is ______.

In interval notation, the solution set is ______.

15. $4 \geq 3x - 5$

In set-builder notation, the solution set is ______.

In interval notation, the solution set is ______.

16. $25 \geq 4x - 3$

In set-builder notation, the solution set is [____].

In interval notation, the solution set is [____].

17. $41 \leq 1 - 4x$

In set-builder notation, the solution set is [____].

In interval notation, the solution set is [____].

18. $33 \leq 8 - 5x$

In set-builder notation, the solution set is [____].

In interval notation, the solution set is [____].

19. $-6x - 4 < -58$

In set-builder notation, the solution set is [____].

In interval notation, the solution set is [____].

20. $-7x - 1 < -29$

In set-builder notation, the solution set is [____].

In interval notation, the solution set is [____].

21. $4 \geq -8x + 4$

In set-builder notation, the solution set is [____].

In interval notation, the solution set is [____].

22. $3 \geq -9x + 3$

In set-builder notation, the solution set is [____].

In interval notation, the solution set is [____].

23. $-5 > 5 - x$

In set-builder notation, the solution set is [____].

In interval notation, the solution set is [____].

24. $-8 > 1 - x$

In set-builder notation, the solution set is [____].

In interval notation, the solution set is [____].

25. $3(x + 4) \geq 24$

In set-builder notation, the solution set is [____].

In interval notation, the solution set is [____].

26. $4(x + 8) \geq 72$

In set-builder notation, the solution set is [____].

In interval notation, the solution set is [____].

27. $7t + 7 < 1t + 37$

In set-builder notation, the solution set is [____].

In interval notation, the solution set is [____].

28. $8t + 4 < 3t + 19$

In set-builder notation, the solution set is [____].

In interval notation, the solution set is [____].

29. $-9z + 6 \leq -z - 58$

In set-builder notation, the solution set is [____].

In interval notation, the solution set is [____].

30. $-8z + 7 \leq -z - 28$

In set-builder notation, the solution set is [____].

In interval notation, the solution set is [____].

31. $a - 9 - 3a > -9 - 4a + 12$

In set-builder notation, the solution set is [____].

In interval notation, the solution set is [____].

32. $a - 9 - 9a > -10 - 10a + 7$

In set-builder notation, the solution set is [____].

In interval notation, the solution set is [____].

33. $-8p + 6 - 8p \geq 2p + 6$

In set-builder notation, the solution set is [____].

In interval notation, the solution set is [____].

34. $-2p + 3 - 6p \geq 3p + 3$

In set-builder notation, the solution set is [   ].

In interval notation, the solution set is [   ].

35. $64 < -4(p - 9)$

In set-builder notation, the solution set is [   ].

In interval notation, the solution set is [   ].

36. $-20 < -5(p - 5)$

In set-builder notation, the solution set is [   ].

In interval notation, the solution set is [   ].

37. $-(x - 2) \geq 8$

In set-builder notation, the solution set is [   ].

In interval notation, the solution set is [   ].

38. $-(x - 8) \geq 13$

In set-builder notation, the solution set is [   ].

In interval notation, the solution set is [   ].

39. $47 \leq 7 - 5(z - 1)$

In set-builder notation, the solution set is [   ].

In interval notation, the solution set is [   ].

40. $20 \leq 8 - 2(z - 8)$

In set-builder notation, the solution set is [   ].

In interval notation, the solution set is [   ].

41. $5 - (y + 9) < 6$

In set-builder notation, the solution set is [   ].

In interval notation, the solution set is [   ].

42. $1 - (y + 7) < -4$

In set-builder notation, the solution set is [   ].

In interval notation, the solution set is [   ].

43. $1 + 7(x - 9) < -15 - (7 - 3x)$

In set-builder notation, the solution set is [   ].

In interval notation, the solution set is [   ].

44. $2 + 9(x - 5) < -20 - (7 - 5x)$

In set-builder notation, the solution set is [   ].

In interval notation, the solution set is [   ].

Applications

45. You are riding in a taxi and can only pay with cash. You have to pay a flat fee of $25, and then pay $2.80 per mile. You have a total of $193 in your pocket.

Let x be the number of miles the taxi will drive you. You want to know how many miles you can afford. Write an inequality to represent this situation in terms of how many miles you can afford:

[   ] [   ]

Solve this inequality. At most how many miles can you afford?

You can afford at most [   ] miles.

Use interval notation to express the number of miles you can afford.

46. You are riding in a taxi and can only pay with cash. You have to pay a flat fee of $30, and then pay $3.50 per mile. You have a total of $135 in your pocket.

Let x be the number of miles the taxi will drive you. You want to know how many miles you can afford. Write an inequality to represent this situation in terms of how many miles you can afford:

Solve this inequality. At most how many miles can you afford?

You can afford at most   miles.

Use interval notation to express the number of miles you can afford.

47. A car rental company offers the following two plans for renting a car:

Plan A: $31 per day and 16 cents per mile

Plan B: $49 dollars per day with free unlimited mileage

How many miles must one drive in order to justify choosing Plan B?

One must drive more than   miles to justify choosing Plan B. In other words, it's more economical to use plan B if your number of miles driven will be in the interval   (answer with interval notation).

48. A car rental company offers the following two plans for renting a car:

Plan A: $29 per day and 17 cents per mile

Plan B: $52 dollars per day with free unlimited mileage

How many miles must one drive in order to justify choosing Plan B?

One must drive more than   miles to justify choosing Plan B. In other words, it's more economical to use plan B if your number of miles driven will be in the interval   (answer with interval notation).

49. You are offered two different sales jobs. The first company offers a straight commission of 9% of the sales. The second company offers a salary of $300 per week *plus* 4% of the sales. How much would you have to sell in a week in order for the straight commission offer to be at least as good?

You'd have to sell more than   worth of goods for the straight commission to be better for you. In other words, the dollar amount of goods sold would have to be in the interval   (answer using interval notation).

50. You are offered two different sales jobs. The first company offers a straight commission of 8% of the sales. The second company offers a salary of $430 per week *plus* 4% of the sales. How much would you have to sell in a week in order for the straight commission offer to be at least as good?

You'd have to sell more than [____] worth of goods for the straight commission to be better for you. In other words, the dollar amount of goods sold would have to be in the interval [____] (answer using interval notation).

3.3 Linear Equations and Inequalities with Fractions

In this section, we will learn how to solve linear equations and inequalities with fractions.

3.3.1 Introduction

So far, in our last step of solving for a variable we have divided each side of the equation by a constant, as in:

$$2x = 10$$

$$\frac{2x}{2} = \frac{10}{2}$$

$$x = 5$$

If we have a coefficient that is a fraction, we *could* proceed in exactly the same manner:

$$\frac{1}{2}x = 10$$

$$\frac{\frac{1}{2}x}{\frac{1}{2}} = \frac{10}{\frac{1}{2}}$$

$$x = 10 \cdot \frac{2}{1} = 20$$

What if our equation or inequality was more complicated though, for example $\frac{1}{4}x + \frac{2}{3} = \frac{1}{6}$? We would have to first do a lot of fraction arithmetic in order to then divide each side by the coefficient of x. An alternate approach is to instead *multiply* each side of the equation by a chosen constant that eliminates the denominator. In the equation $\frac{1}{2}x = 10$, we could simply multiply each side of the equation by 2, which would eliminate the denominator of 2:

$$\frac{1}{2}x = 10$$

$$2 \cdot \left(\frac{1}{2}x\right) = 2 \cdot 10$$

$$x = 20$$

For more complicated equations, we will multiply each side of the equation by the least common denominator (LCD) of all fractions contained in the equation.

3.3.2 Eliminating Denominators

Deshawn planted a sapling in his yard that was 4-feet tall. The tree will grow $\frac{2}{3}$ of a foot every year. How many years will it take for his tree to be 10 feet tall?

Since the tree grows $\frac{2}{3}$ of a foot every year, we can use a table to help write a formula modeling the tree's growth:

Years Passed	Tree's Height (ft)
0	4
1	$4 + \frac{2}{3}$
2	$4 + \frac{2}{3} \cdot 2$
$\vdots$	$\vdots$
y	$4 + \frac{2}{3}y$

Example 3.3.2 From this, we've determined that y years since the tree was planted, the tree's height will be $4 + \frac{2}{3}y$ feet.

To find when Deshawn's tree will be 10 feet tall, we write and solve this equation:

$$4 + \frac{2}{3}y = 10$$

$$3 \cdot \left(4 + \frac{2}{3}y\right) = 3 \cdot 10$$

$$3 \cdot 4 + 3 \cdot \frac{2}{3}y = 30$$

$$12 + 2y = 30$$

$$2y = 18$$

$$y = 9$$

Now we will check the solution 9 in the equation $4 + \frac{2}{3}y = 10$:

$$4 + \frac{2}{3}y = 10$$

$$4 + \frac{2}{3}(9) \stackrel{?}{=} 10$$

$$4 + 6 \stackrel{\checkmark}{=} 10$$

In summary, it will take 9 years for Deshawn's tree to reach 10 feet tall.

Let's look at a few more examples.

Example 3.3.3 Solve for x in $\frac{1}{4}x + \frac{2}{3} = \frac{1}{6}$.

Explanation. To solve this equation, we first need to identify the LCD of all fractions in the equation. On the left side we have $\frac{1}{4}$ and $\frac{2}{3}$. On the right side we have $\frac{1}{6}$. The LCD of $3, 4$, and 6 is 12, so we will multiply each side of the equation by 12 in order to eliminate *all* of the denominators:

$$\frac{1}{4}x + \frac{2}{3} = \frac{1}{6}$$

$$12 \cdot \left(\frac{1}{4}x + \frac{2}{3}\right) = 12 \cdot \frac{1}{6}$$

$$12 \cdot \left(\frac{1}{4}x\right) + 12 \cdot \left(\frac{2}{3}\right) = 12 \cdot \frac{1}{6}$$

$$3x + 8 = 2$$

$$3x = -6$$

$$\frac{3x}{3} = \frac{-6}{3}$$

$$x = -2$$

Checking the solution -2:

$$\frac{1}{4}x + \frac{2}{3} = \frac{1}{6}$$

$$\frac{1}{4}(-2) + \frac{2}{3} \stackrel{?}{=} \frac{1}{6}$$

$$-\frac{2}{4} + \frac{2}{3} \stackrel{?}{=} \frac{1}{6}$$

$$-\frac{6}{12} + \frac{8}{12} \stackrel{?}{=} \frac{1}{6}$$

$$\frac{2}{12} \stackrel{\checkmark}{=} \frac{1}{6}$$

The solution is therefore -2 and the solution set is $\{-2\}$.

Example 3.3.4 Solve for z in $-\frac{2}{5}z - \frac{3}{2} = -\frac{1}{2}z + \frac{4}{5}$.

Explanation.

The first thing we need to do is identify the LCD of all denominators in this equation. Since the denominators are 2 and 5, the LCD is 10. So as our first step, we will multiply each side of the equation by 10 in order to eliminate all denominators:

$$-\frac{2}{5}z - \frac{3}{2} = -\frac{1}{2}z + \frac{4}{5}$$

$$10 \cdot \left(-\frac{2}{5}z - \frac{3}{2}\right) = 10 \cdot \left(-\frac{1}{2}z + \frac{4}{5}\right)$$

$$10\left(-\frac{2}{5}z\right) - 10\left(\frac{3}{2}\right) = 10\left(-\frac{1}{2}z\right) + 10\left(\frac{4}{5}\right)$$

$$-4z - 15 = -5z + 8$$

$$z - 15 = 8$$

$$z = 23$$

Thus the solution is 23 and so the solution set is $\{23\}$.

Checking the solution 23:

$$-\frac{2}{5}z - \frac{3}{2} = -\frac{1}{2}z + \frac{4}{5}$$

$$-\frac{2}{5}(23) - \frac{3}{2} \overset{?}{=} -\frac{1}{2}(23) + \frac{4}{5}$$

$$-\frac{46}{5} - \frac{3}{2} \overset{?}{=} -\frac{23}{2} + \frac{4}{5}$$

$$-\frac{46}{5} \cdot \frac{2}{2} - \frac{3}{2} \cdot \frac{5}{5} \overset{?}{=} -\frac{23}{2} \cdot \frac{5}{5} + \frac{4}{5} \cdot \frac{2}{2}$$

$$-\frac{92}{10} - \frac{15}{10} \overset{?}{=} -\frac{115}{10} + \frac{8}{10}$$

$$-\frac{107}{10} \overset{\checkmark}{=} -\frac{107}{10}$$

Example 3.3.5 Solve for a in the equation $\frac{2}{3}(a + 1) + 5 = \frac{1}{3}$.

Explanation.

$$\frac{2}{3}(a + 1) + 5 = \frac{1}{3}$$

$$3 \cdot \left(\frac{2}{3}(a + 1) + 5\right) = 3 \cdot \frac{1}{3}$$

$$3 \cdot \frac{2}{3}(a + 1) + 3 \cdot 5 = 1$$

$$2(a + 1) + 15 = 1$$

$$2a + 2 + 15 = 1$$

$$2a + 17 = 1$$

$$2a = -16$$

$$a = -8$$

Check the solution -8 in the equation $\frac{2}{3}(a + 1) + 5 = \frac{1}{3}$, we find that:

$$\frac{2}{3}(a + 1) + 5 = \frac{1}{3}$$

$$\frac{2}{3}(-8 + 1) + 5 \overset{?}{=} \frac{1}{3}$$

$$\frac{2}{3}(-7) + 5 \overset{?}{=} \frac{1}{3}$$

$$-\frac{14}{3} + \frac{15}{3} \overset{\checkmark}{=} \frac{1}{3}$$

The solution is therefore -8 and the solution set is $\{-8\}$.

Example 3.3.6 Solve for b in the equation $\frac{2b+1}{3} = \frac{2}{5}$.

Explanation.

$$\frac{2b+1}{3} = \frac{2}{5}$$

$$15 \cdot \frac{2b+1}{3} = 15 \cdot \frac{2}{5}$$

$$5(2b+1) = 6$$

$$10b + 5 = 6$$

$$10b = 1$$

$$b = \frac{1}{10}$$

Checking the solution $\frac{1}{10}$:

$$\frac{2b+1}{3} = \frac{2}{5}$$

$$\frac{2\left(\frac{1}{10}\right)+1}{3} \overset{?}{=} \frac{2}{5}$$

$$\frac{\frac{1}{5}+1}{3} \overset{?}{=} \frac{2}{5}$$

$$\frac{\frac{1}{5}+\frac{5}{5}}{3} \overset{?}{=} \frac{2}{5}$$

$$\frac{\frac{6}{5}}{3} \overset{?}{=} \frac{2}{5}$$

$$\frac{6}{5} \cdot \frac{1}{3} \overset{\checkmark}{=} \frac{2}{5}$$

The solution is $\frac{1}{10}$ and the solution set is $\left\{\frac{1}{10}\right\}$.

Remark 3.3.7. You might know about solving Example 3.3.6 with a technique called **cross-multiplication**. Cross-multiplication is a specialized application of the process of clearing the denominators from an equation. This process will be discussed in Section 3.5.

Example 3.3.8 In a science lab, a container had 21 ounces of water at 9:00 A.M.. Water has been evaporating at the rate of 3 ounces every 5 minutes. When will there be 8 ounces of water left?

Explanation. Since the container has been losing 3 oz of water every 5 minutes, it loses $\frac{3}{5}$ oz every minute. In m minutes since 9:00 A.M., the container would lose $\frac{3}{5}m$ oz of water. Since the container had 21 oz of water at the beginning, the amount of water in the container can be modeled by $21 - \frac{3}{5}m$ (in oz).

To find when there would be 8 oz of water left, we write and solve this equation:

$$21 - \frac{3}{5}m = 8$$

$$5 \cdot \left(21 - \frac{3}{5}x\right) = 5 \cdot 8$$

$$5 \cdot 21 - 5 \cdot \frac{3}{5}x = 40$$

$$105 - 3m = 40$$

$$105 - 3m - 105 = 40 - 105$$

$$-3m = -65$$

$$\frac{-3m}{-3} = \frac{-65}{-3}$$

$$m = \frac{65}{3}$$

Checking the solution $\frac{65}{3}$:

$$21 - \frac{3}{5}m = 8$$

$$21 - \frac{3}{5}\left(\frac{65}{3}\right) \overset{?}{=} 8$$

$$21 - 13 \overset{\checkmark}{=} 8$$

Therefore the solution is $\frac{65}{3}$. As a mixed number, this is $21\frac{2}{3}$. In context, this means that 21 minutes and 40 seconds after 9:00 A.M., at 9:21:40 A.M., the container will have 8 ounces of water left.

Checkpoint 3.3.9. Solve the equation.

$$21 = \frac{x}{5} + \frac{x}{2}$$

Explanation. To clear fractions in an equation, we multiply each term by a common denominator. For this problem, a common denominator is 10.

$$21 = \frac{x}{5} + \frac{x}{2}$$
$$10 \cdot 21 = 10 \cdot \frac{x}{5} + 10 \cdot \frac{x}{2}$$
$$210 = 2x + 5x$$
$$210 = 7x$$
$$\frac{210}{7} = \frac{7x}{7}$$
$$30 = x$$
$$x = 30$$

The solution to this equation is 30. To stress that this is a value assigned to x, some report $x = 30$. We can also say that the solution set is $\{30\}$, or that $x \in \{30\}$. If we substitute 30 in for x in the original equation $21 = \frac{x}{5} + \frac{x}{2}$, the equation will be true. Please check this on your own; it is an important habit.

3.3.3 Solving Inequalities with Fractions

We can also solve linear inequalities involving fractions by multiplying each side of the inequality by the LCD of all fractions within the inequality. Remember that with inequalities, everything works exactly the same except that the inequality sign reverses direction whenever we multiply each side of the inequality by a negative number.

Example 3.3.10 Solve for x in the inequality $\frac{3}{4}x - 2 > \frac{4}{5}x$. Write the solution set in both set-builder notation and interval notation.

Explanation.

$$\frac{3}{4}x - 2 > \frac{4}{5}x$$
$$20 \cdot \left(\frac{3}{4}x - 2\right) > 20 \cdot \frac{4}{5}x$$
$$20 \cdot \frac{3}{4}x - 20 \cdot 2 > 16x$$
$$15x - 40 > 16x$$
$$15x - 40 - 15x > 16x - 15x$$
$$-40 > x$$
$$x < -40$$

The solution set in set-builder notation is $\{x \mid x < -40\}$. Note that it's equivalent to write $\{x \mid -40 > x\}$, but it's easier to understand if we write x first in an inequality.

The solution set in interval notation is $(-\infty, -40)$.

Example 3.3.11 Solve for y in the inequality $\frac{4}{7} - \frac{4}{3}y \leq \frac{2}{3}$. Write the solution set in both set-builder notation and interval notation.

Explanation.

$$\frac{4}{7} - \frac{4}{3}y \leq \frac{2}{3}$$

$$21 \cdot \left(\frac{4}{7} - \frac{4}{3}y\right) \leq 21 \cdot \left(\frac{2}{3}\right)$$

$$21\left(\frac{4}{7}\right) - 21\left(\frac{4}{3}y\right) \leq 21\left(\frac{2}{3}\right)$$

$$12 - 28y \leq 14$$

$$-28y \leq 2$$

$$\frac{-28y}{-28} \geq \frac{2}{-28}$$

$$y \geq -\frac{1}{14}$$

Note that when we divided each side of the inequality by -28, the inequality symbol reversed direction.

The solution set in set-builder notation is $\left\{y \mid y \geq -\frac{1}{14}\right\}$.

The solution set in interval notation is $\left[-\frac{1}{14}, \infty\right)$.

Example 3.3.12 In a certain class, a student's grade is calculated by the average of their scores on 3 tests. Aidan scored 78% and 54% on the first two tests. If he wants to earn at least a grade of C (70%), what's the lowest score he needs to earn on the third exam?

Explanation. Assume Aidan will score x% on the third test. To make his average test score greater than or equal to 70%, we write and solve this inequality:

$$\frac{78 + 54 + x}{3} \geq 70$$

$$\frac{132 + x}{3} \geq 70$$

$$3 \cdot \frac{132 + x}{3} \geq 3 \cdot 70$$

$$132 + x \geq 210$$

$$x \geq 78$$

To earn at least a C grade, Aidan needs to score at least 78% on the third test.

Exercises

Review and Warmup

1. Multiply: $7 \cdot \dfrac{1}{10}$

2. Multiply: $2 \cdot \dfrac{2}{7}$

3. Multiply: $9 \cdot \left(-\dfrac{4}{3}\right)$

4. Multiply: $45 \cdot \left(-\dfrac{5}{9}\right)$

5. Do the following multiplications:

   a. $14 \cdot \dfrac{4}{7} = \boxed{}$

   b. $21 \cdot \dfrac{4}{7} = \boxed{}$

   c. $28 \cdot \dfrac{4}{7} = \boxed{}$

6. Do the following multiplications:

   a. $20 \cdot \dfrac{4}{5} = \boxed{}$

   b. $25 \cdot \dfrac{4}{5} = \boxed{}$

   c. $30 \cdot \dfrac{4}{5} = \boxed{}$

Solving Linear Equations with Fractions   Solve the equation.

7. $\dfrac{n}{9} + 88 = 5n$

8. $\dfrac{p}{6} + 92 = 4p$

9. $\dfrac{x}{3} + 3 = 9$

10. $\dfrac{y}{9} + 10 = 13$

11. $5 - \dfrac{t}{8} = 0$

12. $1 - \dfrac{a}{3} = -3$

13. $2 = 8 - \dfrac{2c}{7}$

14. $-26 = 4 - \dfrac{10A}{3}$

15. $3C = \dfrac{5C}{2} + 4$

16. $3m = \dfrac{9m}{8} + 45$

17. $51 = \dfrac{2}{5}p + 3p$

18. $150 = \dfrac{8}{7}x + 6x$

19. $81 - \dfrac{3}{8}y = 3y$

20. $45 - \dfrac{7}{4}t = 2t$

21. $6a = \dfrac{10}{9}a + 10$

22. $3c = \dfrac{8}{5}c + 7$

23. $\dfrac{9}{4} - 5A = 4$

24. $\dfrac{3}{10} - 2C = 3$

25. $\dfrac{7}{6} - \dfrac{1}{6}m = 9$

26. $\dfrac{3}{4} - \dfrac{1}{4}p = 6$

27. $\dfrac{4x}{9} - 5 = -\dfrac{73}{9}$

28. $\dfrac{8y}{7} - 10 = -\dfrac{102}{7}$

29. $\dfrac{4}{5} + \dfrac{8}{5}t = 3t$

30. $\dfrac{8}{9} + \dfrac{2}{3}a = 3a$

31. $\dfrac{3c}{5} - \dfrac{8}{5} = -\dfrac{1}{5}c$

32. $\dfrac{2A}{7} - \dfrac{25}{7} = -\dfrac{3}{7}A$

33. $\dfrac{8C}{9} + \dfrac{1}{8} = C$

34. $\dfrac{2m}{5} + \dfrac{1}{4} = m$

35. $\dfrac{2p}{3} - 57 = -\dfrac{5}{2}p$

36. $\dfrac{2q}{7} - 19 = -\dfrac{1}{6}q$

37. $-\dfrac{9}{8}y + 81 = \dfrac{9y}{16}$

38. $-\dfrac{1}{2}t + 5 = \dfrac{3t}{4}$

39. $\dfrac{5a}{8} - 5a = \dfrac{9}{16}$

40. $\dfrac{9c}{4} - 5c = \dfrac{7}{8}$

41. $\dfrac{5A}{2} + \dfrac{8}{5} = \dfrac{3}{8}A$

42. $\dfrac{9C}{4} + \dfrac{8}{3} = \dfrac{5}{6}C$

43. $\dfrac{4}{3}m = \dfrac{3}{2} + \dfrac{3m}{7}$

44. $\dfrac{4}{7}p = \dfrac{4}{5} + \dfrac{5p}{3}$

45. $\dfrac{9}{8} = \dfrac{q}{24}$

46. $\dfrac{5}{4} = \dfrac{y}{24}$

47. $-\dfrac{t}{45} = \dfrac{8}{9}$

48. $-\dfrac{a}{28} = \dfrac{2}{7}$

49. $-\dfrac{c}{8} = -\dfrac{7}{4}$

50. $-\dfrac{A}{60} = -\dfrac{3}{10}$

51. $-\dfrac{7}{6} = \dfrac{8C}{9}$

52. $-\dfrac{8}{3} = \dfrac{8m}{5}$

53. $\dfrac{3}{10} = \dfrac{p+9}{50}$

54. $\dfrac{7}{6} = \dfrac{q+8}{36}$

55. $\dfrac{3}{2} = \dfrac{y-9}{7}$

56. $\dfrac{7}{8} = \dfrac{r-9}{3}$

57. $\dfrac{a-10}{4} = \dfrac{a+10}{6}$

58. $\dfrac{c-4}{2} = \dfrac{c+7}{4}$

59. $\dfrac{A+8}{6} - \dfrac{A-4}{12} = \dfrac{13}{6}$

60. $\dfrac{C+2}{4} - \dfrac{C-2}{8} = \dfrac{11}{8}$

61. $\dfrac{m}{3} - 15 = \dfrac{m}{8}$

62. $\dfrac{p}{7} - 4 = \dfrac{p}{9}$

63. $\dfrac{q}{4} - 3 = \dfrac{q}{7} + 6$

64. $\dfrac{y}{2} - 4 = \dfrac{y}{9} + 3$

65. $\dfrac{4}{5}r + \dfrac{3}{5} = \dfrac{2}{5}r + \dfrac{1}{2}$

66. $\dfrac{3}{5}a + \dfrac{7}{5} = \dfrac{8}{5}a + \dfrac{8}{5}$

Solve the equation.

67. $\dfrac{3c+9}{4} - \dfrac{4-c}{8} = \dfrac{2}{3}$

68. $\dfrac{10A+9}{4} - \dfrac{2-A}{8} = \dfrac{5}{7}$

69. $6 = \dfrac{C}{5} + \dfrac{C}{10}$

70. $26 = \dfrac{m}{9} + \dfrac{m}{4}$

71. $p + \dfrac{2}{7} = -\dfrac{4}{5}p - 1$

72. $-2q + \dfrac{9}{4} = -\dfrac{3}{10}q - \dfrac{3}{7}$

Solve the equation.

73. $\dfrac{9}{7}y + \dfrac{3}{5} = -\dfrac{8}{9}y + 1$

74. $-3r + \dfrac{2}{3} = -\dfrac{1}{3}r - 2$

75.
  a. $-\dfrac{a}{6} + 4 = -1$

  b. $\dfrac{-r}{6} + 4 = -1$

  c. $\dfrac{c}{-6} + 4 = -1$

  d. $\dfrac{-q}{-6} + 4 = -1$

76.
  a. $-\dfrac{b}{3} + 8 = 4$

  b. $\dfrac{-n}{3} + 8 = 4$

  c. $\dfrac{C}{-3} + 8 = 4$

  d. $\dfrac{-A}{-3} + 8 = 4$

Applications

77. Nina is jogging in a straight line. She got a head start of 4 meters from the starting line, and she ran 4 meters every 7 seconds. After how many seconds will Nina be 16 meters away from the starting line?

Nina will be 16 meters away from the starting line [______] seconds since she started running.

78. Charlotte is jogging in a straight line. She started at a place 44 meters from the starting line, and ran toward the starting line at the speed of 4 meters every 5 seconds. After how many seconds will Charlotte be 36 meters away from the starting line?

Charlotte will be 36 meters away from the starting line [______] seconds since she started running.

79. Nina had only \$5.00 in her piggy bank, and she decided to start saving more. She saves \$5.00 every 9 days. After how many days will she have \$30.00 in the piggy bank?

Nina will save \$30.00 in her piggy bank after [______] days.

80. Cody has saved \$45.00 in his piggy bank, and he decided to start spending them. He spends \$5.00 every 8 days. After how many days will he have \$25.00 left in the piggy bank?

Cody will have \$25.00 left in his piggy bank after [______] days.

Solving Inequalities with Fractions Solve this inequality.

81. $\dfrac{x}{10} + 98 \geq 5x$

In set-builder notation, the solution set is [______].

In interval notation, the solution set is [______].

82. $\dfrac{x}{2} + 25 \geq 3x$

In set-builder notation, the solution set is [______].

In interval notation, the solution set is [______].

83. $\dfrac{3}{2} - 5y < 5$

In set-builder notation, the solution set is [______].

In interval notation, the solution set is [______].

84. $\dfrac{3}{2} - 2y < 4$

In set-builder notation, the solution set is [______].

In interval notation, the solution set is [______].

85. $-\dfrac{1}{4}t > \dfrac{6}{5}t - 58$

In set-builder notation, the solution set is [______].

In interval notation, the solution set is [______].

86. $-\dfrac{5}{2}t > \dfrac{4}{5}t - 66$

In set-builder notation, the solution set is [______].

In interval notation, the solution set is [______].

87. $\dfrac{3}{8} \geq \dfrac{x}{32}$

In set-builder notation, the solution set is ____.

In interval notation, the solution set is ____.

88. $\dfrac{7}{8} \geq \dfrac{x}{48}$

In set-builder notation, the solution set is ____.

In interval notation, the solution set is ____.

89. $-\dfrac{z}{30} < -\dfrac{3}{10}$

In set-builder notation, the solution set is ____.

In interval notation, the solution set is ____.

90. $-\dfrac{z}{50} < -\dfrac{9}{10}$

In set-builder notation, the solution set is ____.

In interval notation, the solution set is ____.

91. $\dfrac{x}{8} - 5 \leq \dfrac{x}{3}$

In set-builder notation, the solution set is ____.

In interval notation, the solution set is ____.

92. $\dfrac{x}{6} - 2 \leq \dfrac{x}{3}$

In set-builder notation, the solution set is ____.

In interval notation, the solution set is ____.

93. $\dfrac{y - 10}{4} \geq \dfrac{y + 10}{2}$

In set-builder notation, the solution set is ____.

In interval notation, the solution set is ____.

94. $\dfrac{y - 6}{6} \geq \dfrac{y + 4}{4}$

In set-builder notation, the solution set is ____.

In interval notation, the solution set is ____.

Solve this inequality.

95. $\dfrac{3}{2} < \dfrac{x + 3}{4} - \dfrac{x - 8}{8}$

In set-builder notation, the solution set is ____.

In interval notation, the solution set is ____.

96. $\dfrac{13}{8} < \dfrac{x + 9}{4} - \dfrac{x - 2}{8}$

In set-builder notation, the solution set is ____.

In interval notation, the solution set is ____.

Applications

97. Your grade in a class is determined by the average of three test scores. You scored 74 and 88 on the first two tests. To earn at least 83 for this course, how much do you have to score on the third test?

Let x be the score you will earn on the third test. Write an inequality to represent this situation.

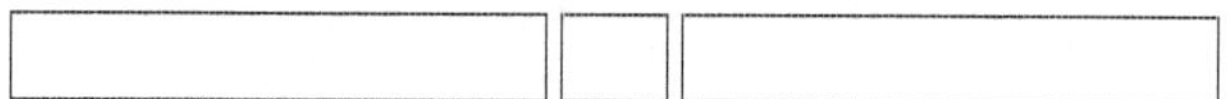

Solve this inequality. What is the minimum that you have to earn on the third test in order to earn a 83 for the course?

You cannot score over 100 on the third test. Use interval notation to represent the range of scores you can earn on the third test in order to earn at least 83 for this course.

98. Your grade in a class is determined by the average of three test scores. You scored 75 and 86 on the first two tests. To earn at least 76 for this course, how much do you have to score on the third test?

Let x be the score you will earn on the third test. Write an inequality to represent this situation.

Solve this inequality. What is the minimum that you have to earn on the third test in order to earn a 76 for the course?

You cannot score over 100 on the third test. Use interval notation to represent the range of scores you can earn on the third test in order to earn at least 76 for this course.

3.4 Isolating a Linear Variable

In this section, we will learn how to solve linear equations and inequalities with more than one variable.

3.4.1 Solving for a Variable

The formula of calculating a rectangle's area is $A = \ell w$, where ℓ stands for the rectangle's length, and w stands for width. When a rectangle's length and width are given, we can easily calculate its area.

What if a rectangle's area and length are given, and we need to calculate its width?

If a rectangle's area is given as $12\,\text{m}^2$, and its length is given as $4\,\text{m}$, we could find its width this way:

$$A = \ell w$$
$$12 = 4w$$
$$\frac{12}{4} = \frac{4w}{4}$$
$$3 = w$$
$$w = 3$$

If we need to do this many times, we would love to have an easier way, without solving an equation each time. We will solve for w in the formula $A = \ell w$:

$$A = \ell w$$
$$\frac{A}{\ell} = \frac{\ell w}{\ell}$$
$$\frac{A}{\ell} = w$$
$$w = \frac{A}{\ell}$$

Now if we want to find the width when $\ell = 4$ is given, we can simply replace ℓ with 4 and simplify.

We solved for w in the formula $A = \ell w$ once, and we could use the new formula $w = \frac{A}{\ell}$ again and again saving us a lot of time down the road. Let's look at a few examples.

Remark 3.4.2. Note that in solving for A, we divided each side of the equation by ℓ. The operations that we apply, and the order in which we do them, are determined by the operations in the original equation. In the original equation $A = \ell w$, we saw that w was *multiplied* by ℓ, and so we knew that in order to "undo" that operation, we would need to *divide* each side by ℓ. We will see this process of "un-doing" the operations throughout this section.

Example 3.4.3 Solve for R in $P = R - C$. (This is the relationship between profit, revenue, and cost.)

To solve for R, we first want to note that C is *subtracted* from R. To "undo" this, we will need to *add* C to each side of the equation:

$$P = \overset{\downarrow}{R} - C$$
$$P + C = \overset{\downarrow}{R} - C + C$$
$$P + C = \overset{\downarrow}{R}$$
$$R = P + C$$

Example 3.4.4 Solve for x in $y = mx + b$. (This is a line's equation in slope-intercept form.)

In the equation $y = mx + b$, we see that x is multiplied by m and then b is added to that. Our first step will be to isolate mx, which we'll do by subtracting b from each side of the equation:

$$y = m\overset{\downarrow}{x} + b$$

$$y - b = m\overset{\downarrow}{x} + b - b$$

$$y - b = m\overset{\downarrow}{x}$$

Now that we have mx on its own, we'll note that x is multiplied by m. To "undo" this, we'll need to divide each side of the equation by m:

$$\frac{y - b}{m} = \frac{m\overset{\downarrow}{x}}{m}$$

$$\frac{y - b}{m} = \overset{\downarrow}{x}$$

$$x = \frac{y - b}{m}$$

Warning 3.4.5. It's important to note in Example 3.4.4 that each *side* was divided by m. We can't simply divide y by m, as the equation would no longer be equivalent.

Example 3.4.6 Solve for b in $A = \frac{1}{2}bh$. (This is the area formula for a triangle.)

To solve for b, we need to determine what operations need to be "undone." The expression $\frac{1}{2}bh$ has multiplication between $\frac{1}{2}$ and b and h. As a first step, we will multiply each side of the equation by 2 in order to eliminate the denominator of 2:

$$A = \frac{1}{2}\overset{\downarrow}{b}h$$

$$2 \cdot A = 2 \cdot \frac{1}{2}\overset{\downarrow}{b}h$$

$$2A = \overset{\downarrow}{b}h$$

As a last step, we will "undo" the multiplication between b and h by dividing each side by h:

$$\frac{2A}{h} = \frac{\overset{\downarrow}{b}h}{h}$$

$$\frac{2A}{h} = \overset{\downarrow}{b}$$

$$b = \frac{2A}{h}$$

Example 3.4.7 Solve for y in $2x + 5y = 10$. (This is a linear equation in standard form.)

To solve for y, we will first have to solve for $5y$ by subtracting $2x$ from each side of the equation. After that, we'll be able to divide each side by 5 to finish solving for y:

$$2x + 5\overset{\downarrow}{y} = 10$$

$$2x + 5\overset{\downarrow}{y} - 2x = 10 - 2x$$

$$5\overset{\downarrow}{y} = 10 - 2x$$

$$\frac{5\overset{\downarrow}{y}}{5} = \frac{10 - 2x}{5}$$

$$y = \frac{10 - 2x}{5}$$

Remark 3.4.8. As we will learn in later sections, the result in Example 3.4.7 can also be written as $y = \frac{10}{5} - \frac{2x}{5}$ which can then be written as $y = 2 - \frac{2}{5}x$.

Example 3.4.9 Solve for F in $C = \frac{5}{9}(F - 32)$. (This represents the relationship between temperature in degrees Celsius and degrees Fahrenheit.)

To solve for F, we first need to see that it is contained inside a set of parentheses. To get the expression $F - 32$ on its own, we'll need to eliminate the $\frac{5}{9}$ outside those parentheses. One way we can "undo" this multiplication is by dividing each side by $\frac{5}{9}$. As we learned in Section 3.3 though, a better approach is to instead multiply each side by the reciprocal of $\frac{9}{5}$:

$$C = \frac{5}{9}(\overset{\downarrow}{F} - 32)$$

$$\frac{9}{5} \cdot C = \frac{9}{5} \cdot \frac{5}{9}(\overset{\downarrow}{F} - 32)$$

$$\frac{9}{5}C = \overset{\downarrow}{F} - 32$$

Now that we have $F - 32$, we simply need to add 32 to each side to finish solving for F:

$$\frac{9}{5}C + 32 = \overset{\downarrow}{F} - 32 + 32$$

$$\frac{9}{5}C + 32 = \overset{\downarrow}{F}$$

$$F = \frac{9}{5}C + 32$$

Exercises

Review and Warmup   Solve the equation.

1. $9q + 2 = 56$ **2.** $6y + 1 = 55$ **3.** $-9r - 8 = 37$ **4.** $-3a - 5 = -17$

5. $-6b + 3 = -b - 12$ **6.** $-2A + 6 = -A - 4$ **7.** $56 = -8(B - 10)$ **8.** $35 = -5(m - 4)$

Solving for a Variable

9.   a. Solve this linear equation for t.

     $t + 1 = 9$

  b. Solve this linear equation for x.

     $x + m = y$

10.   a. Solve this linear equation for t.

     $t + 1 = 5$

  b. Solve this linear equation for r.

     $r + n = y$

11.   a. Solve this linear equation for x.

     $x - 5 = -3$

  b. Solve this linear equation for y.

     $y - B = -3$

12.   a. Solve this linear equation for x.

     $x - 5 = -3$

  b. Solve this linear equation for y.

     $y - c = -3$

13.   a. Solve this linear equation for y.

     $-y + 1 = -5$

  b. Solve this linear equation for r.

     $-r + c = q$

14.   a. Solve this linear equation for y.

     $-y + 7 = 3$

  b. Solve this linear equation for t.

     $-t + C = b$

15.   a. Solve this linear equation for r.

     $7r = 56$

  b. Solve this linear equation for t.

     $ct = x$

16.   a. Solve this linear equation for r.

     $3r = 12$

  b. Solve this linear equation for x.

     $yx = n$

17.   a. Solve this linear equation for r.

     $\frac{r}{7} = 10$

  b. Solve this linear equation for y.

     $\frac{y}{p} = x$

18.   a. Solve this linear equation for t.

     $\frac{t}{3} = 2$

  b. Solve this linear equation for r.

     $\frac{r}{B} = x$

19.   a. Solve this linear equation for t.

     $7t + 5 = 12$

  b. Solve this linear equation for r.

     $qr + a = x$

20.   a. Solve this linear equation for x.

     $6x + 4 = 64$

  b. Solve this linear equation for y.

     $qy + r = n$

21. a. Solve this linear equation for x.

$$xt = m$$

b. Solve this linear equation for t.

$$xt = m$$

22. a. Solve this linear equation for y.

$$yr = c$$

b. Solve this linear equation for r.

$$yr = c$$

23. a. Solve this linear equation for y.

$$y + x = B$$

b. Solve this linear equation for x.

$$y + x = B$$

24. a. Solve this linear equation for r.

$$r + t = x$$

b. Solve this linear equation for t.

$$r + t = x$$

25. a. Solve this linear equation for B.

$$cy + B = C$$

b. Solve this linear equation for c.

$$cy + B = C$$

26. a. Solve this linear equation for m.

$$rt + m = b$$

b. Solve this linear equation for r.

$$rt + m = b$$

27. a. Solve this linear equation for n.

$$x = tn + p$$

b. Solve this linear equation for t.

$$x = tn + p$$

28. a. Solve this linear equation for q.

$$r = cq + n$$

b. Solve this linear equation for c.

$$r = cq + n$$

29. Solve this linear equation for x.

$$y = mx - b$$

30. Solve this linear equation for x.

$$y = -mx + b$$

31. a. Solve this equation for b:

$$12 = \tfrac{1}{2}b \cdot 4$$

b. Solve this equation for b:

$$A = \tfrac{1}{2}b \cdot h$$

32. a. Solve this equation for b:

$$9 = \tfrac{1}{2}b \cdot 6$$

b. Solve this equation for b:

$$A = \tfrac{1}{2}b \cdot h$$

33. Solve this linear equation for r.

$$C = 2\pi r$$

34. Solve this linear equation for h.

$$V = \pi r^2 h$$

35. Solve these linear equations for r.

    a. $\frac{r}{5} + 9 = 12$

    b. $\frac{r}{x} + 9 = B$

36. Solve these linear equations for t.

    a. $\frac{t}{5} + 7 = 8$

    b. $\frac{t}{r} + 7 = x$

37. Solve this linear equation for t.

$$\frac{t}{y} + c = a$$

38. Solve this linear equation for x.

$$\frac{x}{t} + p = m$$

39. Solve this linear equation for x.

$$\frac{x}{9} + r = a$$

40. Solve this linear equation for y.

$$\frac{y}{8} + r = A$$

41. Solve this linear equation for b.

$$t = y - \frac{8b}{q}$$

42. Solve this linear equation for A.

$$C = q - \frac{2A}{B}$$

43. Solve this linear equation for x.

$$Ax + By = C$$

44. Solve this linear equation for y.

$$Ax + By = C$$

Solve the linear equation for y.

45.

$$25x + 5y = -75$$

46.

$$30x - 5y = -65$$

47.

$$4x + 2y = 6$$

48.

$$18x - 2y = 16$$

49.

$$4x - y = 14$$

50.

$$2x - y = -12$$

51.

$$-4x - 6y = -24$$

52.

$$-7x - 6y = -18$$

53.

$$2x + 7y = 2$$

54.

$$6x - 8y = 2$$

55.

$$-87x - 87y = 38$$

56.

$$24y - 46x = 25$$

3.5 Ratios and Proportions

3.5.1 Introduction

A **ratio** is a means of comparing two quantities using division. One common example is a unit price. For example, if a box of cereal costs \$3.99 and weighs 21 ounces then we can write this ratio as:

$$\frac{\$3.99}{21\,\text{oz}}$$

If we want to know the unit price (that is, how much each individual ounce costs), then we can divide \$3.99 by 21 ounces and obtain \$0.19 per ounce. These two ratios, $\frac{\$3.99}{21\,\text{oz}}$ and $0.19\frac{\$}{\text{oz}}$ are equivalent, and the equation showing that they are equal is a **proportion**. In this case, we could write the following proportion:

$$\frac{\$3.99}{21\,\text{oz}} = \frac{\$0.19}{1\,\text{oz}}$$

In this section, we will extend this concept and write proportions where one quantity is unknown and solve for that unknown.

Remark 3.5.2. Sometimes ratios are stated using a colon instead of a fraction. For example, the ratio $\frac{2}{1}$ can be written as $2:1$.

Example 3.5.3 Suppose we want to know the total cost for a box of cereal that weighs 18 ounces, assuming it costs the same per ounce as the 21-ounce box. Letting C be this unknown cost (in dollars), we could set up the following proportion:

$$\frac{\text{cost in dollars}}{\text{weight in oz}} = \frac{\text{cost in dollars}}{\text{weight in oz}}$$

$$\frac{\$3.99}{21\,\text{oz}} = \frac{\$C}{18\,\text{oz}}$$

To solve this proportion, we will first note that it will be easier to solve without units:

$$\frac{3.99}{21} = \frac{C}{18}$$

Next we want to recognize that each side contains a fraction. Our usual approach for solving this type of equation is to multiply each side by the least common denominator (LCD). In this case, the LCD of 21 and 18 is 126. As with many other proportions we solve, it is often easier to just multiply each side by the common denominator of $18 \cdot 21$, which we know will make each denominator cancel:

$$\frac{3.99}{21} = \frac{C}{18}$$

$$18 \cdot 21 \cdot \frac{3.99}{21} = \frac{C}{18} \cdot 18 \cdot 21$$

$$18 \cdot \cancel{21}\frac{3.99}{\cancel{21}} = \frac{C}{\cancel{18}} \cdot \cancel{18} \cdot 21$$

$$71.82 = 21C$$

$$\frac{71.82}{21} = \frac{21C}{21}$$

$$C = 3.42$$

So assuming the cost is proportional to the cost of the 21-ounce box, the cost for an 18-ounce box of cereal would be \$3.42.

3.5.2 Solving Proportions

Solving proportions uses the process of clearing denominators that we covered in Section 3.3. Because a proportion is exactly one fraction equal to another, we can simplify the process of clearing the denominators simply by multiplying both sides of the equation by both denominators. In other words, we don't specifically need the LCD to clear the denominators.

Example 3.5.4 Solve $\frac{x}{8} = \frac{15}{12}$ for x.

Instead of finding the LCD of the two fractions, we'll simply multiply both sides of the equation by 8 and by 12. This will still have the effect of canceling the denominators on both sides of the equation.

$$\frac{x}{8} = \frac{15}{12}$$
$$12 \cdot 8 \cdot \frac{x}{8} = \frac{15}{12} \cdot 12 \cdot 8$$
$$12 \cdot \cancel{8} \cdot \frac{x}{\cancel{8}} = \frac{15}{\cancel{12}} \cdot \cancel{12} \cdot 8$$
$$12 \cdot x = 15 \cdot 8$$
$$12x = 120$$
$$\frac{12x}{12} = \frac{120}{12}$$
$$x = 10$$

Our work indicates 10 is the solution. We can check this as we would for any equation, by substituting 10 for x and verifying we obtain a true statement:

$$\frac{10}{8} \overset{?}{=} \frac{15}{12}$$
$$\frac{5}{4} \overset{\checkmark}{=} \frac{5}{4}$$

Since both fractions reduce to $\frac{5}{4}$, we know the solution to the equation $\frac{x}{8} = \frac{15}{12}$ is 10 and the solution set is $\{10\}$.

When solving proportions, we can use the name **cross-multiplication** to describe the process of what just occurred. Say we have a proportion

$$\frac{a}{b} = \frac{c}{d}$$

To remove fractions, we multiply both sides with the common denominator, bd, and we have:

$$\frac{a}{b} = \frac{c}{d}$$
$$bd \cdot \frac{a}{b} = \frac{c}{d} \cdot bd$$
$$\cancel{b}d \cdot \frac{a}{\cancel{b}} = \frac{c}{\cancel{d}} \cdot b\cancel{d}$$
$$ad = bc$$

Since a and d are diagonally across the equals sign from each other in $\frac{a}{b} = \frac{c}{d}$, as are b and c, we call this approach **cross-multiplication**.

$$\text{If } \frac{a}{b} = \frac{c}{d}, \text{ then } ad = bc.$$

If we understand cross-multiplication, we are able to rewrite a proportion $\frac{a}{b} = \frac{c}{d}$ in an equivalent form that does not have any fractions, $ad = bc$, as our first step of work. If we had used this skill in Example 3.5.4, we would have had:

$$\frac{x}{8} = \frac{15}{12}$$
$$12 \cdot x = 15 \cdot 8$$
$$12x = 120$$

Notice this is the same equation we had in the fifth line of our work in solving Example 3.5.4, but we obtained it without having to contemplate what we need to multiply by to clear the fractions.

We are able to use cross-multiplication when solving proportions, but it is extremely important to note that cross-multiplication only works when we are solving a proportion, an equation that has one ratio or fraction equal to another ratio or fraction. If an equation has anything more than one ratio or fraction on a single side of an equation, we cannot use cross-multiplication. For example, we cannot use cross-multiplication to solve $\frac{3}{4}x - \frac{2}{5} = \frac{9}{4}$, unless we first manipulate the equation to have exactly one fraction and nothing else on each side of the equation.

It is also important to be aware of the fact that cross-multiplication is a special version of our general process of clearing fractions: multiplying both sides of an equation by a common denominator of all the fractions in an equation.

Example 3.5.5 Solve $\frac{t}{5} = \frac{t+2}{3}$ for t.

Explanation. Again this equation is a proportion, so we are able to multiply both sides of the equation by both denominators to clear the fractions:

$$\frac{t}{5} = \frac{t+2}{3}$$
$$5 \cdot 3 \cdot \frac{t}{5} = \frac{t+2}{3} \cdot 5 \cdot 3$$
$$\cancel{5} \cdot 3 \cdot \frac{t}{\cancel{5}} = \frac{t+2}{\cancel{3}} \cdot 5 \cdot \cancel{3}$$
$$3 \cdot t = 5 \cdot (t+2)$$

It is critical that we include the parentheses around $t+2$, so that we are multiplying 5 against the entire numerator.

$$3t = 5(t+2)$$
$$3t = 5t + 10$$
$$3t - 5t = 5t + 10 - 5t$$
$$-2t = 10$$
$$\frac{-2t}{-2} = \frac{10}{-2}$$
$$t = -5$$

We should check that this value -5 is actually the solution of the equation:

$$\frac{-5}{5} \overset{?}{=} \frac{-5+2}{3}$$
$$-1 \overset{?}{=} \frac{-3}{3}$$
$$-1 \overset{\checkmark}{=} -1$$

Since we have verified that -5 is the solution for $\frac{t}{5} = \frac{t+2}{3}$, we know that the solution set is $\{-5\}$.

Example 3.5.6 Solve $\frac{r+7}{8} = -\frac{9}{4}$ for r.

Explanation. This proportion is a bit different in the fact that one fraction is negative. The key to working with a negative fraction is to attach the negative sign to either the numerator or denominator, but not both:

$$\frac{-9}{4} = -\frac{9}{4} \quad \text{and} \quad \frac{9}{-4} = -\frac{9}{4}, \quad \text{but} \quad \frac{-9}{-4} = +\frac{9}{4}$$

Since we're trying to eliminate the fractions, it will likely make the work a bit easier to attach the negative to the numerator.

We'll work with the equation in the form $\frac{r+7}{8} = \frac{-9}{4}$

$$\frac{r+7}{8} = \frac{-9}{4}$$
$$8 \cdot 4 \cdot \frac{r+7}{8} = \frac{-9}{4} \cdot 8 \cdot 4$$
$$\cancel{8} \cdot 4 \cdot \frac{r+7}{\cancel{8}} = \frac{-9}{\cancel{4}} \cdot 8 \cdot \cancel{4}$$
$$4 \cdot (r+7) = 8 \cdot (-9)$$
$$4r + 28 = -72$$
$$4r + 28 - 28 = -72 - 28$$
$$4r = -100$$
$$\frac{4r}{4} = \frac{-100}{4}$$
$$r = -25$$

We should check that this value -25 is actually the solution of the equation:

$$\frac{-25 + 7}{8} \overset{?}{=} -\frac{9}{4}$$
$$\frac{-18}{8} \overset{?}{=} -\frac{9}{4}$$
$$-\frac{9}{4} \overset{\checkmark}{=} -\frac{9}{4}$$

Since we have verified that -25 is the solution for $\frac{r+7}{8} = -\frac{9}{4}$, we know that the solution set is $\{-25\}$.

Example 3.5.7 Solve $\frac{x}{15} = \frac{40}{25}$ for x.

Explanation. To solve this proportion, begin by multiplying both sides by both denominators.

$$\frac{x}{15} = \frac{40}{25}$$
$$15 \cdot 25 \cdot \frac{x}{15} = \frac{40}{25} \cdot 15 \cdot 25$$
$$\cancel{15} \cdot 25 \cdot \frac{x}{\cancel{15}} = \frac{40}{\cancel{25}} \cdot 15 \cdot \cancel{25}$$
$$25 \cdot x = 40 \cdot 15$$
$$25x = 600$$
$$\frac{25x}{25} = \frac{600}{25}$$
$$x = 24$$

You can easily verify that this value 24 is actually the solution of the equation:

$$\frac{24}{15} \overset{?}{=} \frac{40}{25}$$
$$\frac{8}{5} \overset{\checkmark}{=} \frac{8}{5}$$

Since we have verified that 24 is the solution for $\frac{x}{15} = \frac{40}{25}$, we know that the solution set is $\{24\}$.

Example 3.5.8 Solve $\frac{x-4}{6} = \frac{x+3}{4}$ for x.

Explanation. To solve this proportion, begin by multiplying both sides by both denominators.

$$\frac{x-4}{6} = \frac{x+3}{4}$$
$$6 \cdot 4 \cdot \frac{x-4}{6} = \frac{x+3}{4} \cdot 6 \cdot 4$$
$$\cancel{6} \cdot 4 \cdot \frac{x-4}{\cancel{6}} = \frac{x+3}{\cancel{4}} \cdot 6 \cdot \cancel{4}$$
$$4 \cdot (x-4) = (x+3) \cdot 6$$
$$4x - 16 = 6x + 18$$
$$4x - 16 + 16 = 6x + 18 + 16$$
$$4x = 6x + 34$$
$$4x - 6x = 6x + 34 - 6x$$
$$-2x = 34$$
$$\frac{-2x}{-2} = \frac{34}{-2}$$
$$x = -17$$

We can check that this value is correct by substituting it back into the original equation:

$$\frac{x-4}{6} = \frac{x+3}{4}$$
$$\frac{-17-4}{6} \overset{?}{=} \frac{-17+3}{4}$$
$$\frac{-21}{6} \overset{?}{=} \frac{-14}{4}$$
$$\frac{-7}{2} \overset{\checkmark}{=} \frac{-7}{2}$$

Since we have verified that -17 is the solution for $\frac{x-4}{6} = \frac{x+3}{4}$, we know that the solution set is $\{-17\}$.

3.5.3 Proportionality in Similar Triangles

One really useful example of ratios and proportions involves similar triangles. Two triangles are considered **similar** if they have the same angles and their side lengths are proportional, as shown in Figure 3.5.9:

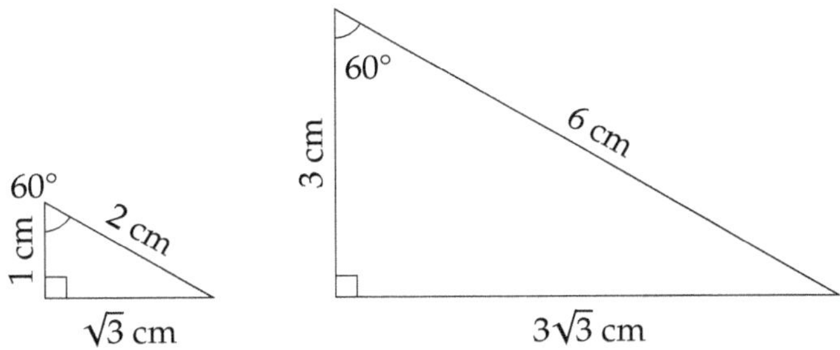

Figure 3.5.9: Similar Triangles

In the first triangle in Figure 3.5.9, the ratio of the left side length to the hypotenuse length is $\frac{1\,\text{cm}}{2\,\text{cm}}$; in the second triangle, the ratio of the left side length to the hypotenuse length is $\frac{3\,\text{cm}}{6\,\text{cm}}$. Since both reduce to $\frac{1}{2}$, we can write the following proportion:

$$\frac{1\,\text{cm}}{2\,\text{cm}} = \frac{3\,\text{cm}}{6\,\text{cm}}$$

If we extend this concept, we can use it to solve for an unknown side length. Consider the two similar triangles in the next example.

Example 3.5.10

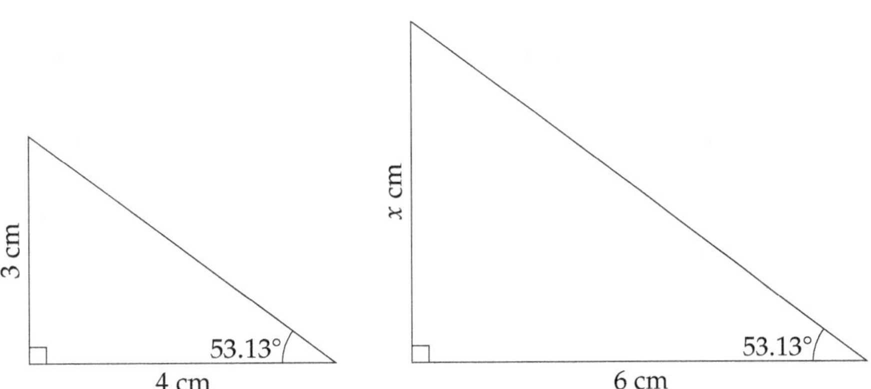

Figure 3.5.11: Similar Triangles

Since the two triangles are similar, we know that their side length should be proportional. To determine the unknown length, we can set up a proportion and solve for x:

$$\frac{\text{bigger triangle's left side length in cm}}{\text{bigger triangle's bottom side length in cm}} = \frac{\text{smaller triangle's left side length in cm}}{\text{smaller triangle's bottom side length in cm}}$$

$$\frac{x\,\text{cm}}{6\,\text{cm}} = \frac{3\,\text{cm}}{4\,\text{cm}}$$

$$\frac{x}{6} = \frac{3}{4}$$

$$12 \cdot \frac{x}{6} = 12 \cdot \frac{3}{4} \qquad \text{(12 is the least common denominator)}$$

$$2x = 9$$

$$\frac{2x}{2} = \frac{9}{2}$$

$$x = \frac{9}{2} \text{ or } 4.5$$

The unknown side length is then 4.5 cm.

Remark 3.5.12. Looking at the triangles in Figure 3.5.9, you may notice that there are many different proportions you could set up, such as:

$$\frac{2\,\text{cm}}{1\,\text{cm}} = \frac{6\,\text{cm}}{3\,\text{cm}}$$

$$\frac{2\,\text{cm}}{6\,\text{cm}} = \frac{1\,\text{cm}}{3\,\text{cm}}$$

$$\frac{6\,\text{cm}}{2\,\text{cm}} = \frac{3\,\text{cm}}{1\,\text{cm}}$$

$$\frac{3\sqrt{3}\,\text{cm}}{\sqrt{3}\,\text{cm}} = \frac{3\,\text{cm}}{1\,\text{cm}}$$

This is often the case when we set up ratios and proportions.

If we take a second look at Figure 3.5.11, there are also several other proportions we could have used to find the value of x.

$$\frac{\text{bigger triangle's left side length}}{\text{smaller triangle's left side length}} = \frac{\text{bigger triangle's bottom side length}}{\text{smaller triangle's bottom side length}}$$

$$\frac{\text{smaller triangle's bottom side length}}{\text{bigger triangle's bottom side length}} = \frac{\text{smaller triangle's left side length}}{\text{bigger triangle's left side length}}$$

$$\frac{\text{bigger triangle's bottom side length}}{\text{smaller triangle's bottom side length}} = \frac{\text{bigger triangle's left side length}}{\text{smaller triangle's left side length}}$$

Written as algebraic proportions, these three equations would, respectively, be

$$\frac{x\,\text{cm}}{3\,\text{cm}} = \frac{6\,\text{cm}}{4\,\text{cm}}, \qquad \frac{4\,\text{cm}}{6\,\text{cm}} = \frac{3\,\text{cm}}{x\,\text{cm}}, \qquad \frac{6\,\text{cm}}{4\,\text{cm}} = \frac{x\,\text{cm}}{3\,\text{cm}}$$

While these are only a few of the possibilities, if we clear the denominators from any properly designed proportion, every one is equivalent to $x = 4.5$.

3.5.4 Creating and Solving Proportions

Proportions can be used to solve many real-life applications. The key to using proportions to solve such applications is to first set up a ratio where all values are known. We then set up a second ratio that will be proportional to the first, but has one value in the ratio unknown. Let's look at a few examples.

Example 3.5.13 Property taxes for a residential property are proportional to the assessed value of the property. Assume that a certain property in a given neighborhood is assessed at \$234,100 and its annual property taxes are \$2,518.92. What are the annual property taxes for a house that is assessed at \$287,500?

Explanation. Let T be the annual property taxes (in dollars) for a property assessed at \$287,500. We can write and solve this proportion:

$$\frac{\text{tax}}{\text{property value}} = \frac{\text{tax}}{\text{property value}}$$

$$\frac{2518.92}{234100} = \frac{T}{287500}$$

The least common denominator of this proportion is rather large, so we will instead multiply each side by 234100 and 287500 and simplify from there:

$$\frac{2518.92}{234100} = \frac{T}{287500}$$

$$234100 \cdot 287500 \cdot \frac{2518.92}{234100} = \frac{T}{287500} \cdot 234100 \cdot 287500$$

$$287500 \cdot 2518.92 = T \cdot 234100$$

$$\frac{287500 \cdot 2518.92}{234100} = \frac{234100T}{234100}$$

$$T \approx 3093.50$$

The property taxes for a property assessed at $\$287{,}500$ are $\$3{,}093.50$.

Example 3.5.14 Tagging fish is a means of estimating the size of the population of fish in a body of water (such as a lake). A sample of fish is taken, tagged, and then redistributed into the lake. When another sample is taken, the proportion of fish that are tagged out of that sample are assumed to be proportional to the total number of fish tagged out of the entire population of fish in the lake.

$$\frac{\text{number of tagged fish in sample}}{\text{number of fish in sample}} = \frac{\text{number of tagged fish total}}{\text{number of fish total}}$$

Assume that 90 fish are caught and tagged. Once they are redistributed, a sample of 200 fish is taken. Of these, 7 are tagged. Estimate how many fish total are in the lake.

Explanation. Let n be the number of fish in the lake. We can set up a proportion for this scenario:

$$\frac{7}{200} = \frac{90}{n}$$

To solve for n, which is in a denominator, we'll need to multiply each side by both 200 and n:

$$\frac{7}{200} = \frac{90}{n}$$

$$200 \cdot n \cdot \frac{7}{200} = \frac{90}{n} \cdot 200 \cdot n$$

$$200 \cdot n \cdot \frac{7}{200} = \frac{90}{n} \cdot 200 \cdot n$$

$$7n = 1800$$

$$\frac{7n}{7} = \frac{1800}{7}$$

$$n \approx 2471.4286$$

According to this sample, we can estimate that there are about $2{,}471$ fish in the lake.

Example 3.5.15 Infant Tylenol contains $160\,\text{mg}$ of acetaminophen in each $5\,\text{mL}$ of liquid medicine. If Bao's baby is prescribed $60\,\text{mg}$ of acetaminophen, how many milliliters of liquid medicine should he give them?

Explanation. Assume Bao should give q milliliters of liquid medicine, and we can set up the following proportion:

$$\frac{\text{amount of liquid medicine in mL}}{\text{amount of acetaminophen in mg}} = \frac{\text{amount of liquid medicine in mL}}{\text{amount of acetaminophen in mg}}$$

$$\frac{5\,\text{mL}}{160\,\text{mg}} = \frac{q\,\text{mL}}{60\,\text{mg}}$$

$$\frac{5}{160} = \frac{q}{60}$$

$$160 \cdot 60 \cdot \frac{5}{160} = \frac{q}{60} \cdot 160 \cdot 60$$

$$60 \cdot 5 = q \cdot 160$$

$$300 = 160q$$

$$\frac{300}{160} = \frac{160q}{160}$$

$$q = 1.875$$

So to give $60\,\text{mg}$ of acetaminophen to his baby, Bao should give $1.875\,\text{mL}$ of liquid medicine.

Example 3.5.16 Sarah is an architect and she's making a scale model of a building. The actual building will be $30\,\text{ft}$ tall. In the model, the height of the building will be $2\,\text{in}$. How tall should she make the model of a person who is $5\,\text{ft}\,6\,\text{in}$ tall so that the model is to scale?

Explanation. Let h be the height of the person in Sarah's model, which we'll measure in inches. We'll create a proportion that compares the building and person's heights in the model to their heights in real life:

$$\frac{\text{height of model building in inches}}{\text{height of actual building in feet}} = \frac{\text{height of model person in inches}}{\text{height of actual person in feet}}$$

$$\frac{2\,\text{in}}{30\,\text{ft}} = \frac{h\,\text{in}}{5\,\text{ft}\,6\text{in}}$$

Before we can just eliminate the units, we'll need to convert $5\,\text{ft}\,6\,\text{in}$ to feet:

$$\frac{2\,\text{in}}{30\,\text{ft}} = \frac{h\,\text{in}}{5.5\,\text{ft}}$$

Now we can remove the units and continue solving:

$$\frac{2}{30} = \frac{h}{5.5}$$

$$30 \cdot 5.5 \cdot \frac{2}{30} = \frac{h}{5.5} \cdot 30 \cdot 5.5$$

$$5.5 \cdot 2 = h \cdot 30$$

$$11 = 30h$$

$$\frac{11}{30} = \frac{30h}{30}$$

$$\frac{11}{30} = h$$

$$h \approx 0.3667$$

Sarah should make the model of a person who is 5 ft 6 in tall be $\frac{11}{30}$ inches (about 0.3667 inches) tall.

Exercises

Review and Warmup

1. Reduce the fraction $\frac{3}{15}$.

2. Reduce the fraction $\frac{2}{10}$.

3. Reduce the fraction $\frac{9}{21}$.

4. Reduce the fraction $\frac{8}{14}$.

5. Reduce the fraction $\frac{60}{105}$.

6. Reduce the fraction $\frac{84}{147}$.

7. Reduce the fraction $\frac{252}{105}$.

8. Reduce the fraction $\frac{420}{245}$.

Setting Up Ratios and Proportions

9. Ibuprofen for infants comes in a liquid form and contains 30 milligrams of ibuprofen for each 0.75 milliliters of liquid. If a child is to receive a dose of 50 milligrams of ibuprofen, how many milliliters of liquid should they be given?

Assume l milliliters of liquid should be given. Write an equation to model this scenario. There is no need to solve it.

10. Ibuprofen for infants comes in a liquid form and contains 35 milligrams of ibuprofen for each 0.875 milliliters of liquid. If a child is to receive a dose of 45 milligrams of ibuprofen, how many milliliters of liquid should they be given?

Assume l milliliters of liquid should be given. Write an equation to model this scenario. There is no need to solve it.

11. The property taxes on a 2400-square-foot house are \$2,904.00 per year. Assuming these taxes are proportional, what are the property taxes on a 1200-square-foot house?

Assume property taxes on a 1200-square-foot house is t dollars. Write an equation to model this scenario. There is no need to solve it.

12. The property taxes on a 1900-square-foot house are \$2,527.00 per year. Assuming these taxes are proportional, what are the property taxes on a 2500-square-foot house?

Assume property taxes on a 2500-square-foot house is t dollars. Write an equation to model this scenario. There is no need to solve it.

Solving Proportions

13. Solve $\frac{x}{48} = \frac{15}{40}$ for x.

14. Solve $\frac{x}{63} = \frac{10}{35}$ for x.

15. Solve $\frac{10}{x} = \frac{35}{63}$ for x.

16. Solve $\frac{12}{x} = \frac{16}{24}$ for x.

17. Solve $\frac{x}{6} = \frac{x-15}{9}$ for x.

18. Solve $\frac{x}{5} = \frac{x+24}{9}$ for x.

19. Solve $\frac{x}{7} = \frac{x-3}{6}$ for x.

20. Solve $\frac{x}{7} = \frac{x-20}{11}$ for x.

21. Solve $\frac{x+3}{5} = \frac{x-5}{9}$ for x.

22. Solve $\frac{x-10}{5} = \frac{x+16}{7}$ for x.

23. Solve $\frac{x-16}{6} = \frac{x-12}{14}$ for x.

24. Solve $\frac{x-8}{7} = \frac{x-8}{11}$ for x.

25. Solve $\frac{x}{24} = -\frac{45}{27}$ for x.

26. Solve $\frac{x}{21} = -\frac{18}{14}$ for x.

27. Solve $\frac{x+2}{42} = -\frac{24}{18}$ for x.

28. Solve $\frac{x-2}{6} = -\frac{45}{10}$ for x.

Applications

29. The following two triangles are similar to each other. Find the length of the missing side.

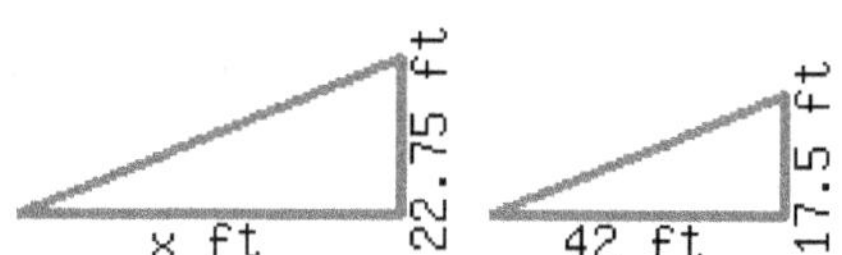

The missing side's length is ☐

30. The following two triangles are similar to each other. Find the length of the missing side.

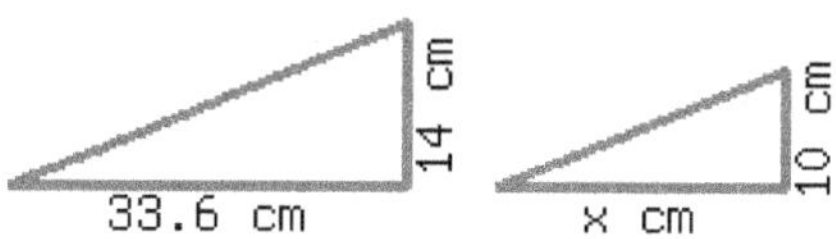

The missing side's length is ☐

31. The following two triangles are similar to each other. Find the length of the missing side.

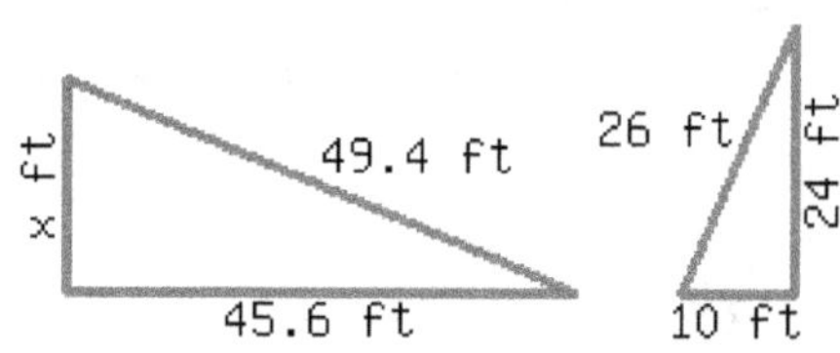

The missing side's length is ☐

32. The following two triangles are similar to each other. Find the length of the missing side.

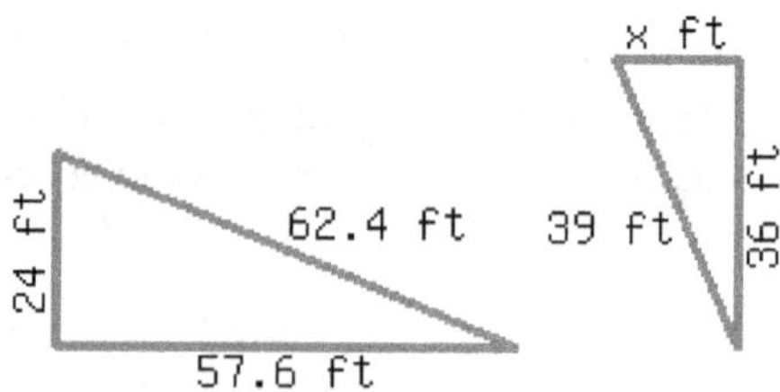

The missing side's length is ☐

33. According to a salad recipe, each serving requires 2 teaspoons of vegetable oil and 12 teaspoons of vinegar. If 12 teaspoons of vegetable oil were used, how many teaspoons of vinegar should be used?

    If 12 teaspoons of vegetable oil were used, ______ teaspoons of vinegar should be used.

34. According to a salad recipe, each serving requires 5 teaspoons of vegetable oil and 35 teaspoons of vinegar. If 119 teaspoons of vinegar were used, how many teaspoons of vegetable oil should be used?

    If 119 teaspoons of vinegar were used, ______ teaspoons of vegetable oil should be used.

35. Laurie makes $105 every six hours she works. How much will she make if she works twenty-two hours this week?

    If Laurie works twenty-two hours this week, she will make ______.

36. Corey makes $81 every six hours he works. How much will he make if he works twenty-six hours this week?

    If Corey works twenty-six hours this week, he will make ______.

37. A mutual fund consists of 23% stock and 77% bond. In other words, for each 23 dollars of stock, there are 77 dollars of bond. For a mutual fund with $2,850.00 of stock, how many dollars of bond are there?

    For a mutual fund with $2,850.00 of stock, there are approximately ______ of bond.

38. A mutual fund consists of 32% stock and 68% bond. In other words, for each 32 dollars of stock, there are 68 dollars of bond. For a mutual fund with $2,510.00 of bond, how many dollars of stock are there?

    For a mutual fund with $2,510.00 of bond, there are approximately ______ of stock.

39. Farshad jogs every day. Last month, he jogged 14.5 hours for a total of 17.4 miles. At this speed, if Farshad runs 35 hours, how far can he run?

    At this speed, Farshad can run ______ in 35 hours.

40. Scot jogs every day. Last month, he jogged 5.5 hours for a total of 9.9 miles. At this speed, how long would it take Scot to run 90 miles?

    At this speed, Scot can run 90 mi in ______.

41. Blake purchased 3.7 pounds of apples at the total cost of $16.28. If he purchases 9.4 pounds of apples at this store, how much would it cost?

    It would cost ______ to purchase 9.4 pounds of apples.

42. Jay purchased 5.4 pounds of apples at the total cost of $22.14. If the price doesn't change, how many pounds of apples can Jay purchase with $40.18?

    With $40.18, Jay can purchase ______ of apples.

43. Timothy collected a total of 2261 stamps over the past 17 years. At this rate, how many stamps would he collect in 23 years?

    At this rate, Timothy would collect ______ stamps in 23 years.

44. Tiffany collected a total of 2016 stamps over the past 14 years. At this rate, how many years would it take she to collect 3888 stamps?

    At this rate, Tiffany can collect 3888 stamps in ______ years.

45. In a city, the owner of a house valued at 380 thousand dollars needs to pay $425.60 in property tax. At this tax rate, how much property tax should the owner pay if a house is valued at 890 thousand dollars?

The owner of a 890-thousand-dollar house should pay ⬚ in property tax.

46. In a city, the owner of a house valued at 300 thousand dollars needs to pay $525.00 in property tax. At this tax rate, if the owner of a house paid $1,540.00 of property tax, how much is the house worth?

If the owner of a house paid $1,540.00 of property tax, the house is worth ⬚ thousand dollars.

47. To try to determine the health of the Rocky Mountain elk population in the Wenaha Wildlife Area, the Oregon Department of Fish and Wildlife caught, tagged, and released 39 Rocky Mountain elk. A week later, they returned and observed 42 Rocky Mountain elk, 9 of which had tags. Approximately how many Rocky Mountain elk are in the Wenaha Wildlife Area?

There are approximately ⬚ elk in the wildlife area.

48. To try to determine the health of the black-tailed deer population in the Jewell Meadow Wildlife Area, the Oregon Department of Fish and Wildlife caught, tagged, and released 28 black-tailed deer. A week later, they returned and observed 63 black-tailed deer, 18 of which had tags. Approximately how many black-tailed deer are in the Jewell Meadow Wildlife Area?

There are approximately ⬚ deer in the wildlife area.

49. A restaurant used 1105.2 lb of vegetable oil in 36 days. At this rate, how many pounds of vegetable oil will be used in 50 days?

The restaurant will use ⬚ of vegetable oil in 50 days.

50. A restaurant used 777.6 lb of vegetable oil in 24 days. At this rate, 1458 lb of oil will last how many days?

The restaurant will use 1458 lb of vegetable oil in ⬚ days.

Challenge

51. The ratio of girls to boys in a preschool is 6 to 7. If there are 104 kids in the school, how many girls are there in the preschool?

3.6 Special Solution Sets

Most of the time, a linear equation's final equivalent equation looks like $x = 3$, and the solution set is written to show that there is only one solution: $\{3\}$. Similarly, a linear inequality's final equivalent equation looks like $x < 5$, and the solution set is represented with either $(-\infty, 5)$ in interval notation or $\{x | x < 5\}$ in set-builder notation. It's possible that both linear equations and inequalities have all real numbers as possible solutions, and it's possible that no real numbers are solutions to each. In this section, we will explore these special solution sets.

3.6.1 Special Solution Sets

Recall that for the equation $x + 2 = 5$, there is only one number which will make the equation true: 3. This means that our solution is 3, and we write the **solution set** as $\{3\}$. We say the equation's solution set has one **element**, 3.

We'll now explore equations that have all real numbers as possible solutions or no real numbers as possible solutions.

Example 3.6.2 Solve for x in $3x = 3x + 4$.

To solve this equation, we need to move all terms containing x to one side of the equals sign:

$$3x = 3x + 4$$
$$3x - 3x = 3x + 4 - 3x$$
$$0 = 4$$

Notice that x is no longer present in the equation. What value can we substitute into x to make $0 = 4$ true? Nothing! We say this equation has no solution. Or, the equation has an empty solution set. We can write this as $\emptyset$, which is the symbol for the empty set.

The equation $0 = 4$ is known a **false statement** since it is false no matter what x is. It indicates there is no solution to the original equation.

Example 3.6.3 Solve for x in $2x + 1 = 2x + 1$.

We will move all terms containing x to one side of the equals sign:

$$2x + 1 = 2x + 1$$
$$2x + 1 - 2x = 2x + 1 - 2x$$
$$1 = 1$$

At this point, x is no longer contained in the equation. What value can we substitute into x to make $1 = 1$ true? Any number! This means that all real numbers are possible solutions to the equation $2x + 1 = 2x + 1$. We say this equation's solution set contains *all real numbers*. We can write this set using set-builder notation as $\{x \mid x \text{ is a real number}\}$ or using interval notation as $(-\infty, \infty)$.

The equation $1 = 1$ is known as an **identity** since it is true no matter what x is. It indicates that all real numbers are solutions to the original linear equation.

Remark 3.6.4. What would have happened if we had continued solving after we obtained $1 = 1$ in Example 3.6.3?

$$1 = 1$$
$$1 - 1 = 1 - 1$$
$$0 = 0$$

As we can see, all we found was another identity — a different equation that is true for all values of x.

Warning 3.6.5. Note that there is a very important difference between ending with an equivalent equation of $0 = 0$ and $x = 0$. The first holds true for all real numbers, and the solution set is $\{x \mid x$ is a real number$\}$. The second has only one solution: 0. We write that solution set to show that only the number zero is the solution: $\{0\}$.

Example 3.6.6 Solve for t in the inequality $4t + 5 > 4t + 2$.

To solve for t, we will first subtract $4t$ from each side to get all terms containing t on one side:

$$4t + 5 > 4t + 2$$
$$4t + 5 - 4t > 4t + 2 - 4t$$
$$5 > 2$$

Notice that again, the variable t is no longer contained in the inequality. We then need to consider which values of t make the inequality true. The answer is *all values*, so our solution set is all real numbers, which we can write as $\{t \mid t$ is a real number$\}$.

Example 3.6.7 Solve for x in the inequality $-5x + 1 \leq -5x$.

To solve for x, we will first add $5x$ to each side to get all terms containing x on one side:

$$-5x + 1 \leq -5x$$
$$-5x + 1 + 5x \leq -5x + 5x$$
$$1 \leq 0$$

Once more, the variable x is absent. So we can ask ourselves, "For which values of x is $1 \leq 0$ true?" The answer is *none*, and so there is no solution to this inequality. We can write the solution set using $\emptyset$.

Remark 3.6.8. Again consider what would have happened if we had continued solving after we obtained $1 \leq 0$ in Example 3.6.7.

$$1 \leq 0$$
$$1 - 1 \leq 0 - 1$$
$$0 \leq -1$$

As we can see, all we found was another false statement—a different equation that is not true for any real number.

Let's summarize the two special cases when solving linear equations and inequalities:

> **All Real Numbers** When the equivalent equation or inequality is an *identity* such as $2 = 2$ or $0 < 2$, all real numbers are solutions. We write this solution set as either $(-\infty, \infty)$ or $\{x \mid x \text{ is a real number}\}$.
>
> **No Solution** When the equivalent equation or inequality is a *false statement* such as $0 = 2$ or $0 > 2$, no real number is a solution. We write this solution set as either $\{\ \}$ or $\emptyset$ or write the words "no solution exists."

List 3.6.9: Special Solution Sets for Equations and Inequalities

3.6.2 Solving Equations and Inequalities with Special Solution Sets

Example 3.6.10 Solve for a in $\frac{2}{3}(a + 1) - \frac{5}{6} = \frac{2}{3}a$.

To solve this equation for a, we'll want to recall the technique of multiplying each side of the equation by the LCD of all fractions. Here, this means that we will multiply each side by 6 as our first step. After that, we'll be able to simplify each side of the equation and continue solving for a:

$$\frac{2}{3}(a + 1) - \frac{5}{6} = \frac{2}{3}a$$

$$6 \cdot \left(\frac{2}{3}(a + 1) - \frac{5}{6}\right) = 6 \cdot \frac{2}{3}a$$

$$6 \cdot \frac{2}{3}(a + 1) - 6 \cdot \frac{5}{6} = \frac{2}{3}a$$

$$4(a + 1) - 5 = 4a$$

$$4a + 4 - 5 = 4a$$

$$4a - 1 = 4a$$

$$4a - 1 - 4a = 4a - 4a$$

$$-1 = 0$$

The statement $-1 = 0$ is false, so the equation has no solution. We can write the empty set as: $\emptyset$.

Example 3.6.11 Solve for x in the equation $3(x + 2) - 8 = (5x + 4) - 2(x + 1)$.

To solve for x, we will first need to simplify the left side and right side of the equation as much as possible by distributing and combining like terms:

$$3(x + 2) - 8 = (5x + 4) - 2(x + 1)$$

$$3x + 6 - 8 = 5x + 4 - 2x - 2$$

$$3x - 2 = 3x + 2$$

From here, we'll want to subtract $3x$ from each side:

$$3x - 2 - 3x = 3x + 2 - 3x$$

$$-2 = 2$$

As the equation $-2 = 2$ is not true for any value of x, there is no solution to this equation. We write the solution set as: $\emptyset$.

Example 3.6.12 Solve for z in the inequality $\frac{3z}{5} + \frac{1}{2} \leq \left(\frac{z}{10} + \frac{3}{4}\right) + \left(\frac{z}{2} - \frac{1}{4}\right)$.

To solve for z, we will first need to multiply each side of the inequality by the LCD, which is 40. After that, we'll finish solving by putting all terms containing a variable on one side of the inequality:

$$\frac{3z}{5} + \frac{1}{2} \leq \left(\frac{z}{10} + \frac{3}{4}\right) + \left(\frac{z}{2} - \frac{1}{4}\right)$$

$$40 \cdot \left(\frac{3z}{5} + \frac{1}{2}\right) \leq 40 \cdot \left(\left(\frac{z}{10} + \frac{3}{4}\right) + \left(\frac{z}{2} - \frac{1}{4}\right)\right)$$

$$40 \cdot \left(\frac{3z}{5}\right) + 40 \cdot \left(\frac{1}{2}\right) \leq 40 \cdot \left(\frac{z}{10} + \frac{3}{4}\right) + 40 \cdot \left(\frac{z}{2} - \frac{1}{4}\right)$$

$$40 \cdot \left(\frac{3z}{5}\right) + 40 \cdot \left(\frac{1}{2}\right) \leq 40 \cdot \left(\frac{z}{10}\right) + 40 \cdot \left(\frac{3}{4}\right) + 40 \cdot \left(\frac{z}{2}\right) - 40 \cdot \left(\frac{1}{4}\right)$$

$$24z + 20 \leq 4z + 30 + 20z - 10$$

$$24z + 20 \leq 24z + 20$$

$$24z + 20 - 24z \leq 24z + 20 - 24z$$

$$20 \leq 20$$

As the equation $20 \leq 20$ is true for all values of z, all real numbers are solutions to this inequality. Thus the solution set is $\{z \mid z \text{ is a real number}\}$.

Exercises

Review and Warmup Solve the equation.

1. $7n + 4 = 18$ **2.** $4q + 3 = 27$ **3.** $-2x - 2 = -18$ **4.** $-5r - 9 = 11$

5. $-4t + 8 = -t - 10$ **6.** $-10b + 3 = -b - 24$ **7.** $96 = -6(c - 6)$ **8.** $15 = -3(B - 10)$

Solving Equations with Special Solution Sets Solve the equation.

9. $10C = 10C + 4$ **10.** $6n = 6n + 8$ **11.** $2p + 2 = 2p + 2$

12. $10x + 6 = 10x + 6$ **13.** $6r - 2 - 7r = -4 - r + 2$ **14.** $3t - 6 - 4t = -7 - t + 1$

15. $-9 - 10b + 8 = -b + 13 - 9b$ **16.** $-7 - 6c + 4 = -c + 12 - 5c$ **17.** $2(B - 9) = 2(B - 1)$

18. $8(C - 5) = 8(C - 4)$

19.
  a. $7n + 7 = 4n + 7$
  b. $7n + 7 = 7n + 7$
  c. $7n + 7 = 7n + 11$

20.
  a. $5p + 10 = 2p + 10$
  b. $5p + 10 = 5p + 10$
  c. $5p + 10 = 5p + 13$

Solve the equation.

21. $4(4 - 2x) - (6x - 8) = 23 - 2(8 + 7x)$

22. $3(8 - 10r) - (6r - 3) = 19 - 2(9 + 18r)$

23. $19 - 6(4 + 5t) = -31t - (5 - t)$

24. $30 - 5(7 + 4b) = -21b - (5 - b)$

Solving Inequalities with Special Solution Sets Solve this inequality. Answer using interval notation.

25. $6x > 6x + 3$

26. $6x > 6x + 9$

27. $-8x \leq -8x - 5$

28. $-10x \leq -10x - 8$

29. $-8 + 10x + 18 \geq 10x + 10$

30. $-2 + 2x + 8 \geq 2x + 6$

31. $-6 + 2x + 9 < 2x + 3$

32. $-10 + 4x + 19 < 4x + 9$

33. $-7 - 8z + 3 > -z + 16 - 7z$

34. $-7 - 5z + 6 > -z + 10 - 4z$

35. $6(k - 9) \leq 6(k - 1)$

36. $8(k - 7) \leq 8(k - 3)$

37. $10x \leq 10x + 2$

38. $10x \leq 10x + 7$

Solve this inequality. Answer using interval notation.

39. $2(4 - 10m) - (2m - 4) > 8 - 2(8 + 11m)$

40. $2(1 - 4m) - (10m - 4) > 10 - 2(3 + 9m)$

Challenge

41. Fill in the right side of the equation to create a linear equation with the properties listed.

  a. Create a linear equation with *infinitely many solutions.*

$$6(x + 4) = \boxed{}$$

  b. Create a linear equation with the solution $x = 2$.

$$6(x + 4) = \boxed{}$$

3.7 Linear Equations and Inequalities Chapter Review

3.7.1 Solving Multistep Linear Equations

In Section 3.1 we covered the steps to solve a linear equation and the differences among simplifying expressions, evaluating expressions and solving equations.

Example 3.7.1 Solve for a in $4 - (3 - a) = -2 - 2(2a + 1)$.

Explanation.   To solve this equation, we will simplify each side of the equation, manipulate it so that all variable terms are on one side and all constant terms are on the other, and then solve for a:

$$
\begin{aligned}
4 - (3 - a) &= -2 - 2(2a + 1) \\
4 - 3 + a &= -2 - 4a - 2 \\
1 + a &= -4 - 4a \\
1 + a + 4a &= -4 - 4a + 4a \\
1 + 5a &= -4 \\
1 + 5a - 1 &= -4 - 1 \\
5a &= -5 \\
\frac{5a}{5} &= \frac{-5}{5} \\
a &= -1
\end{aligned}
$$

Checking the solution -1 in the original equation, we get:

$$
\begin{aligned}
4 - (3 - a) &= -2 - 2(2a + 1) \\
4 - (3 - (-1)) &\overset{?}{=} -2 - 2(2(-1) + 1) \\
4 - (4) &\overset{?}{=} -2 - 2(-1) \\
0 &\overset{\checkmark}{=} 0
\end{aligned}
$$

Therefore the solution to the equation is -1 and the solution set is $\{-1\}$.

3.7.2 Solving Multistep Linear Inequalities

In Section 3.2 we covered how solving inequalities is very much like how we solve equations, except that if we multiply or divide by a negative we switch the inequality sign.

Example 3.7.2 Solve for x in $-2 - 2(2x + 1) > 4 - (3 - x)$. Write the solution set in both set-builder notation and interval notation.

Explanation.

$$
\begin{aligned}
-2 - 2(2x + 1) &> 4 - (3 - x) \\
-2 - 4x - 2 &> 4 - 3 + x \\
-4x - 4 &> x + 1
\end{aligned}
$$

$$-4x - 4 - x > x + 1 - x$$
$$-5x - 4 > 1$$
$$-5x - 4 + 4 > 1 + 4$$
$$-5x > 5$$
$$\frac{-5x}{-5} < \frac{5}{-5}$$
$$x < -1$$

Note that when we divided both sides of the inequality by -5, we had to switch the direction of the inequality symbol.

The solution set in set-builder notation is $\{x \mid x < -1\}$. The solution set in interval notation is $(-\infty, -1)$.

3.7.3 Linear Equations and Inequalities with Fractions

In Section 3.3 we covered how to eliminate denominators in an equation with the LCD to help solve the equation.

Example 3.7.3 Solve for x in $\frac{1}{4}x + \frac{2}{3} = \frac{1}{6}$.

Explanation.

We'll solve by multiplying each side of the equation by 12:

$$\frac{1}{4}x + \frac{2}{3} = \frac{1}{6}$$
$$12 \cdot \left(\frac{1}{4}x + \frac{2}{3}\right) = 12 \cdot \frac{1}{6}$$
$$12 \cdot \left(\frac{1}{4}x\right) + 12 \cdot \left(\frac{2}{3}\right) = 12 \cdot \frac{1}{6}$$
$$3x + 8 = 2$$
$$3x = -6$$
$$\frac{3x}{3} = \frac{-6}{3}$$
$$x = -2$$

Checking the solution:

$$\frac{1}{4}x + \frac{2}{3} = \frac{1}{6}$$
$$\frac{1}{4}(-2) + \frac{2}{3} \overset{?}{=} \frac{1}{6}$$
$$-\frac{2}{4} + \frac{2}{3} \overset{?}{=} \frac{1}{6}$$
$$-\frac{6}{12} + \frac{8}{12} \overset{?}{=} \frac{1}{6}$$
$$\frac{2}{12} \overset{?}{=} \frac{1}{6}$$
$$\frac{1}{6} \overset{\checkmark}{=} \frac{1}{6}$$

The solution is therefore -2. We write the solution set s $\{-2\}$.

3.7.4 Isolating a Linear Variable

In Section 3.4 we covered how to solve an equation when there are multiple variables in the equation.

Example 3.7.4 Solve for x in $y = mx + b$.

Explanation.

$$y = mx + b$$
$$y - b = mx + b - b$$
$$y - b = mx$$
$$\frac{y - b}{m} = \frac{mx}{m}$$
$$\frac{y - b}{m} = x$$

3.7.5 Ratios and Proportions

In Section 3.5 we covered the definitions of a ratio and a proportion and how to solve a proportion. We learned about cross multiplication, did problems about similar triangles, and used proportions to solve word problems.

Example 3.7.5 Solve $\frac{6-x}{5} = \frac{x}{4}$ for x.

Explanation. To solve this proportion, begin by multiplying both sides by both denominators.

$$\frac{6 - x}{5} = \frac{x}{4}$$
$$5 \cdot 4 \cdot \frac{6 - x}{5} = \frac{x}{4} \cdot 5 \cdot 4$$
$$\cancel{5} \cdot 4 \cdot \frac{6 - x}{\cancel{5}} = \frac{x}{\cancel{4}} \cdot 5 \cdot \cancel{4}$$
$$4 \cdot (6 - x) = x \cdot 5$$
$$24 - 4x = 5x$$
$$24 - 4x + 4x = 5x + 4x$$
$$24 = 9x$$
$$\frac{24}{9} = \frac{9x}{9}$$
$$\frac{24}{9} = x$$

So, the solution set is $\left\{\frac{24}{9}\right\}$.

Example 3.7.6 Property taxes for a residential property are proportional to the assessed value of the property. Assume that a certain property in a given neighborhood is assessed at \$234,100 and its annual property taxes are \$2,518.92. What are the annual property taxes for a house that is assessed at \$287,500?

Explanation. Let T be the annual property taxes (in dollars) for a property assessed at \$287,500. We can write and solve this proportion:

$$\frac{\text{tax}}{\text{property value}} = \frac{\text{tax}}{\text{property value}}$$
$$\frac{2518.92}{234100} = \frac{T}{287500}$$

$$234100 \cdot 287500 \cdot \frac{2518.92}{234100} = \frac{T}{287500} \cdot 234100 \cdot 287500$$

$$287500 \cdot 2518.92 = T \cdot 234100$$

$$\frac{287500 \cdot 2518.92}{234100} = \frac{234100T}{234100}$$

$$T \approx 3093.50$$

The property taxes for a property assessed at \$287,500 are \$3,093.50.

Example 3.7.7

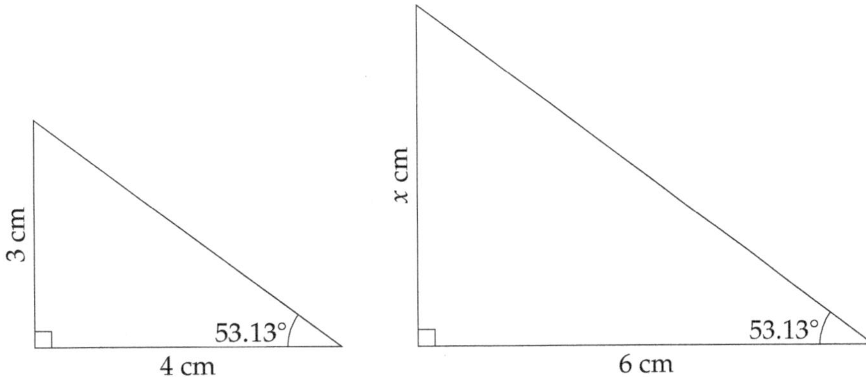

Figure 3.7.8: Similar Triangles

Since the two triangles are similar, we know that their side length should be proportional. To determine the unknown length, we can set up a proportion and solve for x:

$$\frac{\text{bigger triangle's left side length in cm}}{\text{bigger triangle's bottom side length in cm}} = \frac{\text{smaller triangle's left side length in cm}}{\text{smaller triangle's bottom side length in cm}}$$

$$\frac{x\,\text{cm}}{6\,\text{cm}} = \frac{3\,\text{cm}}{4\,\text{cm}}$$

$$\frac{x}{6} = \frac{3}{4}$$

$$12 \cdot \frac{x}{6} = 12 \cdot \frac{3}{4} \qquad \text{(12 is the least common denominator)}$$

$$2x = 9$$

$$\frac{2x}{2} = \frac{9}{2}$$

$$x = \frac{9}{2} \text{ or } 4.5$$

The unknown side length is then 4.5 cm.

3.7.6 Special Solution Sets

In Section 3.6 we covered linear equations that have no solutions and also linear equations that have infinitely many solutions. When solving linear inequalities, it's also possible that no solution exists or that all real numbers are solutions.

Example 3.7.9

    a. Solve for x in the equation $3x = 3x + 4$.

    b. Solve for t in the inequality $4t + 5 > 4t + 2$.

Explanation.

    a. To solve this equation, we need to move all terms containing x to one side of the equals sign:

$$3x = 3x + 4$$
$$3x - 3x = 3x + 4 - 3x$$
$$0 = 4$$

This equation has no solution. We write the solution set as $\emptyset$, which is the symbol for the empty set.

    b. To solve for t, we will first subtract $4t$ from each side to get all terms containing t on one side:

$$4t + 5 > 4t + 2$$
$$4t + 5 - 4t > 4t + 2 - 4t$$
$$5 > 2$$

All values of the variable t make the inequality true. The solution set is all real numbers, which we can write as $\{t \mid t \text{ is a real number}\}$ in set notation, or $(-\infty, \infty)$ in interval notation.

Exercises

1.   a. Solve the following linear equation:

    $3(y - 7) - 4 = -13$

  b. Evaluate the following expression when $y = 4$:

    $3(y - 7) - 4 = \boxed{}$

  c. Simplify the following expression:

    $3(y - 7) - 4 = \boxed{}$

2.   a. Solve the following linear equation:

    $2(y + 1) - 9 = 1$

  b. Evaluate the following expression when $y = 4$:

    $2(y + 1) - 9 = \boxed{}$

  c. Simplify the following expression:

    $2(y + 1) - 9 = \boxed{}$

3. Solve the equation.

    $-55 = -8B - 10 - B$

4. Solve the equation.

    $1 = -5C - 5 - C$

5. Solve the equation.

    $5 + 10(n - 4) = -73 - (2 - 2n)$

6. Solve the equation.

    $4 + 7(p - 10) = -84 - (7 - 2p)$

7. Solve the equation.

    $-6 - 8x + 4 = -x + 4 - 7x$

8. Solve the equation.

    $-10 - 5y + 4 = -y + 0 - 4y$

9. Solve the equation.

$$13 = \frac{t}{7} + \frac{t}{6}$$

10. Solve the equation.

$$9 = \frac{a}{5} + \frac{a}{10}$$

11. Solve the equation.

$$\frac{c-4}{6} = \frac{c+6}{8}$$

12. Solve the equation.

$$\frac{B-8}{4} = \frac{B+3}{6}$$

13. Solve this inequality.

$$4 - (y + 7) < 4$$

In set-builder notation, the solution set is ☐.

In interval notation, the solution set is ☐.

14. Solve this inequality.

$$4 - (y + 10) < -7$$

In set-builder notation, the solution set is ☐.

In interval notation, the solution set is ☐.

15. Solve this inequality. Answer using interval notation.

$$10(k - 8) \leq 10(k - 4)$$

16. Solve this inequality. Answer using interval notation.

$$2(k - 6) \leq 2(k - 2)$$

17. Solve this inequality.

$$1 + 10(x - 9) < -23 - (2 - 2x)$$

In set-builder notation, the solution set is ☐.

In interval notation, the solution set is ☐.

18. Solve this inequality.

$$2 + 8(x - 4) < -21 - (1 - 4x)$$

In set-builder notation, the solution set is ☐.

In interval notation, the solution set is ☐.

19. Solve this inequality.

$$-\frac{1}{4}t > \frac{2}{5}t - 13$$

In set-builder notation, the solution set is ☐.

In interval notation, the solution set is ☐.

20. Solve this inequality.

$$-\frac{1}{2}t > \frac{6}{5}t - 17$$

In set-builder notation, the solution set is ☐.

In interval notation, the solution set is ☐.

21. Solve this linear equation for x.

$$Ax + By = C$$

22. Solve this linear equation for y.

$$Ax + By = C$$

23. Solve this linear equation for n.

$$r = a - \frac{2n}{m}$$

24. Solve this linear equation for p.

$$q = m - \frac{2p}{b}$$

25.   Carly has $70 in her piggy bank. She plans to purchase some Pokemon cards, which costs $1.15 each. She plans to save $52.75 to purchase another toy. At most how many Pokemon cards can he purchase?

Write an equation to solve this problem.

Carly can purchase at most ⬚ Pokemon cards.

26.   Maygen has $72 in her piggy bank. She plans to purchase some Pokemon cards, which costs $2.55 each. She plans to save $43.95 to purchase another toy. At most how many Pokemon cards can he purchase?

Write an equation to solve this problem.

Maygen can purchase at most ⬚ Pokemon cards.

27.   Use a linear equation to solve the word problem.

Evan has $85.00 in his piggy bank, and he spends $2.50 every day.

Bobbi has $31.00 in her piggy bank, and she saves $2.00 every day.

If they continue to spend and save money this way, how many days later would they have the same amount of money in their piggy banks?

⬚ days later, Evan and Bobbi will have the same amount of money in their piggy banks.

28.   Use a linear equation to solve the word problem.

Will has $100.00 in his piggy bank, and he spends $4.00 every day.

Ross has $34.00 in his piggy bank, and he saves $2.00 every day.

If they continue to spend and save money this way, how many days later would they have the same amount of money in their piggy banks?

⬚ days later, Will and Ross will have the same amount of money in their piggy banks.

29.   A hockey team played a total of 167 games last season. The number of games they won was 17 more than five times of the number of games they lost.

Write and solve an equation to answer the following questions.

The team lost ⬚ games. The team won ⬚ games.

30.   A hockey team played a total of 117 games last season. The number of games they won was 13 more than three times of the number of games they lost.

Write and solve an equation to answer the following questions.

The team lost ⬚ games. The team won ⬚ games.

31.   A rectangle's perimeter is 278 ft. Its length is 4 ft longer than four times its width. Use an equation to find the rectangle's length and width.

Its width is ⬚.

Its length is ⬚.

32.   A rectangle's perimeter is 226 ft. Its length is 1 ft longer than three times its width. Use an equation to find the rectangle's length and width.

Its width is [   ].

Its length is [   ].

33. Briana has saved $45.00 in her piggy bank, and she decided to start spending them. She spends $5.00 every 7 days. After how many days will she have $30.00 left in the piggy bank?

Briana will have $30.00 left in her piggy bank after [   ] days.

34. Maygen has saved $49.00 in her piggy bank, and she decided to start spending them. She spends $5.00 every 6 days. After how many days will she have $29.00 left in the piggy bank?

Maygen will have $29.00 left in her piggy bank after [   ] days.

35. The following two triangles are similar to each other. Find the length of the missing side.

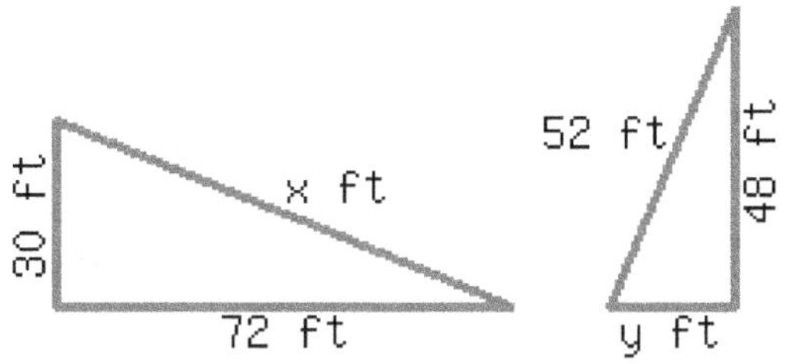

The length of the side labeled x is [   ] and the length of the side labeled y is [   ].

36. The following two triangles are similar to each other. Find the length of the missing side.

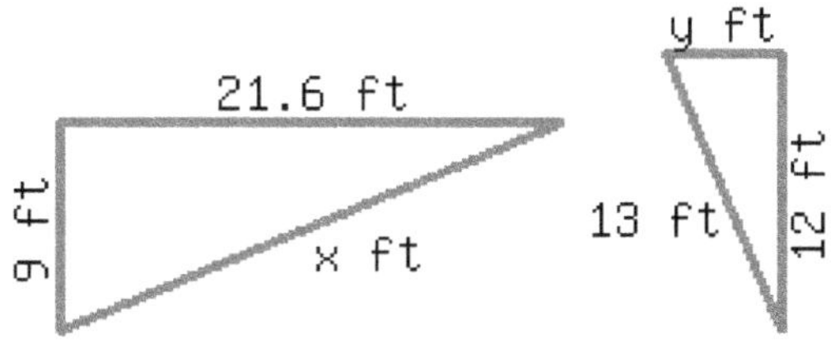

The length of the side labeled x is [   ] and the length of the side labeled y is [   ].

37. A restaurant used 639.4 lb of vegetable oil in 23 days. At this rate, 1306.6 lb of oil will last how many days?

The restaurant will use 1306.6 lb of vegetable oil in [   ] days.

38. A restaurant used 914.5 lb of vegetable oil in 31 days. At this rate, 1209.5 lb of oil will last how many days?

The restaurant will use 1209.5 lb of vegetable oil in [   ] days.

39.   Use a linear equation to solve the word problem.

Massage Heaven and Massage You are competitors. Massage Heaven has 6500 registered customers, and it gets approximately 550 newly registered customers every month. Massage You has 8500 registered customers, and it gets approximately 450 newly registered customers every month. How many months would it take Massage Heaven to catch up with Massage You in the number of registered customers?

These two companies would have approximately the same number of registered customers [____________] months later.

40.   Use a linear equation to solve the word problem.

Two truck rental companies have different rates. V-Haul has a base charge of $70.00, plus $0.70 per mile. W-Haul has a base charge of $60.40, plus $0.80 per mile. For how many miles would these two companies charge the same amount?

If a driver drives [____________] miles, those two companies would charge the same amount of money.

41.   A rectangle's perimeter is 134 ft. Its length is 5 ft shorter than three times its width. Use an equation to find the rectangle's length and width.

Its width is [____________].

Its length is [____________].

42.   A rectangle's perimeter is 178 ft. Its length is 2 ft longer than two times its width. Use an equation to find the rectangle's length and width.

Its width is [____________].

Its length is [____________].

CHAPTER 4

Graphing Lines

4.1 Cartesian Coordinates

When we model relationships with graphs, we use the **Cartesian coordinate system**. This section covers the basic vocabulary and ideas that come with the Cartesian coordinate system.

The Cartesian coordinate system identifies the location of every point in a plane. Basically, the system gives every point in a plane its own "address" in relation to a starting point. We'll use a street grid as an analogy. Here is a map with Carl's home at the center. The map also shows some nearby businesses. Assume each unit in the grid represents one city block.

René Descartes. Several conventions used in mathematics are attributed to (or at least named after) René Descartes[a]. The Cartesian coordinate system is one of these.

---

[a]en.wikipedia.org/wiki/René_Descartes

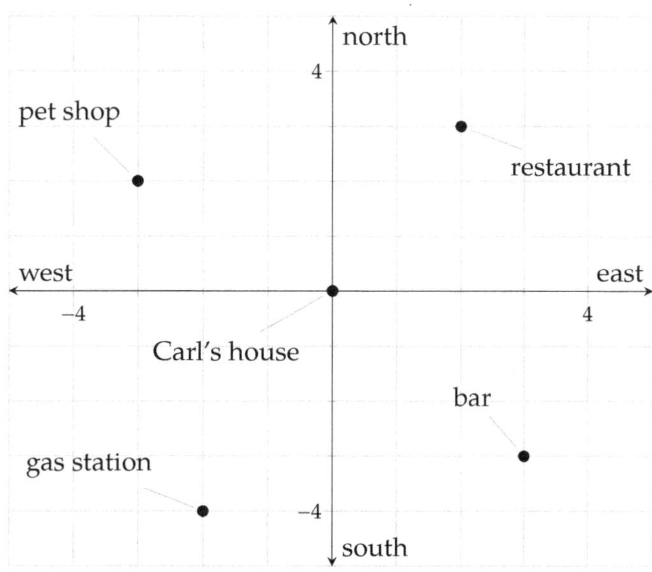

Figure 4.1.2: Carl's neighborhood

If Carl has an out-of-town guest who asks him how to get to the restaurant, Carl could say:

"First go 2 blocks east, then go 3 blocks north."

Carl uses two numbers to locate the restaurant. In the Cartesian coordinate system, these numbers are called **coordinates** and they are written as the **ordered pair** $(2,3)$. The first coordinate, 2, represents distance traveled from Carl's house to the east (or to the right horizontally on the graph). The second coordinate, 3, represents distance to the north (up vertically on the graph).

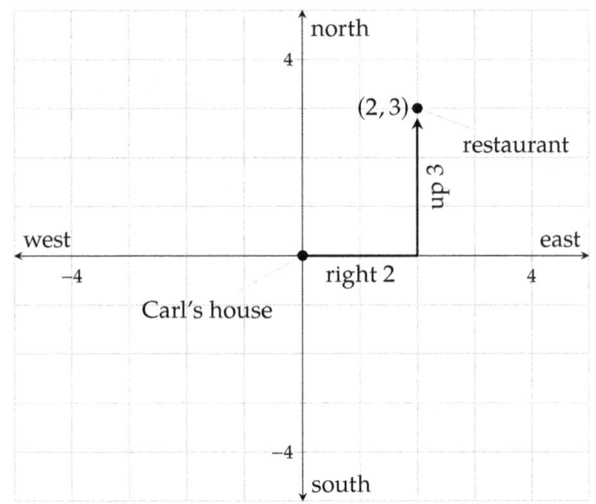

Figure 4.1.3: Carl's path to the restaurant

Alternatively, to travel from Carl's home to the pet shop, he would go 3 blocks west, and then 2 blocks north.

In the Cartesian coordinate system, the *positive* directions are to the *right* horizontally and *up* vertically. The *negative* directions are to the *left* horizontally and *down* vertically. So the pet shop's Cartesian coordinates are $(-3,2)$.

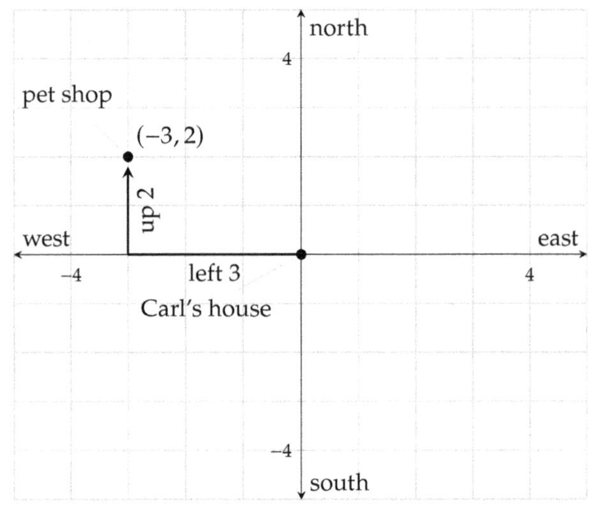

Figure 4.1.4: Carl's path to the pet shop

Remark 4.1.5. It's important to know that the order of Cartesian coordinates is (horizontal, vertical). This idea of communicating horizontal information *before* vertical information is consistent throughout most of mathematics.

Checkpoint 4.1.6. Use Figure 4.1.2 to answer the following questions.

    a. What are the coordinates of the bar?

    b. What are the coordinates of the gas station?

    c. What are the coordinates of Carl's house?

Traditionally, the variable x represents numbers on the horizontal axis, so it is called the x-**axis**. The variable y represents numbers on the vertical axis, so it is called the y-**axis**. The axes meet at the point $(0,0)$, which is called the **origin**. Every point in the plane is represented by an **ordered pair**, (x, y).

In a Cartesian coordinate system, the map of Carl's neighborhood would look like this:

Notation Issue: Coordinates or Interval?. Unfortunately, the notation for an ordered pair looks exactly like interval notation for an open interval. *Context* will help you understand if $(2, 3)$ indicates the point 2 units right of the origin and 3 units up, or if $(2, 3)$ indicates the interval of all real numbers between 2 and 3.

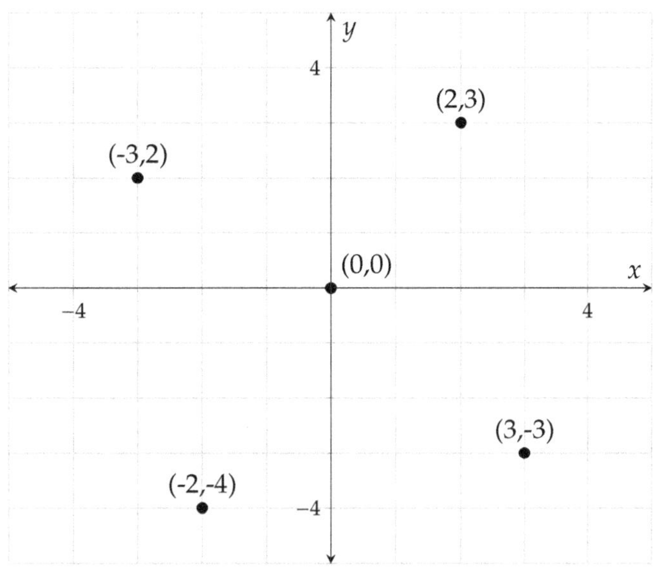

Figure 4.1.7: Carl's Neighborhood in a Cartesian Coordinate System

Definition 4.1.8 Cartesian Coordinate System. A Cartesian coordinate system[1] is a coordinate system that specifies each point uniquely in a plane by a pair of numerical coordinates, which are the signed (positive/negative) distances to the point from two fixed perpendicular directed lines, measured in the same unit of length. Those two reference lines are called the **horizontal axis** and **vertical axis**, and the point where they meet is the **origin**. The horizontal and vertical axes are often called the x-**axis** and y-**axis**.

The plane based on the x-axis and y-axis is called a **coordinate plane**. The ordered pair used to locate a point is called the point's **coordinates**, which consists of an x-**coordinate** and a y-**coordinate**. For example, for the point $(1, 2)$, its x-coordinate is 1, and its y-coordinate is 2. The origin has coordinates $(0, 0)$.

A Cartesian coordinate system is divided into four **quadrants**, as shown in Figure 4.1.9. The quadrants are traditionally labeled with Roman numerals.

---

[1]en.wikipedia.org/wiki/Cartesian_coordinate_system

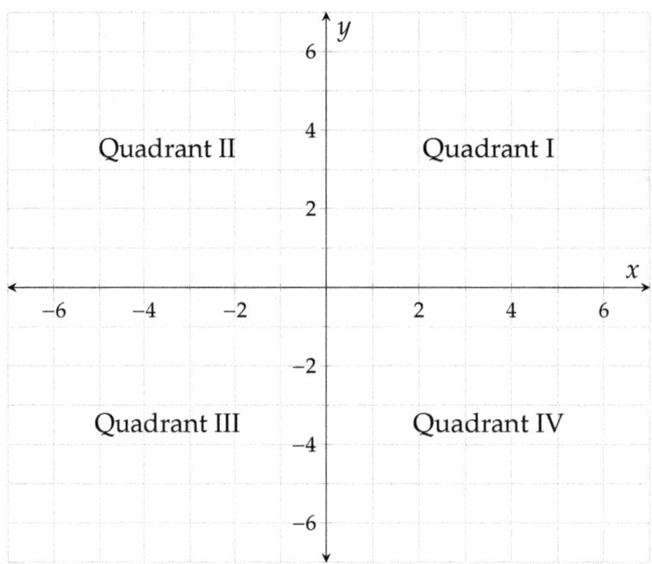

Figure 4.1.9: A Cartesian grid with four quadrants marked

Example 4.1.10 On paper, sketch a Cartesian coordinate system with units, and then plot the following points: $(3, 2), (-5, -1), (0, -3), (4, 0)$.

Explanation.

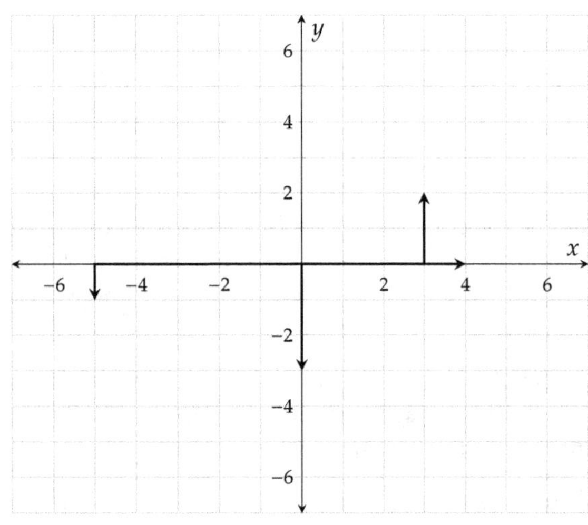
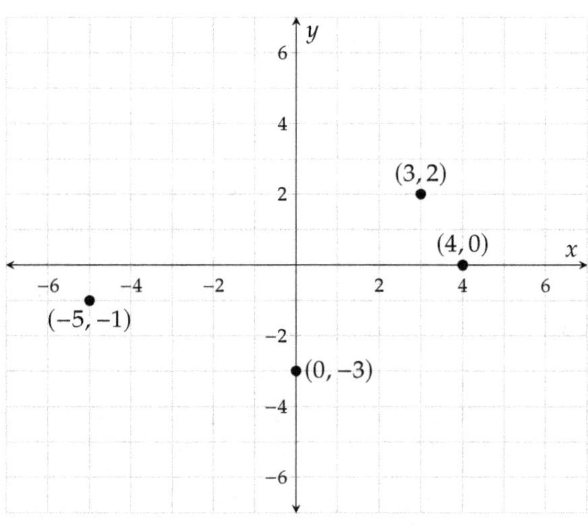

Exercises

Identifying Coordinates   Locate each point in the graph:

1.

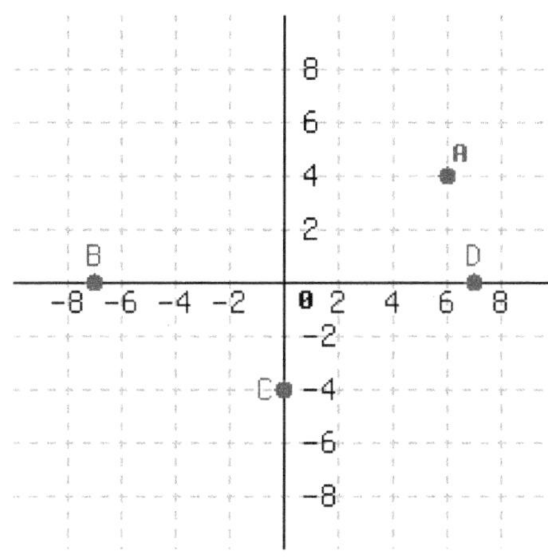

Write each point's position as an ordered pair, like $(1, 2)$.

$A =$ _______ $B =$ _______
$C =$ _______ $D =$ _______

2.

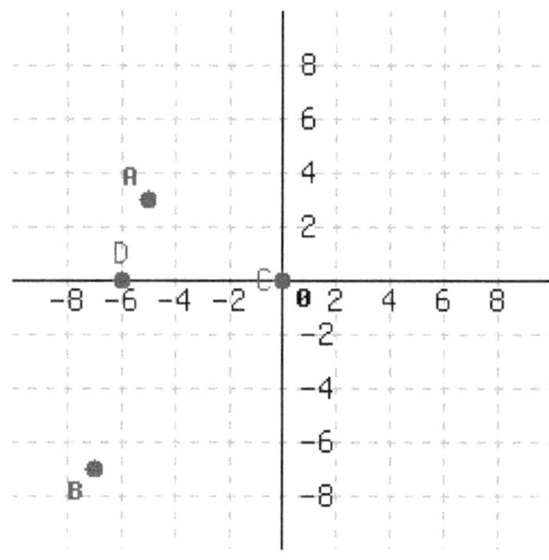

Write each point's position as an ordered pair, like $(1, 2)$.

$A =$ _______ $B =$ _______
$C =$ _______ $D =$ _______

Creating Sketches of Graphs

3. Sketch the points $(8, 2)$, $(5, 5)$, $(-3, 0)$, and $(2, -6)$ on a Cartesian plane.

4. Sketch the points $(1, -4)$, $(-3, 5)$, $(0, 4)$, and $(-2, -6)$ on a Cartesian plane.

5. Sketch the points $(208, -50)$, $(97, 112)$, $(-29, 103)$, and $(-80, -172)$ on a Cartesian plane.

6. Sketch the points $(110, 38)$, $(-205, 52)$, $(-52, 125)$, and $(-172, -80)$ on a Cartesian plane.

7. Sketch the points $(5.5, 2.7)$, $(-7.3, 2.75)$, $\left(-\frac{10}{3}, \frac{1}{2}\right)$, and $\left(-\frac{28}{5}, -\frac{29}{4}\right)$ on a Cartesian plane.

8. Sketch the points $(1.9, -3.3)$, $(-5.2, -8.11)$, $\left(\frac{7}{11}, \frac{15}{2}\right)$, and $\left(-\frac{16}{3}, \frac{19}{5}\right)$ on a Cartesian plane.

9. Sketch a Cartesian plane and shade the quadrants where the x-coordinate is negative.

10. Sketch a Cartesian plane and shade the quadrants where the y-coordinate is positive.

11. Sketch a Cartesian plane and shade the quadrants where the x-coordinate has the same sign as the y-coordinate.

12. Sketch a Cartesian plane and shade the quadrants where the x-coordinate and the y-coordinate have opposite signs.

Cartesian Plots in Context

13. This graph gives the minimum estimates of the wolf population in Washington from 2008 through 2015.
(Source: http://wdfw.wa.gov/publications/01793/ wdfw01793.pdf)

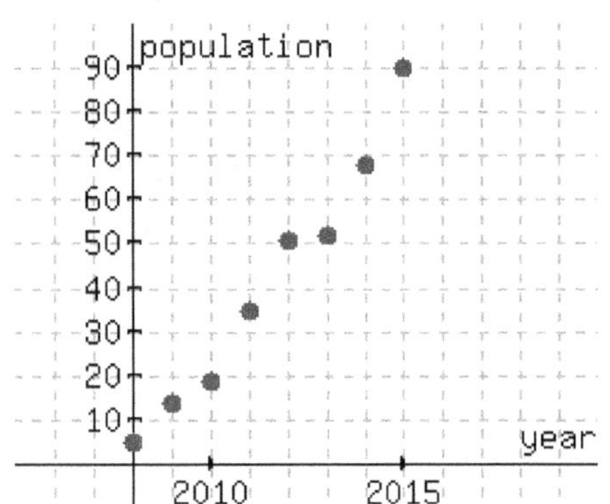

What are the Cartesian coordinates for the point representing the year 2010?

Between 2010 and 2011, the wolf population grew by ☐ wolves.

List at least three ordered pairs in the graph.

14. Here is a graph of the foreign-born US population (in millions) during Census years 1960 to 2010.
(Source: http://www.pewhispanic.org/2015/09/28/ chapter-5-u-s-foreign-born-population-trends/.)

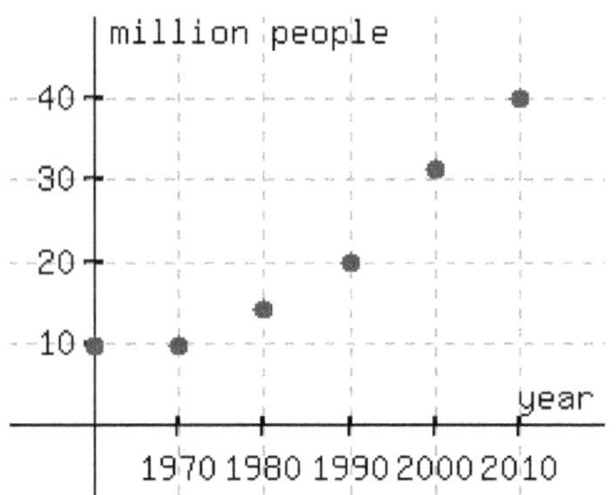

What are the Cartesian coordinates for the point representing the year 1970?

Between 1970 and 1990, the US population that is foreign-born increased by ☐ million people.

List at least three ordered pairs in the graph.

Regions in the Cartesian Plane

15. The point $(1, -10)$ is in Quadrant   (☐ I   ☐ II   ☐ III   ☐ IV) .

   The point $(4, 2)$ is in Quadrant   (☐ I   ☐ II   ☐ III   ☐ IV) .

   The point $(-6, 8)$ is in Quadrant   (☐ I   ☐ II   ☐ III   ☐ IV) .

   The point $(-10, -2)$ is in Quadrant   (☐ I   ☐ II   ☐ III   ☐ IV) .

16. The point $(4, 4)$ is in Quadrant   (☐ I   ☐ II   ☐ III   ☐ IV) .

   The point $(-7, -10)$ is in Quadrant   (☐ I   ☐ II   ☐ III   ☐ IV) .

   The point $(4, -9)$ is in Quadrant   (☐ I   ☐ II   ☐ III   ☐ IV) .

   The point $(-9, 10)$ is in Quadrant   (☐ I   ☐ II   ☐ III   ☐ IV) .

17. Assume the point (x, y) if in Quadrant II, locate the following points:

   The point $(-x, y)$ is in Quadrant   (☐ I   ☐ II   ☐ III   ☐ IV) .

   The point $(x, -y)$ is in Quadrant   (☐ I   ☐ II   ☐ III   ☐ IV) .

   The point $(-x, -y)$ is in Quadrant   (☐ I   ☐ II   ☐ III   ☐ IV) .

18. Assume the point (x, y) if in Quadrant IV, locate the following points:

The point $(-x, y)$ is in Quadrant   ($\square$ I $\square$ II $\square$ III $\square$ IV)   .

The point $(x, -y)$ is in Quadrant   ($\square$ I $\square$ II $\square$ III $\square$ IV)   .

The point $(-x, -y)$ is in Quadrant   ($\square$ I $\square$ II $\square$ III $\square$ IV)   .

19. Answer the following questions on the coordinate system:

For the point (x, y), if $x > 0$ and $y > 0$, then the point is in/on   ($\square$ Quadrant I $\square$ Quadrant II $\square$ Quadrant III $\square$ Quadrant IV $\square$ the x-axis $\square$ the y-axis)   .

For the point (x, y), if $x > 0$ and $y < 0$, then the point is in/on   ($\square$ Quadrant I $\square$ Quadrant II $\square$ Quadrant III $\square$ Quadrant IV $\square$ the x-axis $\square$ the y-axis)   .

For the point (x, y), if $x < 0$ and $y < 0$, then the point is in/on   ($\square$ Quadrant I $\square$ Quadrant II $\square$ Quadrant III $\square$ Quadrant IV $\square$ the x-axis $\square$ the y-axis)   .

For the point (x, y), if $x < 0$ and $y > 0$, then the point is in/on   ($\square$ Quadrant I $\square$ Quadrant II $\square$ Quadrant III $\square$ Quadrant IV $\square$ the x-axis $\square$ the y-axis)   .

For the point (x, y), if $y = 0$, then the point is in/on   ($\square$ Quadrant I $\square$ Quadrant II $\square$ Quadrant III $\square$ Quadrant IV $\square$ the x-axis $\square$ the y-axis)   .

For the point (x, y), if $x = 0$, then the point is in/on   ($\square$ Quadrant I $\square$ Quadrant II $\square$ Quadrant III $\square$ Quadrant IV $\square$ the x-axis $\square$ the y-axis)   .

Plotting Points and Choosing a Scale

20. What would be the difficulty with trying to plot $(12, 4)$, $(13, 5)$, and $(310, 208)$ all on the same graph?

21. The points $(3, 5)$, $(5, 6)$, $(7, 7)$, and $(9, 8)$ all lie on a straight line. What can go wrong if you make a plot of a Cartesion plane with these points marked, and you don't have tick marks that are evenly spaced apart?

4.2 Graphing Equations

We have graphed *points* in a coordinate system, and now we will graph *lines* and *curves*.

A **graph** of an equation is a picture of that equation's solution set. For example, the graph of $y = -2x + 3$ is shown in Figure 4.1c. The graph plots the ordered pairs whose coordinates make $y = -2x + 3$ true. Table 4.2.2 shows a few points that make the equation true.

$$\begin{array}{ll} y = -2x + 3 & (x, y) \\ 5 \overset{\checkmark}{=} -2(-1) + 3 & (-1, 5) \\ 3 \overset{\checkmark}{=} -2(0) + 3 & (0, 3) \\ 1 \overset{\checkmark}{=} -2(1) + 3 & (1, 1) \\ -1 \overset{\checkmark}{=} -2(2) + 3 & (2, -1) \\ -3 \overset{\checkmark}{=} -2(3) + 3 & (3, -3) \\ -5 \overset{\checkmark}{=} -2(4) + 3 & (4, -5) \end{array}$$

Table 4.2.2

Table 4.2.2 tells us that the points $(-1, 5)$, $(0, 3)$, $(1, 1)$, $(2, -1)$, $(3, -3)$, and $(4, -5)$ are all solutions to the equation $y = -2x + 3$, and so they should all be shaded as part of that equation's graph. You can see them in Figure 4.1a. But there are many more points that make the equation true. More points are plotted in Figure 4.1b. Even more points are plotted in Figure 4.1c — so many, that together the points look like a straight line.

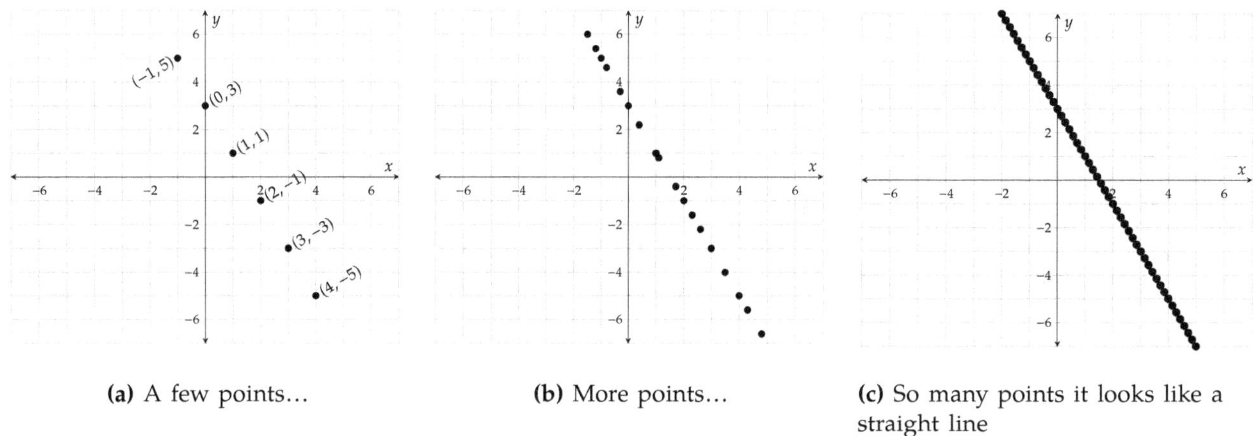

(a) A few points...

(b) More points...

(c) So many points it looks like a straight line

Figure 4.2.2: Graphs of the Equation $y = -2x + 3$

Remark 4.2.3. The graph of an equation shades all the points (x, y) that make the equation true once the x- and y-values are substituted in. Typically, there are *so many* points shaded, that the final graph appears to be a continuous line or curve that you could draw with one stroke of a pen.

Checkpoint 4.2.4. The point $(4, -5)$ is on the graph in Figure 4.2.3.(c). What happens when you substitute these values into the equation $y = -2x + 3$?

$$\begin{array}{rcl} y & = & -2x + 3 \\ \underline{} & = & \underline{} \end{array}$$

This equation is (□ true   □ false) .

Checkpoint 4.2.5. Decide whether $(5, -2)$ and $(-10, -7)$ are on the graph of the equation $y = -\frac{3}{5}x + 1$.

At $(5, -2)$:

$$y \quad = \quad -\frac{3}{5}x + 1$$
$$\underline{} \quad = \quad \underline{}$$

This equation is　(□ true　□ false)　and $(5, -2)$ is　(□ part of　□ not part of)　the graph of $y = -\frac{3}{5}x + 1$.

At $(-10, -7)$:

$$y \quad = \quad -\frac{3}{5}x + 1$$
$$\underline{} \quad = \quad \underline{}$$

This equation is　(□ true　□ false)　and $(-10, -7)$ is　(□ part of　□ not part of)　the graph of $y = -\frac{3}{5}x + 1$.

Explanation.　If the point $(5, -2)$ is on $y = -\frac{3}{5}x + 1$, once we substitute $x = 5$ and $y = -2$ into the line's equation, the equation should be true. Let's try:

$$y = -\frac{3}{5}x + 1$$
$$-2 \stackrel{?}{=} -\frac{3}{5}(5) + 1$$
$$-2 \stackrel{\checkmark}{=} -3 + 1$$

Because this last equation is true, we can definitively say that $(5, -2)$ is on the graph of $y = -\frac{3}{5}x + 1$.

However if we substitute $x = -10$ and $y = -7$ into the equation, it leads to $-7 = 7$, which is false. This definitively tells us that $(-10, -7)$ is not on the graph.

So to make our own graph of an equation with two variables x and y, we can choose some reasonable x-values, then calculate the corresponding y-values, and then plot the (x, y)-pairs as points. For many (not-so-complicated) algebraic equations, connecting those points with a smooth curve will produce an excellent graph.

Example 4.2.6 Let's plot a graph for the equation $y = -2x + 5$. We use a table to organize our work:

x	$y = -2x + 5$	Point
-2		
-1		
0		
1		
2		

(a) Set up the table

x	$y = -2x + 5$	Point
-2	$-2(-2) + 5 = 9$	$(-2, 9)$
-1	$-2(-1) + 5 = 7$	$(-1, 7)$
0	$-2(0) + 5 = 5$	$(0, 5)$
1	$-2(1) + 5 = 3$	$(1, 3)$
2	$-2(2) + 5 = 1$	$(2, 1)$

(b) Complete the table

Figure 4.2.6: Making a table for $y = -2x + 5$

We use points from the table to graph the equation. First, plot each point carefully. Then, connect the points with a smooth curve. Here, the curve is a straight line. Lastly, we can communicate that the graph extends further by sketching arrows on both ends of the line.

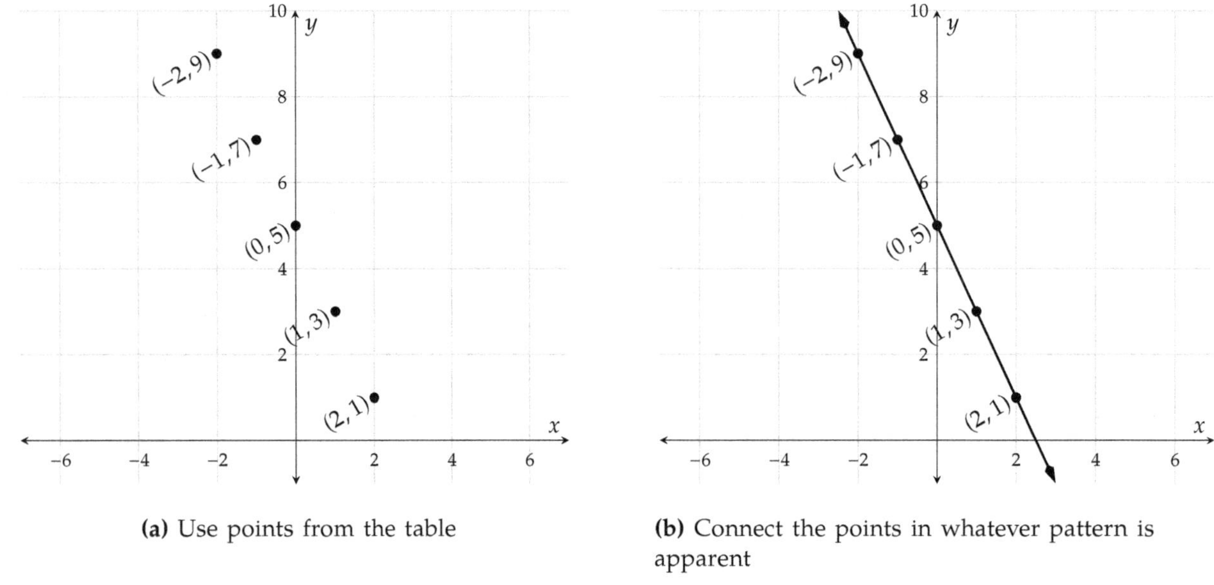

(a) Use points from the table

(b) Connect the points in whatever pattern is apparent

Figure 4.2.6: Graphing the Equation $y = -2x + 5$

Remark 4.2.7. Note that our choice of x-values is arbitrary. As long as we determine the coordinates of enough points to indicate the behavior of the graph, we may choose whichever x-values we like. For simpler calculations, people often start with the integers from -2 to 2. However sometimes the equation has context that suggests using other x-values, as in the next examples.

Example 4.2.8 The gas tank in Sofia's car holds $14\,\text{gal}$ of fuel. Over the course of a long road trip, her car uses fuel at an average rate of $0.032\,\frac{\text{gal}}{\text{mi}}$. If Sofia fills the tank at the beginning of a long trip, then the amount of fuel remaining in the tank, y, after driving x miles is given by the equation $y = 14 - 0.032x$. Make a suitable table of values and graph this equation.

Explanation. Choosing x-values from -2 to 2, as in our previous example, wouldn't make sense here. Sofia cannot drive a negative number of miles, and any long road trip is longer than 2 miles. So in this context, choose x-values that reflect the number of miles Sofia might drive in a day.

x	$y = 14 - 0.032x$	Point
20	13.36	$(20, 13.36)$
50	12.4	$(50, 12.4)$
80	11.44	$(80, 11.44)$
100	10.8	$(100, 10.8)$
200	7.6	$(200, 7.6)$

Table 4.2.9: Make the table

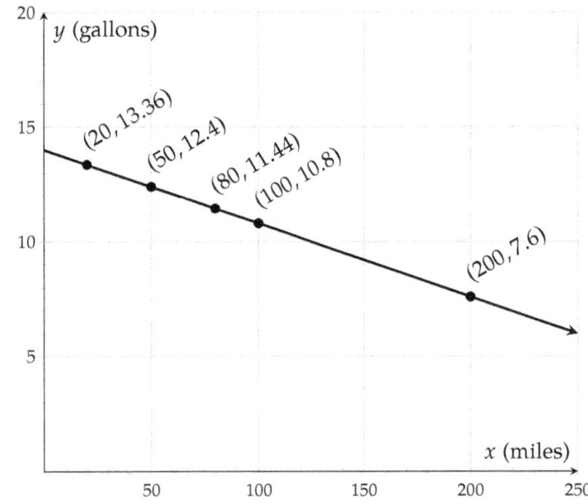

Figure 4.2.10: Make the graph

In the graph from Example 4.2.8, notice how both axes indicate units that help describe the meaning of each variable. Whenever a graph has real-world context, be sure to label both axes clearly with both variable name (like x) and units.

Example 4.2.11 Plot a graph for the equation $y = \frac{4}{3}x - 4$.

Explanation. This equation doesn't have any context to help us choose x-values for a table. We could use x-values like -2, -1, and so on. But note the fraction in the equation. If we use an x-value like -2, we will have to multiply by the fraction $\frac{4}{3}$ which will leave us still holding a fraction. And then we will have to subtract 4 from that fraction. Since we know that everyone can make mistakes with that kind of arithmetic, maybe we can avoid it with a more wise selection of x-values.

If we use onnly multiples of 3 for the x-values, then multiplying by $\frac{4}{3}$ will leave us with an integer, which will be easy to subtract 4 from. So we decide to use -6, -3, 0, 3, and 6 for x.

x	$y = \frac{4}{3}x - 4$	Point
-6		
-3		
0		
3		
6		

(a) Set up the table

x	$y = \frac{4}{3}x - 4$	Point
-6	$\frac{4}{3}(-6) - 2 = -12$	$(-6, -12)$
-3	$\frac{4}{3}(-3) - 2 = -8$	$(-3, -8)$
0	$\frac{4}{3}(0) - 2 = -4$	$(0, -4)$
3	$\frac{4}{3}(3) - 2 = 0$	$(3, 0)$
6	$\frac{4}{3}(6) - 2 = 4$	$(6, 4)$

(b) Complete the table

Figure 4.2.11: Making a table for $y = \frac{4}{3}x - 4$

We use points from the table to graph the equation. First, plot each point carefully. Then, connect the points with a smooth curve. Here, the curve is a straight line. Lastly, we can communicate that the graph extends further by sketching arrows on both ends of the line.

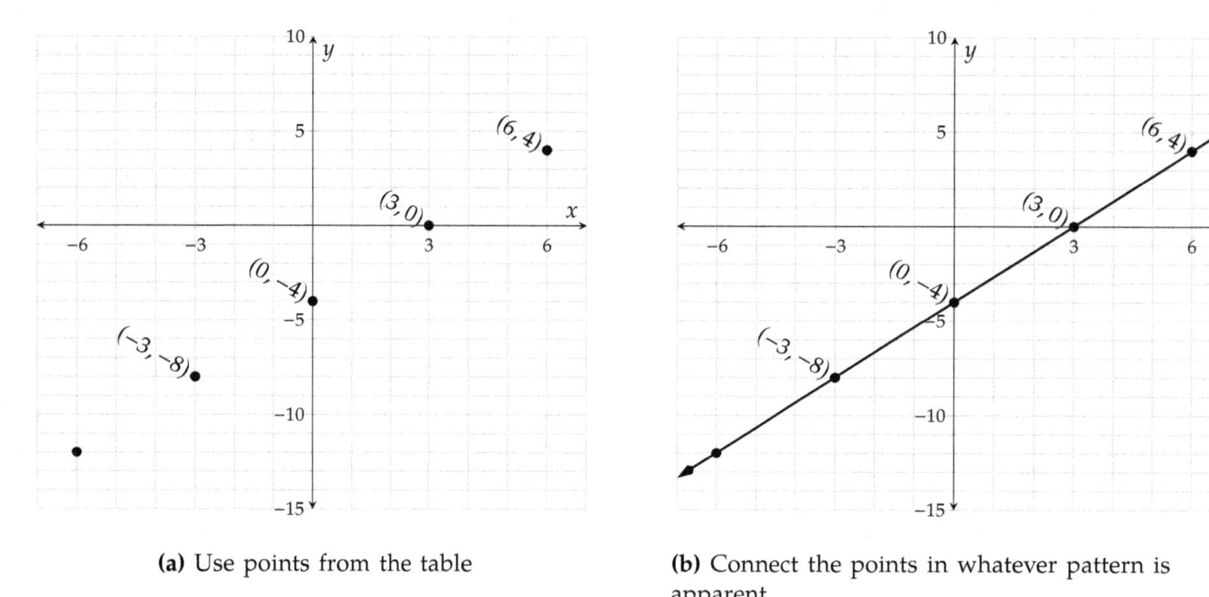

(a) Use points from the table

(b) Connect the points in whatever pattern is apparent

Figure 4.2.11: Graphing the Equation $y = \frac{4}{3}x - 4$

Not all equations make a straight line once they are plotted.

Example 4.2.12 Build a table and graph the equation $y = x^2$. Use x-values from -3 to 3.

Explanation.

x	$y = x^2$	Point
-3	$(-3)^2 = 9$	$(-3, 9)$
-2	$(-2)^2 = 4$	$(-2, 4)$
-1	$(-1)^2 = 1$	$(-1, 1)$
0	$(0)^2 = 0$	$(0, 0)$
1	$(1)^2 = 1$	$(0, 1)$
2	$(2)^2 = 4$	$(2, 4)$
3	$(3)^2 = 9$	$(3, 9)$

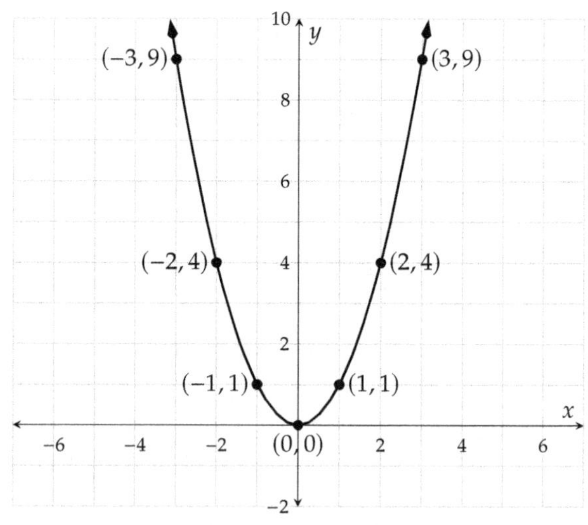

In this example, the points do not fall on a straight line. Many algebraic equations have graphs that are non-linear, where the points do not fall on a straight line. Since each x-value corresponds to a single y-value (the square of x) we connected the points with a smooth curve, sketching from left to right.

Exercises

Testing Points as Solutions   Consider the equation

1. $y = 8x + 7$

Which of the following ordered pairs are solutions to the given equation? There may be more than one correct answer.

☐ $(10, 90)$ ☐ $(-4, -25)$
☐ $(0, 10)$ ☐ $(-5, -33)$

2. $y = 9x + 3$

Which of the following ordered pairs are solutions to the given equation? There may be more than one correct answer.

☐ $(0, 4)$ ☐ $(-5, -42)$
☐ $(-3, -24)$ ☐ $(4, 41)$

3. $y = -2x - 2$

Which of the following ordered pairs are solutions to the given equation? There may be more than one correct answer.

☐ $(8, -18)$ ☐ $(-6, 13)$
☐ $(-2, 2)$ ☐ $(0, -2)$

4. $y = -10x - 5$

Which of the following ordered pairs are solutions to the given equation? There may be more than one correct answer.

☐ $(3, -35)$ ☐ $(0, -5)$
☐ $(-6, 56)$ ☐ $(-4, 35)$

5. $y = \frac{2}{3}x - 3$

Which of the following ordered pairs are solutions to the given equation? There may be more than one correct answer.

☐ $(-15, -13)$ ☐ $(0, 0)$
☐ $(6, 1)$ ☐ $(-9, -6)$

6. $y = \frac{2}{3}x - 5$

Which of the following ordered pairs are solutions to the given equation? There may be more than one correct answer.

☐ $(3, -3)$ ☐ $(-15, -15)$
☐ $(0, 0)$ ☐ $(-9, -10)$

7. $y = -\frac{3}{4}x - 3$

Which of the following ordered pairs are solutions to the given equation? There may be more than one correct answer.

☐ $(-12, 6)$ ☐ $(20, -15)$
☐ $(0, -3)$ ☐ $(-16, 14)$

8. $y = -\frac{3}{4}x - 5$

Which of the following ordered pairs are solutions to the given equation? There may be more than one correct answer.

☐ $(12, -11)$ ☐ $(0, -5)$
☐ $(-20, 10)$ ☐ $(-16, 9)$

Tables for Equations   Make a table for the equation.

9. The first row is an example.

x	$y = -x + 6$	Points
-3	9	$(-3, 9)$
-2	_____	_____
-1	_____	_____
0	_____	_____
1	_____	_____
2	_____	_____

10. The first row is an example.

x	$y = -x + 7$	Points
-3	10	$(-3, 10)$
-2	_____	_____
-1	_____	_____
0	_____	_____
1	_____	_____
2	_____	_____

11. The first row is an example.

x	$y = 5x + 1$	Points
−3	−14	$(-3, -14)$
−2	______	____________
−1	______	____________
0	______	____________
1	______	____________
2	______	____________

12. The first row is an example.

x	$y = 6x + 8$	Points
−3	−10	$(-3, -10)$
−2	______	____________
−1	______	____________
0	______	____________
1	______	____________
2	______	____________

13. The first row is an example.

x	$y = -5x + 4$	Points
−3	19	$(-3, 19)$
−2	______	____________
−1	______	____________
0	______	____________
1	______	____________
2	______	____________

14. The first row is an example.

x	$y = -5x + 1$	Points
−3	16	$(-3, 16)$
−2	______	____________
−1	______	____________
0	______	____________
1	______	____________
2	______	____________

15. The first row is an example.

x	$y = \frac{3}{8}x + 5$	Points
−24	−4	$(-24, -4)$
−16	______	____________
−8	______	____________
0	______	____________
8	______	____________
16	______	____________

16. The first row is an example.

x	$y = \frac{3}{4}x - 7$	Points
−12	−16	$(-12, -16)$
−8	______	____________
−4	______	____________
0	______	____________
4	______	____________
8	______	____________

17. The first row is an example.

x	$y = -\frac{5}{2}x + 2$	Points
−6	17	$(-6, 17)$
−4	______	____________
−2	______	____________
0	______	____________
2	______	____________
4	______	____________

18. The first row is an example.

x	$y = -\frac{5}{8}x + 10$	Points
−24	25	$(-24, 25)$
−16	______	____________
−8	______	____________
0	______	____________
8	______	____________
16	______	____________

19. x $y = 8x$

20. x $y = 12x$

21. x $y = 8x + 2$

22. x $y = 10x - 4$

23. x $y = \frac{19}{2}x - 8$

24. x $y = \frac{3}{4}x - 6$

25. x $y = -\frac{11}{19}x - 4$

26. x $y = \frac{15}{8}x - 1$

Cartesian Plots in Context

27. A certain water heater will cost you \$900 to buy and have installed. This water heater claims that its operating expense (money spent on electricity or gas) will be about \$31 per month. According to this information, the equation $y = 900 + 31x$ models the total cost of the water heater after x months, where y is in dollars. Make a table of at least five values and plot a graph of this equation.

28. You bought a new Toyota Corolla for \$18,600 with a zero interest loan over a five-year period. That means you'll have to pay \$310 each month for the next five years (sixty months) to pay it off. According to this information, the equation $y = 18600 - 310x$ models the loan balance after x months, where y is in dollars. Make a table of at least five values and plot a graph of this equation. Make sure to include a data point representing when you will have paid off the loan.

29. The pressure inside a full propane tank will rise and fall if the ambient temperature rises and falls. The equation $P = 0.1963(T + 459.67)$ models this relationship, where the temperature T is measured in °F and the pressure and the pressure P is measured in $\frac{lb}{in^2}$. Make a table of at least five values and plot a graph of this equation. Make sure to use T-values that make sense in context.

30. A beloved coworker is retiring and you want to give her a gift of week-long vacation rental at the coast that costs \$1400 for the week. You might end up paying for it yourself, but you ask around to see if the other 29 office coworkers want to split the cost evenly. The equation $y = \frac{1400}{x}$ models this situation, where x people contribute to the gift, and y is the dollar amount everyone contributes. Make a table of at least five values and plot a graph of this equation. Make sure to use x-values that make sense in context.

Graphs of Equations

31. Create a table of ordered pairs and then make a plot of the equation $y = 2x + 3$.

32. Create a table of ordered pairs and then make a plot of the equation $y = 3x + 5$.

33. Create a table of ordered pairs and then make a plot of the equation $y = -4x + 1$.

34. Create a table of ordered pairs and then make a plot of the equation $y = -x - 4$.

35. Create a table of ordered pairs and then make a plot of the equation $y = \frac{5}{2}x$.

36. Create a table of ordered pairs and then make a plot of the equation $y = \frac{4}{3}x$.

37. Create a table of ordered pairs and then make a plot of the equation $y = -\frac{2}{5}x - 3$.

38. Create a table of ordered pairs and then make a plot of the equation $y = -\frac{3}{4}x + 2$.

39. Create a table of ordered pairs and then make a plot of the equation $y = x^2 + 1$.

40. Create a table of ordered pairs and then make a plot of the equation $y = (x-2)^2$. Use x-values from 0 to 4.

41. Create a table of ordered pairs and then make a plot of the equation $y = -3x^2$.

42. Create a table of ordered pairs and then make a plot of the equation $y = -x^2 - 2x - 3$.

4.3 Exploring Two-Variable Data and Rate of Change

This section is about examining data that has been plotted on a Cartesian coordinate system, and then making observations. In some cases, we'll be able to turn those observations into useful mathematical calculations.

4.3.1 Modeling data with two variables

Using mathematics, we can analyze real data from the world around us. We can use what we discover to better understand the world, and sometimes to make predictions. Here's an example of data about the economic situation in the US:

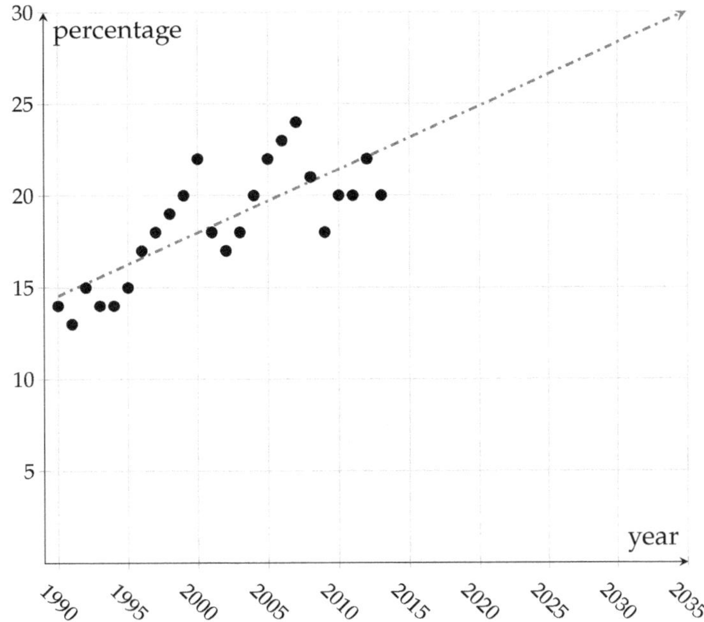

Figure 4.3.2: Share of all income held by the top 1 %, United States, 1990–2013 (www.epi.org)

If this trend continues, what percentage of all income will the top 1 % have in the year 2030? If we model data in the chart with the trend line, we can estimate the value to be 28.6 %. This is one way math is used in real life.

Does that trend line have an equation like those we looked at in Section 4.2? Is it even correct to look at this data set and decide that a straight line is a good model? These are some of the questions we want to consider as we begin this section. The answers will evolve through the next several sections.

4.3.2 Patterns in Tables

Example 4.3.3 Find a pattern in each table. What is the missing entry in each table? Can you describe each pattern in words and/or mathematics?

black	white
big	small
short	tall
few	

USA	Washington
UK	London
France	Paris
Mexico	

1	2
2	4
3	6
5	

Figure 4.3.4: Patterns in 3 tables

Explanation.

black	white
big	small
short	tall
few	*many*

USA	Washington
UK	London
France	Paris
Mexico	*Mexico City*

1	2
2	4
3	6
5	*10*

Figure 4.3.5: Patterns in 3 tables

First table Each word on the right has the opposite meaning of the word to its left.

Second table Each city on the right is the capital of the country to its left.

Third table Each number on the right is double the number to its left.

We can view each table as assigning each input in the left column a corresponding output in the right column. In the first table, for example, when the input "big" is on the left, the output "small" is on the right. The first table's function is to output a word with the opposite meaning of each input word. (This is not a numerical example.)

The third table *is* numerical. And its function is to take a number as input, and give twice that number as its output. Mathematically, we can describe the pattern as "$y = 2x$," where x represents the input, and y represents the output. Labeling the table mathematically, we have Table 4.3.6.

x (input)	y (output)
1	2
2	4
3	6
5	10
10	20
Pattern: $y = 2x$	

Table 4.3.6: Table with a mathematical pattern

The equation $y = 2x$ summarizes the pattern in the table. For each of the following tables, find an equation that describes the pattern you see. Numerical pattern recognition may or may not come naturally for you. Either way, pattern recognition is an important mathematical skill that anyone can develop. Solutions for these exercises provide some ideas for recognizing patterns.

Checkpoint 4.3.7. Write an equation in the form $y = \ldots$ suggested by the pattern in the table.

x	y
0	10
1	11
2	12
3	13

Explanation. Looking for a similar relationship in each row is one approach to pattern recognition. Here, the y-value in each row is 10 greater than its corresponding x-value. So the equation $y = x + 10$ describes the pattern. Of course, there are more complicated patterns to explore, as we'll see in the next exercise.

Checkpoint 4.3.8. Write an equation in the form $y = \ldots$ suggested by the pattern in the table.

x	y
0	-1
1	2
2	5
3	8

Explanation. The relationship between x and y in each row is not as clear here. Another popular approach for finding patterns: in each column, consider how the values change from one row to the next. From row to row, the x-value increases by 1. Also, the y-value increases by 3 from row to row.

$$
\begin{array}{rccl}
 & x & y & \\
 & 0 & -1 & \\
+1 \rightarrow & 1 & 2 & \leftarrow +3 \\
+1 \rightarrow & 2 & 5 & \leftarrow +3 \\
+1 \rightarrow & 3 & 8 & \leftarrow +3
\end{array}
$$

Since row-to-row change is always 1 for x and is always 3 for y, the rate of change from one row to another row is always the same: 3 units of y for every 1 unit of x.

We know that the output for $x = 0$ is $y = -1$. And our observation about the constant rate of change tells us that if we increase the input by x units from 0, the output should increase by $\overbrace{3 + 3 + \cdots + 3}^{x \text{ times}}$, which is $3x$. So the output would be $-1 + 3x$. So the equation is $y = 3x - 1$.

Checkpoint 4.3.9. Write an equation in the form $y = \ldots$ suggested by the pattern in the table.

x	y
0	0
1	1
2	4
3	9

Explanation.   Looking for a relationship in each row here, we see that each y-value is the square of the corresponding x-value. So the equation is $y = x^2$.

What if we had tried the approach we used in the previous exercise, comparing change from row to row in each column?

$$
\begin{array}{rcccl}
 & x & y & \\
 & 0 & 0 & \\
+1 \rightarrow & 1 & 1 & \leftarrow +1 \\
+1 \rightarrow & 2 & 4 & \leftarrow +3 \\
+1 \rightarrow & 3 & 9 & \leftarrow +5 \\
\end{array}
$$

Here, the rate of change is *not* constant from one row to the next. While the x-values are increasing by 1 from row to row, the y-values increase more and more from row to row. Notice that there is a pattern there as well? Mathematicians are fascinated by relationships that produce more complicated patterns. (We'll study more complicated patterns later.)

4.3.3 Rate of Change

For an hourly wage-earner, the amount of money they earn depends on how many hours they work. If a worker earns \$15 per hour, then 10 hours of work corresponds to \$150 of pay. Working *one* additional hour will change 10 hours to 11 hours; and this will cause the \$150 in pay to rise by *fifteen* dollars to \$165 in pay. Any time we compare how one amount changes (dollars earned) as a consequence of another amount changing (hours worked), we are talking about a **rate of change**.

Given a table of two-variable data, between any two rows we can compute a **rate of change**.

Example 4.3.10 The following data, given in both table and graphed form, gives the counts of invasive cancer diagnoses in Oregon over a period of time. (wonder.cdc.gov)

Year	Invasive Cancer Incidents
1999	17,599
2000	17,446
2001	17,847
2002	17,887
2003	17,559
2004	18,499
2005	18,682
2006	19,112
2007	19,376
2008	20,370
2009	19,909
2010	19,727
2011	20,636
2012	20,035
2013	20,458

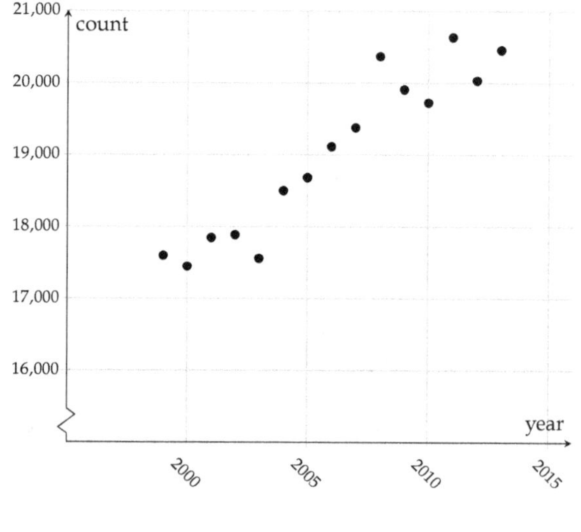

What is the **rate of change** in Oregon invasive cancer diagnoses between 2000 and 2010? The total (net) change in diagnoses over that timespan is

$$19727 - 17446 = 2281.$$

Since 10 years passed (which you can calculate as $2010 - 2000$), the rate of change is 2281 diagnoses per 10 years, or

$$\frac{2281 \text{ diagnoses}}{10 \text{ year}} = 228.1 \frac{\text{diagnoses}}{\text{year}}.$$

We read that last quantity as "228.1 diagnoses per year." This rate of change means that between the years 2000 and 2010, there were 228.1 more diagnoses each year, on average. (Notice that there was no single year in that span when diagnoses increased by 228.1.)

Let's practice calculating rates of change over different timespans:

Checkpoint 4.3.11. Use the data in Example 4.3.10 to find the rate of change in Oregon invasive cancer diagnoses between 1999 and 2002. Just give the numerical value; the units are provided.

$$\boxed{} \frac{\text{diagnoses}}{\text{year}}$$

And what was the rate of change between 2003 and 2011?

$$\boxed{} \frac{\text{diagnoses}}{\text{year}}$$

Explanation. To find the rate of change between 1999 and 2002, calculate

$$\frac{17887 - 17599}{2002 - 1999} = 96.$$

To find the rate of change between 2003 and 2011, calculate

$$\frac{20636 - 17559}{2011 - 2003} = 384.625.$$

We are ready to give a formal definition for **rate of change**. Considering our work from Example 4.3.10 and Checkpoint 4.3.11, we settle on:

Definition 4.3.12 Rate of Change. If (x_1, y_1) and (x_2, y_2) are two data points from a set of two-variable data, then the **rate of change** between them is

$$\frac{\text{change in } y}{\text{change in } x} = \frac{\Delta y}{\Delta x} = \frac{y_2 - y_1}{x_2 - x_1}.$$

(The Greek letter delta, Δ, is used to represent "change in" since it is the first letter of the Greek word for "difference.")

In Example 4.3.10 and 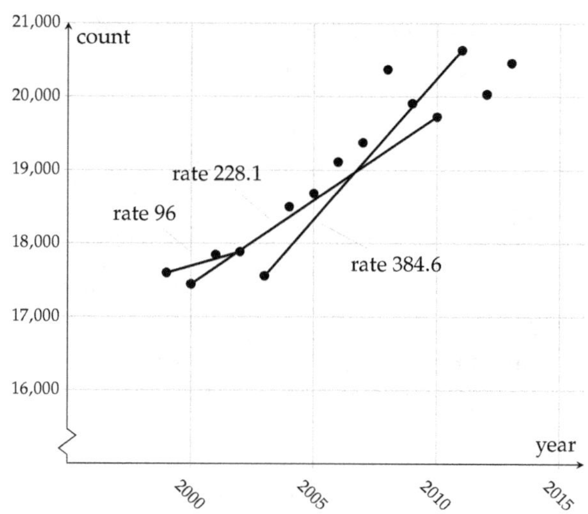 Checkpoint 4.3.11 we found three rates of change. Figure 4.3.13 highlights the three pairs of points that were used to make these calculations.

Figure 4.3.13

Note how the larger the numerical rate of change between two points, the steeper the line is that connects them. This is such an important observation, we'll put it in an official remark.

Remark 4.3.14. The rate of change between two data points is intimately related to the steepness of the line segment that connects those points.

1. The steeper the line, the larger the rate of change, and vice versa.

2. If one rate of change between two data points equals another rate of change between two different data points, then the corresponding line segments will have the same steepness.

3. When a line segment between two data points slants down from left to right, the rate of change between those points will be negative.

In the solution to Checkpoint 4.3.8, the key observation was that the **rate of change** from one row to the next was constant: 3 units of increase in y for every 1 unit of increase in x. Graphing this pattern in Figure 4.3.15, we see that every line segment here has the same steepness, so the entire graph is a line.

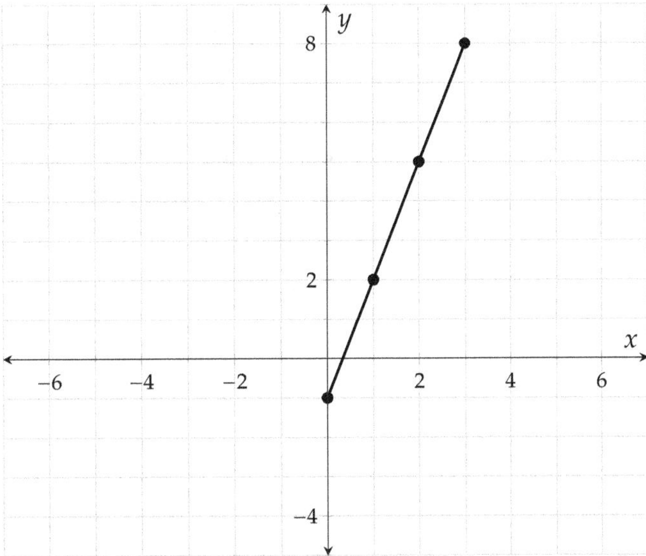

Figure 4.3.15

Whenever the rate of change is constant no matter which two (x, y)-pairs (or data pairs) are chosen from a data set, then you can conclude the graph will be a straight line *even without making the graph*. We call this kind of relationship a **linear** relationship. We'll study linear relationships in more detail throughout this chapter. Right now in this section, we feel it is important to simply identify if data has a linear relationship or not.

Checkpoint 4.3.16. Is there a linear relationship in the table?

x	y
-8	3.1
-5	2.1
-2	1.1
1	0.1

($\square$ The relationship is linear $\square$ The relationship is not linear)

Explanation.   From one x-value to the next, the change is always 3. From one y-value to the next, the change is alwasy -1. So the rate of change is always $\frac{-1}{3} = -\frac{1}{3}$. Since the rate of change is constant, the data have a linear relationship.

Checkpoint 4.3.17. Is there a linear relationship in the table?

x	y
11	208
13	210
15	214
17	220

($\square$ The relationship is linear $\square$ The relationship is not linear)

Explanation. The rate of change between the first two points is $\frac{210-208}{13-11} = 1$. The rate of change between the last two points is $\frac{220-214}{17-15} = 3$. This is one way to demonstrate that the rate of change differs for different pairs of points, so this pattern is not linear.

Checkpoint 4.3.18. Is there a linear relationship in the table?

x	y
3	-2
6	-8
8	-12
12	-20

($\square$ The relationship is linear $\quad$ $\square$ The relationship is not linear)

Explanation. The changes in x from one row to the next are $+3, +2$, and $+8$. That's not a consistent pattern, but we need to consider rates of change between points. The rate of change between the first two points is $\frac{-8-(-2)}{6-3} = -2$. The rate of change between the next two points is $\frac{-12-(-8)}{8-6} = -2$. And the rate of change between the last two points is $\frac{-20-(-12)}{12-8} = -2$. So the rate of change, -2, is constant regardless of which pairs we choose. That means these pairs describe a linear relationship.

Let's return to the data that we opened the section with, in Figure 4.3.2. Is that data linear? Well, yes and no. To be completely honest, it's not linear. It's easy to pick out pairs of points where the steepness changes from one pair to the next. In other words, the points do not all fall into a single line.

However if we stand back, there does seem to be an overall upward trend that is captured by the line someone has drawn over the data. Points *on this line* do have a linear pattern. Let's estimate the rate of change between some points on this line. We are free to use any points to do this, so let's make this calculation easier by choosing points we can clearly identify on the graph: $(1991, 15)$ and $(2020, 25)$.

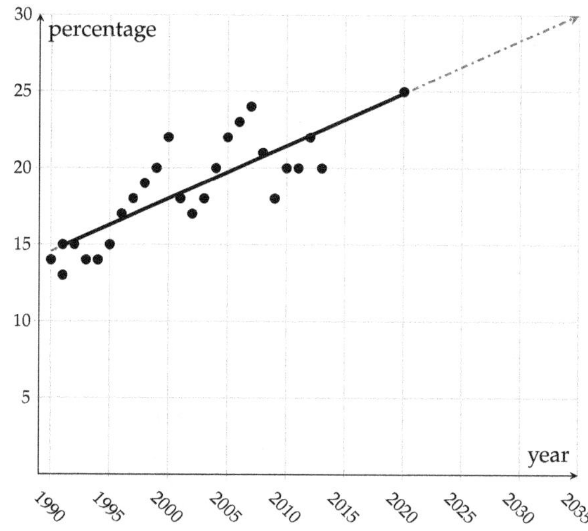

Figure 4.3.19: Share of all income held by the top 1 %, United States, 1990–2013 (www.epi.org)

The rate of change between those two points is

$$\frac{25-15}{2020-1991} = \frac{10}{29} \approx 0.3448.$$

So we might say that *on average* the rate of change expressed by this data is $0.3448\,\frac{\%}{\text{yr}}$.

Exercises

Finding Patterns   Write an equation in the form $y = \ldots$ suggested by the pattern in the table.

1. x	y
-2	-6
-1	-3
0	0
1	3
2	6

2. x	y
3	12
4	16
5	20
6	24
7	28

3. x	y
5	11
6	12
7	13
8	14
9	15

4. x	y
6	10
7	11
8	12
9	13
10	14

5. x	y
15	23
13	21
6	14
4	12
1	9

6. x	y
17	16
6	5
14	13
19	18
1	0

7. x	y
25	5
1	1
4	2
16	4
9	3

8. x	y
-5	5
-2	2
-3	3
-2	2
-4	4

9. x	y
2	4
3	9
4	16
5	25
6	36

10. x	y
7	49
9	81
11	121
13	169
15	225

11. x	y
43	$\frac{1}{43}$
62	$\frac{1}{62}$
84	$\frac{1}{84}$
58	$\frac{1}{58}$
1	1

12. x	y
54	$\frac{1}{54}$
27	$\frac{1}{27}$
25	$\frac{1}{25}$
33	$\frac{1}{33}$
99	$\frac{1}{99}$

Linear Relationships   Does the following table show that x and y have a linear relationship?   ($\square$ yes $\square$ no)

13. x	y
0	93
1	100
2	107
3	114
4	121
5	128

14. x	y
0	62
1	70
2	78
3	86
4	94
5	102

15. x	y
5	51
6	49
7	47
8	45
9	43
10	41

16. x	y
10	74
11	72
12	70
13	68
14	66
15	64

17. x	y
3	19
4	27
5	43
6	75
7	139
8	267

18. x	y
8	260
9	516
10	1028
11	2052
12	4100
13	8196

19. x	y
0	17
1	18
2	25
3	44
4	81
5	142

20. x	y
1	11
2	18
3	37
4	74
5	135
6	226

21. x	y
-10	35.92
-9	36.32
-8	36.72
-7	37.12
-6	37.52
-5	37.92

22. x	y
-2	82.57
-1	84.08
0	85.59
1	87.1
2	88.61
3	90.12

23. x	y
5	85
10	125
12	141
16	173
17	181
18	189

24. x	y
1	18
5	50
7	66
13	114
16	138
19	162

Calculating Rate of Change

25. This table gives population estimates for Portland, Oregon from 1990 through 2014.

Year	Population	Year	Population
1990	487849	2003	539546
1991	491064	2004	533120
1992	493754	2005	534112
1993	497432	2006	538091
1994	497659	2007	546747
1995	498396	2008	556442
1996	501646	2009	566143
1997	503205	2010	585261
1998	502945	2011	593859
1999	503637	2012	602954
2000	529922	2013	609520
2001	535185	2014	619360
2002	538803		

Find the rate of change in Portland population between 2005 and 2006. Just give the numerical value; the units are provided.

$\dfrac{\text{people}}{\text{year}}$

And what was the rate of change between 2008 and 2014?

$\dfrac{\text{people}}{\text{year}}$

List all the years where there is a negative rate of change between that year and the next year.

26. This table and graph gives population estimates for Portland, Oregon from 1990 through 2014.

Year	Population	Year	Population
1990	487849	2003	539546
1991	491064	2004	533120
1992	493754	2005	534112
1993	497432	2006	538091
1994	497659	2007	546747
1995	498396	2008	556442
1996	501646	2009	566143
1997	503205	2010	585261
1998	502945	2011	593859
1999	503637	2012	602954
2000	529922	2013	609520
2001	535185	2014	619360
2002	538803		

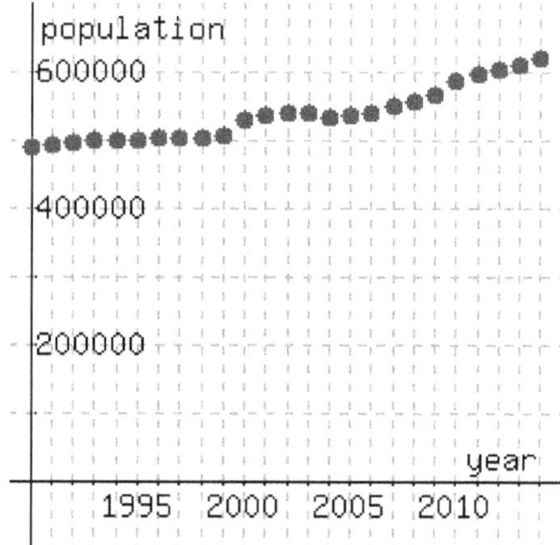

Between what two years that are two years apart was the rate of change highest?

What was that rate of change? Just give the numerical value; the units are provided.

$\dfrac{\text{people}}{\text{year}}$

4.4 Slope

In Section 4.3, we observed that a constant rate of change between points produces a linear relationship, whose graph is a straight line. Such a constant rate of change has a special name, **slope**, and we'll explore slope in more depth here.

4.4.1 What is slope?

When the **rate of change** from point to point never changes, those points must fall on a straight line, as in Figure 4.4.2, and there is a **linear relationship** between the variables x and y.

Rather than say "constant rate of change" in every such situation, mathematicians call that common rate of change **slope.**

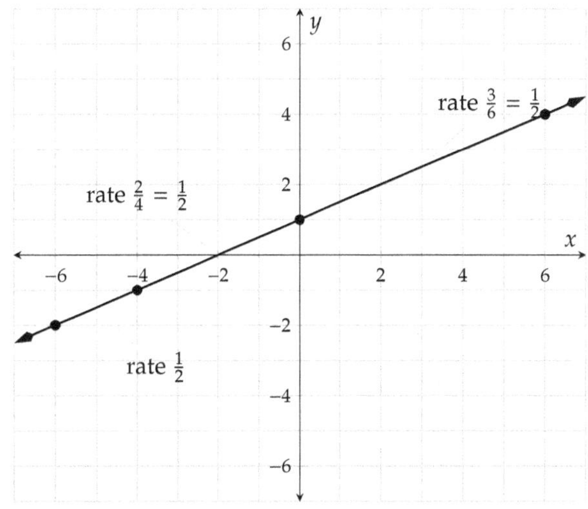

Figure 4.4.2: Between successive points, the rate of change is always 1/2.

Definition 4.4.3 Slope. When x and y are two variables where the rate of change between any two points is always the same, we call this common rate of change the **slope**. Since having a constant rate of change means the graph will be a straight line, its also called the **slope of the line**.

Considering the definition for Definition 4.3.12, this means that when x and y are two variables where the rate of change between any two points is always the same, then you can calculate slope, m, by finding two distinct data points (x_1, y_1) and (x_2, y_2), and calculating

$$m = \frac{\text{change in } y}{\text{change in } x} = \frac{\Delta y}{\Delta x} = \frac{y_2 - y_1}{x_2 - x_1}. \tag{4.4.1}$$

A slope is a rate of change. So if there are units for the horizontal and vertical variables, then there will be units for the slope. The slope will be measured in $\frac{\text{vertical units}}{\text{horizontal units}}$.

If the slope is nonzero, we say that there is a **linear relationship** between x and y. When the slope is 0, we say that y is **constant** with respect to x.

Slope m. Why is the letter m commonly used as the symbol for "slope?" Some believe that it comes from the French word "monter" which means "to climb."

Here are some scenarios with different slopes. Note that a slope is more meaningful with units.

- If a tree grows 2.5 feet every year, its rate of change in height is the same from year to year. So the height and time have a linear relationship where the **slope** is 2.5 $\frac{\text{ft}}{\text{yr}}$.

- If a company loses 2 million dollars every year, its rate of change in reserve funds is the same from year to year. So the company's reserve funds and time have a linear relationship where the **slope** is -2 million dollars per year.

- If Sakura is an adult who has stopped growing, her rate of change in height is the same from year to year—it's zero. So the **slope** is $0\,\frac{\text{in}}{\text{yr}}$. Sakura's height is **constant** with respect to time. In a statistics course, you would say that height and time don't have a relationship at all, in the sense that information about Sakura's height tells you nothing about her age.

Remark 4.4.4. A useful phrase for remembering the definition of **slope** is "rise over run." Here, "rise" refers to "change in y," Δy, and "run" refers to "change in x," Δx. Be careful though. As we have learned, the horizontal direction comes *first* in mathematics, followed by the vertical direction. The phrase "rise over run" reverses this. (It's a bit awkward to say, but the phrase "run under rise" puts the horizontal change first.)

Example 4.4.5 Yara's Savings. On Dec. 31, Yara had only \$50 in her savings account. For the the new year, she resolved to deposit \$20 into her savings account each week, without withdrawing any money from the account.

Yara keeps her resolution, and her account balance increases steadily by \$20 each week. That's a constant rate of change, so her account balance and time have a linear relationship with slope $20\,{}^{\text{dollars}}/\text{wk}$.

We can model the balance, y, in Yara's savings account after x weeks with an equation. Since Yara started with \$50 and adds \$20 each week, the account balance y after x weeks is

$$y = 50 + 20x \qquad (4.4.2)$$

where y is a dollar amount. Notice that the slope, $20\,{}^{\text{dollars}}/\text{wk}$, serves as the multiplier for x weeks.

We can also consider Yara's savings using a table.

	x, weeks since DEC. 31	y, savings account balance (dollars)	
	0	50	
x increases by 1 ⟶	1	70	⟵ y increases by 20
x increases by 1 ⟶	2	90	⟵ y increases by 20
x increases by 2 ⟶	4	130	⟵ y increases by 40
x increases by 3 ⟶	7	190	⟵ y increases by 60
x increases by 5 ⟶	12	290	⟵ y increases by 100

Table 4.4.6: Yara's savings

In first few rows of the table, we see that when the number of weeks x increases by 1, the balance y increases by 20. The row-to-row rate of change is $\frac{20}{1} = 20$, the slope. In any table for a linear relationship, whenever x increases by 1 unit, y will increase by the slope.

In further rows, notice that as row-to-row change in x increases, row-to-row change in y increases proportionally to preserve the constant rate of change. Looking at the change in the last two rows of the table, we see x increases by 5 and y increases by 100, which gives a rate of change of $\frac{100}{5} = 20$, the value of the slope again.

We can "see" the rates of change between consecutive rows of the table on a graph of Yara's savings by including **slope triangles**.

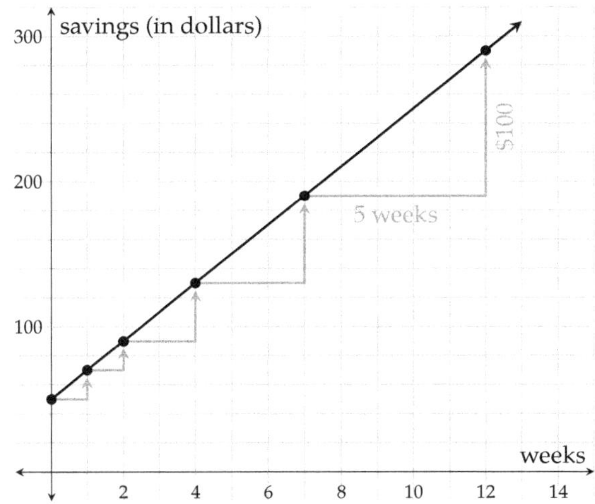

Figure 4.4.7: Yara's savings

The large, labeled slope triangle indicates that when 5 weeks pass, Yara saves \$100. This is the rate of change between the last two rows of the table, $\frac{100}{5} = 20$ dollars/wk.

The smaller slope triangles indicate, from left to right, the rates of change $\frac{20}{1}$, $\frac{20}{1}$, $\frac{40}{2}$, and $\frac{60}{3}$ respectively. All of these rates simplify to the slope, 20 dollars/wk.

Every slope triangle on the graph of Yara's savings has the same shape (geometrically, they are called similar triangles) since the ratio of vertical change to horizontal change is always 20 dollars/wk. On any graph of any line, we can draw a slope triangle and compute slope as "rise over run."

Of course, we could draw a slope triangle on the other side of the line:

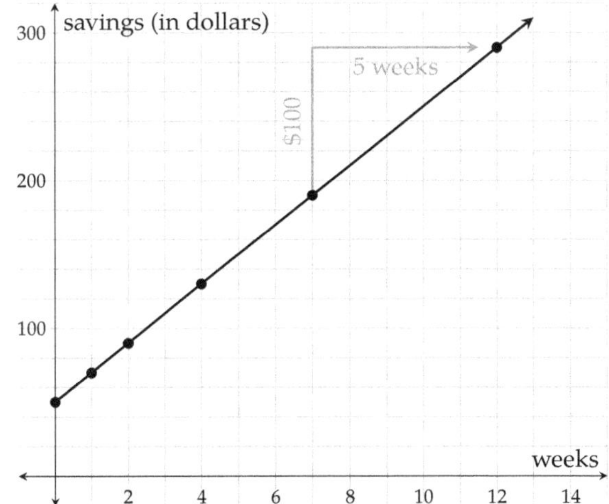

Figure 4.4.8: Yara's savings

This slope triangle works just as well for identifying "rise" and "run," but it focuses on vertical change before horizontal change. For consistency with mathematical conventions, we will generally draw slope triangles showing horizontal change followed by vertical change, as in Figure 4.4.7.

Example 4.4.9 The following graph of a line models the amount of gas, in gallons, in Kiran's gas tank as they drive their car. Find the line's slope, and interpret its meaning in this context.

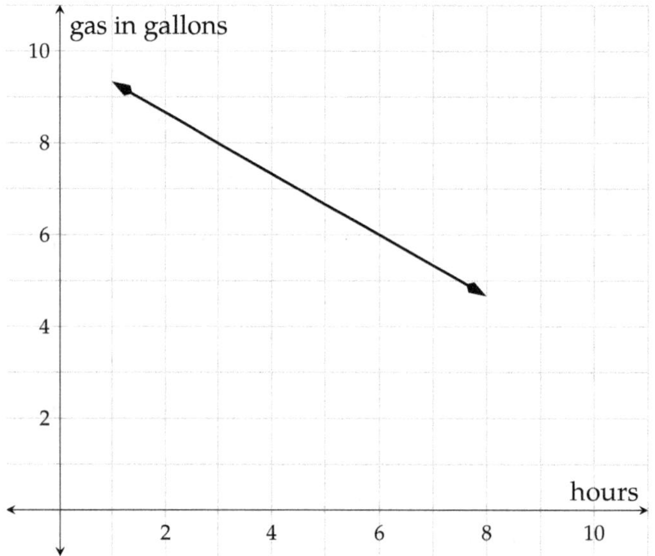

Figure 4.4.10: Amount of gas in Kiran's gas tank

Explanation. To find a line's slope using its graph, we first identify two points on it, and then draw a slope triangle. Naturally, we would want to choose two points whose x- and y-coordinates are easy to identify exactly based on the graph. We will pick the two points where $x = 3$ and $x = 6$, because they are right on the grid lines:

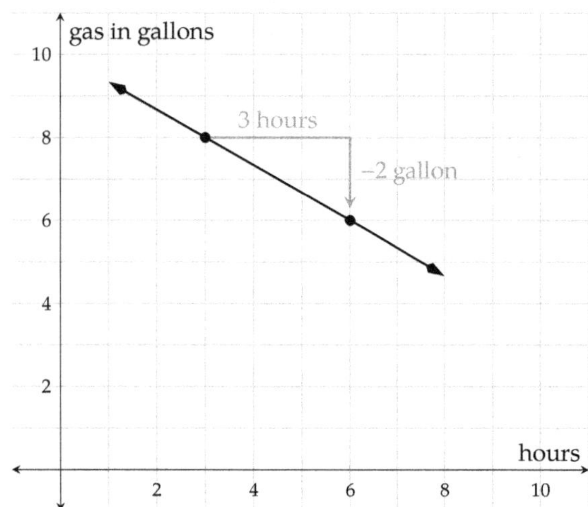

Notice that the *change* in y is negative, because the amount of gas is decreasing. Since we chose points with integer coordinates, we can easily calculate the slope:

$$\text{slope} = \frac{-2}{3} = -\frac{2}{3}.$$

Figure 4.4.11: A Good Slope Triangle

With units, the slope is $-\frac{2}{3}\,\frac{\text{gal}}{\text{h}}$. In the given context, this slope implies gas in the tank is *decreasing* at the rate of $\frac{2}{3}\,\frac{\text{gal}}{\text{h}}$. Since this slope is written as a fraction, there is another way to understand it: the gas in Kiran's tank is decreasing by 2 gallons every 3 hours.

Checkpoint 4.4.12. Below is a line's graph.

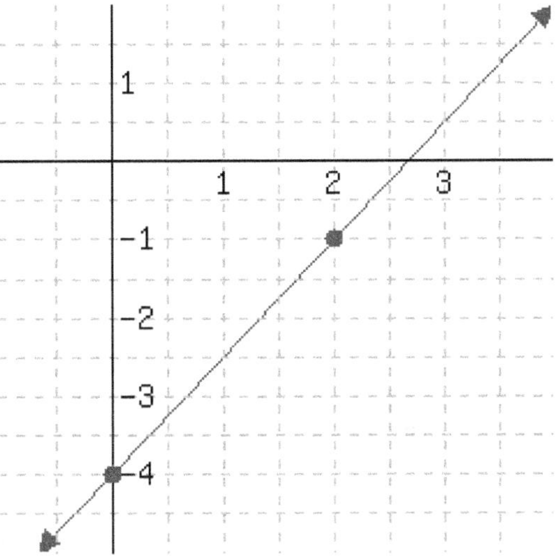

The slope of this line is [   ].

Explanation. To find the slope of a line from its graph, we first need to identify two points that the line passes through. It is wise to choose points with integer coordinates. For this problem, we choose $(0, -4)$ and $(2, -1)$.

Next, we sketch a slope triangle and find the *rise* and *run*. In the sketch below, the rise is 3 and the run is 2.

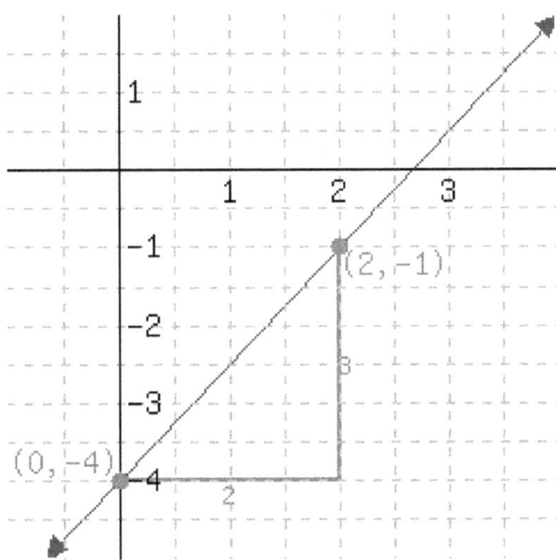

$$\text{slope} = \frac{\text{rise}}{\text{run}}$$
$$= \frac{3}{2}$$

This line's slope is $\frac{3}{2}$.

Checkpoint 4.4.13. Make a table and plot the equation $y = \frac{3}{4}x + 2$, which makes a straight line. Use the plot to determine the slope of this line.

Explanation. First, we choose some x-values to make a table, and compute the corresponding y-values.

x	$y = \frac{3}{4}x + 2$	Point
-2	$\frac{3}{4}(-2) + 2 = 0.5$	$(-2, 0.5)$
-1	$\frac{3}{4}(-1) + 2 = 1.25$	$(-1, 1.25)$
0	$\frac{3}{4}(0) + 2 = 2$	$(0, 2)$
1	$\frac{3}{4}(1) + 2 = 2.75$	$(1, 2.75)$
2	$\frac{3}{4}(2) + 2 = 3.5$	$(1, 3.5)$
3	$\frac{3}{4}(3) + 2 = 4.25$	$(1, 4.25)$
4	$\frac{3}{4}(4) + 2 = 5$	$(1, 5)$
5	$\frac{3}{4}(5) + 2 = 5.75$	$(2, 5.75)$

This table lets us plot the graph and identify a slope triangle that is easy to work with.

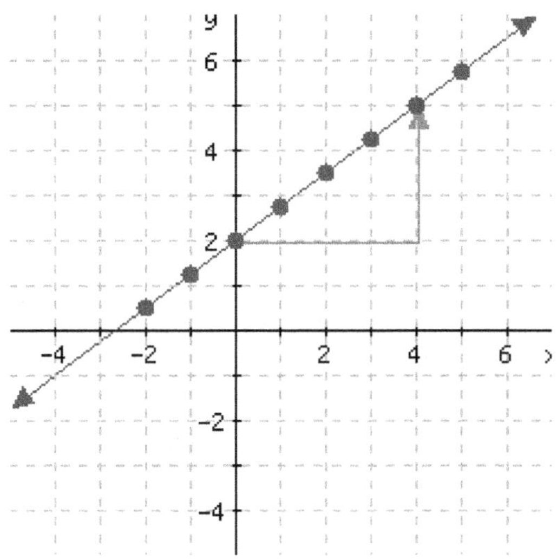

Since the slope triangle runs 4 units and then rises 3 units, the slope is $\frac{3}{4}$.

4.4.2 Comparing Slopes

It's useful to understand what it means for different slopes to appear on the same coordinate system.

Example 4.4.14 Effie, Ivan and Cleo are in a foot race. Figure 4.4.15 models the distance each has traveled in the first few seconds. Each runner takes a second to accelerate up to their running speed, but then runs at a constant speed. So they are then traveling with a constant rate of change, and the straight line portions of their graphs have a slope. Find each line's slope, and interpret its meaning in this context. What comparisons can you make with these runners?

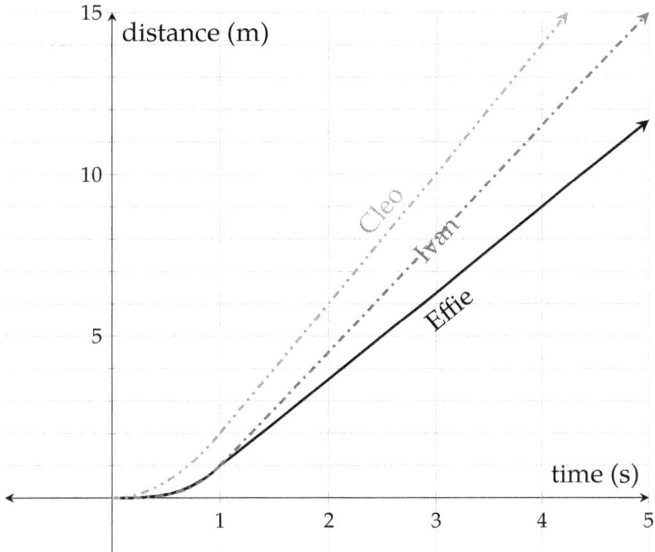

Figure 4.4.15: A three-way foot race

We will draw slope triangles to find each line's slope.

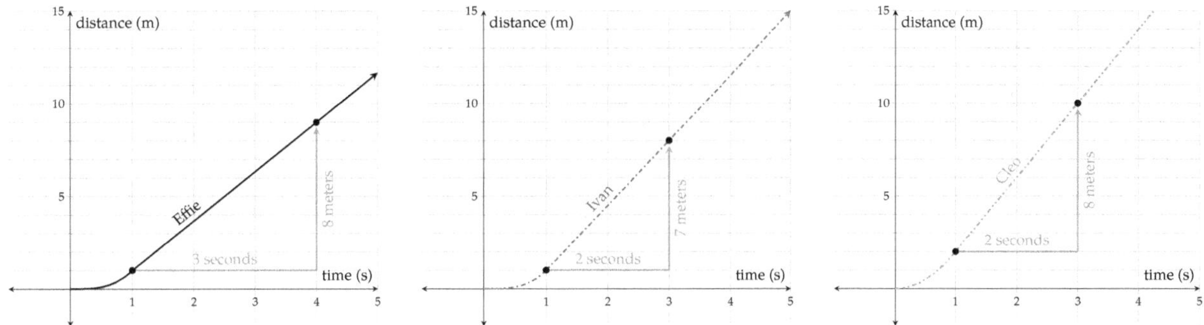

Figure 4.4.16: Find the Slope of Each Line

Using Formula (4.4.1), we have:

- Effie's slope is $\frac{8}{3} \approx 2.666$ meters per second

- Ivan's slope is $\frac{7}{2} = 3.5$ meters per second

- Cleo's slope is $\frac{8}{2} = 4$ meters per second

In a time-distance graph, the slope of a line represents speed. The slopes in these examples and the running speeds of these runners are both measured in $\frac{m}{s}$. Another important relationship we can see is that, the more sharply a line is slanted, the bigger the slope is. This should make sense because for each passing second, the faster person travels longer, making a slope triangle's height taller. This means that, numerically, we can tell that Cleo is the fastest runner (and Effie is the slowest) just by comparing the slopes $4 > 3.5 > 2.666$.

Checkpoint 4.4.17 Jogging on Mt. Hood. Kato is training for a race up the slope of Mt. Hood, from Sandy to Government Camp, and then back. The graph below models his elevation from his starting point as time passes. Find the slopes of the three line segments, and interpret their meanings in this context.

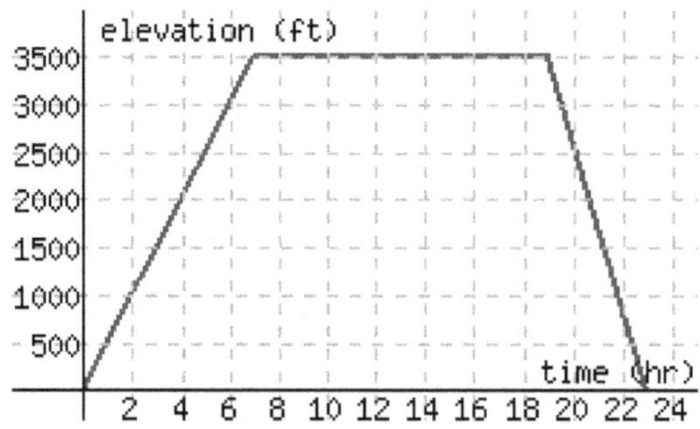

a. The first segment has slope [＿＿＿＿].

b. The second segment has slope [＿＿＿＿].

c. The third segment has slope [＿＿＿＿].

Explanation. The first segment started at $(0, 0)$ and stopped at $(7, 3500)$. This implies, Kato started at the starting point, traveled 7 hours and reached a point 3500 feet higher in elevation from the starting point. The slope of the line is

$$\frac{\Delta y}{\Delta x} = \frac{3500}{7} = 500$$

and with units, that is 500 ft/hr. In context, Kato was running, gaining 500 feet in elevation per hour.

The third segment started from $(19, 3500)$ and stopped at $(23, 0)$. This implies, Kato started this part of his trip from a spot 3500 feet higher in elevation from the starting point, traveled for 4 hours and returned to the starting elevation. The slope of the line is

$$\frac{\Delta y}{\Delta x} = \frac{-3500}{4} = -875$$

and with units, that is -875 ft/hr. In context, Kato was running, dropping in elevation by 875 feet per hour.

What happened in the second segment, which started at $(7, 3500)$ and ended at $(19, 3500)$? This implies he started this portion 3500 feet higher in elevation from the starting point, and didn't change elevation for 19 hours. The slope of the line is

$$\frac{\Delta y}{\Delta x} = \frac{0}{19} = 0$$

and with units, that is 0 ft/hr. In context, Kato was running but neither gaining nor losing elevation.

Some important properties are demonstrated in Exercise 4.4.17.

Fact 4.4.18 The Relationship Between Slope and Increase/Decrease. *In a linear relationship, as the x-value increases (in other words as you read its graph from left to right):*

- *if the y-values increase (in other words, the line goes upward), its slope is positive.*

- *if the y-values decrease (in other words, the line goes downward), its slope is negative.*

- *if the y-values don't change (in other words, the line is flat, or horizontal), its slope is 0.*

These properties are summarized graphically in Figure 4.4.18.

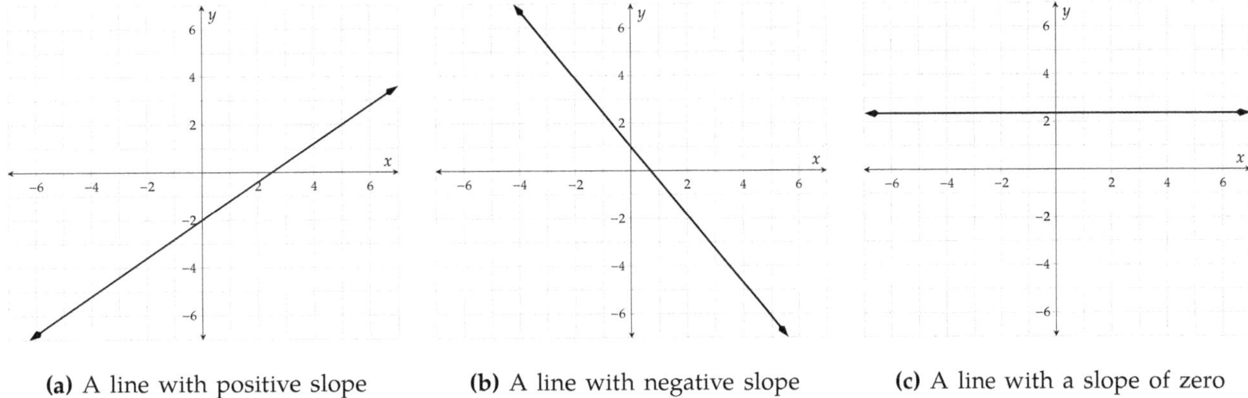

(a) A line with positive slope   **(b)** A line with negative slope   **(c)** A line with a slope of zero

Figure 4.4.18

4.4.3 Finding Slope by Two Given Points

Several times in this section we computed a slope by drawing a slope triangle. That's not really necessary if you have coordinates for two points that a line passes through. In fact, sometimes it's impractical to draw a slope triangle.[1] Here we will stress how to find a line's slope without drawing a slope triangle.

Example 4.4.19 Your neighbor planted a sapling from Portland Nursery in his front yard. Ever since, for several years now, it has been growing at a constant rate. By the end of the third year, the tree was 15 ft tall; by the end of the sixth year, the tree was 27 ft tall. What's the tree's rate of growth (i.e.the slope)?

We *could* sketch a graph for this scenario, and include a slope triangle. If we did that, it would look like:

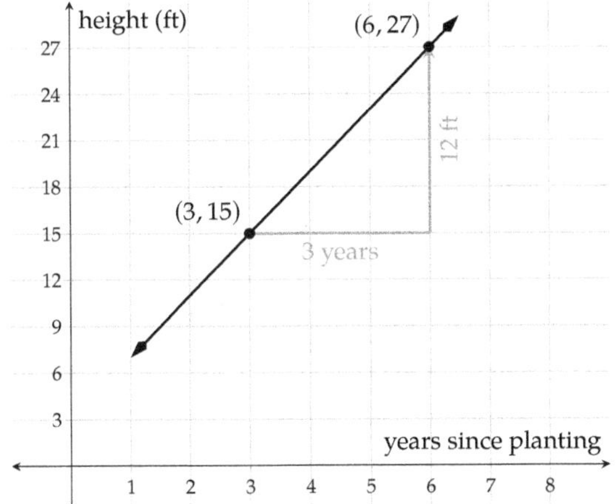

Figure 4.4.20: Height of a Tree

By the slope triangle and Equation (4.4.1) we have:

$$\text{slope} = m = \frac{\Delta y}{\Delta x}$$
$$= \frac{12}{3}$$
$$= 4$$

So the tree is growing at a rate of $4\,\frac{\text{ft}}{\text{yr}}$.

---

[1]For instance if you only have specific information about two points that are too close together to draw a triangle, or if you cannot clearly see precise coordinates where you might start and stop your slope triangle.

But hold on. Did we really *need* this picture? The "rise" of 12 came from a subtraction of two y-values: $27 - 15$. And the "run" of 3 came from a subtraction of two x-values: $6 - 3$.

Here is a picture-free approach. We know that after 3 yr, the height is 15 ft. As an ordered pair, that information gives us the point $(3, 15)$ which we can label as $(\overset{x_1}{3}, \overset{y_1}{15})$. Similarly, the background information tells us to consider $(6, 27)$, which we label as $(\overset{x_2}{6}, \overset{y_2}{27})$. Here, x_1 and y_1 represent the first point's x-value and y-value, and x_2 and y_2 represent the second point's x-value and y-value.

Now we can write an alternative to Equation (4.4.1):

$$\text{slope} = m = \frac{\Delta y}{\Delta x} = \frac{y_2 - y_1}{x_2 - x_1} \tag{4.4.3}$$

This is known as the **slope formula**. The following graph will help you understand why this formula works. Basically, we are still using a slope triangle to calculate the slope.

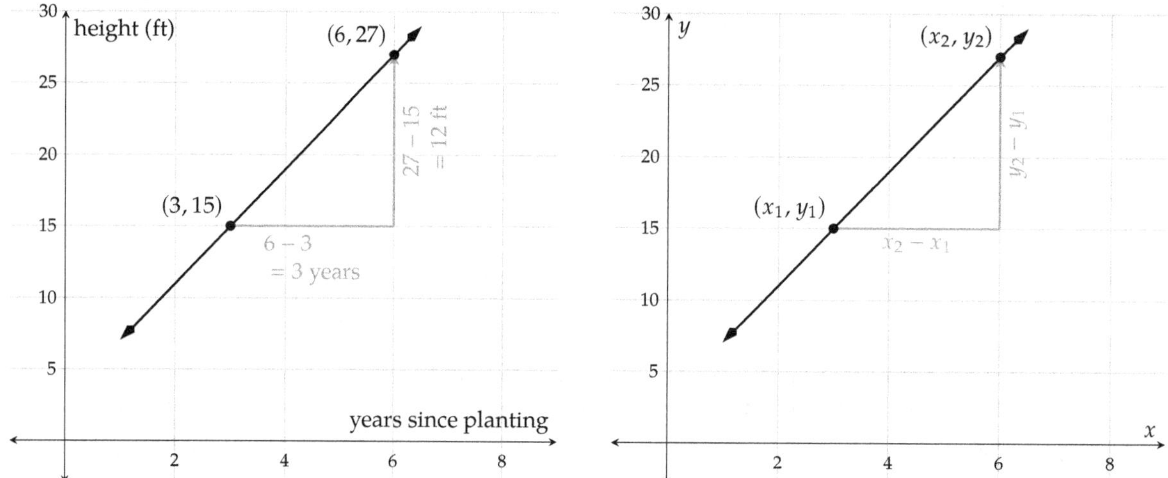

Figure 4.4.21: Understanding the slope formula

It's important to use subscript instead of superscript in the slope equation, because y^2 means to take the number y and square it. Whereas y_2 tells you that there are at least two y-values in the conversation, and y_2 is the second of them.

The beauty of the slope formula (4.4.3) is that to find a line's slope, we don't need to draw a slope triangle any more. Let's look at an example.

Example 4.4.22 A line passes the points $(-5, 25)$ and $(4, -2)$. Find this line's slope.

Explanation. If you are new to this formula, it's important to label each number before using the formula. The two given points are:

$$(\overset{x_1}{-5}, \overset{y_1}{25}) \text{ and } (\overset{x_2}{4}, \overset{y_2}{-2})$$

Now apply the slope formula (4.4.3):

$$\text{slope} = \frac{y_2 - y_1}{x_2 - x_1}$$
$$= \frac{-2 - 25}{4 - (-5)}$$
$$= \frac{-27}{9}$$
$$= -3$$

Note that we used parentheses when substituting in x_1 and y_1. This is a good habit to protect yourself from making errors with subtraction and double negatives.

Checkpoint 4.4.23. A line passes through the points $(-6, 26)$ and $(6, -16)$. Find this line's slope.

Explanation. To find a line's slope, we can use the slope formula:

$$\text{slope} = \frac{y_2 - y_1}{x_2 - x_1}$$

First, we mark which number corresponds to which variable in the formula:

$$(-6, 26) \longrightarrow (x_1, y_1)$$

$$(6, -16) \longrightarrow (x_2, y_2)$$

Now we substitute these numbers into the corresponding variables in the slope formula:

$$\text{slope} = \frac{y_2 - y_1}{x_2 - x_1}$$
$$= \frac{-16 - 26}{6 - (-6)}$$
$$= \frac{-42}{12}$$
$$= -\frac{7}{2}$$

So the line's slope is $-\dfrac{7}{2}$.

Exercises

Review and Warmup

1. Reduce the fraction $\dfrac{5}{40}$.

2. Reduce the fraction $\dfrac{3}{27}$.

3. Reduce the fraction $\dfrac{15}{18}$.

4. Reduce the fraction $\dfrac{15}{27}$.

5. Reduce the fraction $\dfrac{35}{210}$.

6. Reduce the fraction $\dfrac{42}{189}$.

7. Reduce the fraction $\dfrac{135}{75}$.

8. Reduce the fraction $\dfrac{100}{30}$.

9. Reduce the fraction $\dfrac{245}{35}$.

10. Reduce the fraction $\dfrac{280}{35}$.

Slope and Points

11. A line passes through the points $(2, 1)$ and $(7, 21)$. Find this line's slope.

12. A line passes through the points $(4, 27)$ and $(6, 37)$. Find this line's slope.

13. A line passes through the points $(1, -1)$ and $(9, -9)$. Find this line's slope.

14. A line passes through the points $(3, -22)$ and $(8, -47)$. Find this line's slope.

15. A line passes through the points $(-4, 2)$ and $(-8, -2)$. Find this line's slope.

16. A line passes through the points $(-3, -7)$ and $(-5, -11)$. Find this line's slope.

17. A line passes through the points $(-1, -7)$ and $(3, -11)$. Find this line's slope.

18. A line passes through the points $(-3, -14)$ and $(3, 4)$. Find this line's slope.

19. A line passes through the points $(-2, 2)$ and $(-8, 14)$. Find this line's slope.

20. A line passes through the points $(-4, -1)$ and $(-5, 1)$. Find this line's slope.

21. A line passes through the points $(14, 16)$ and $(-7, -8)$. Find this line's slope.

22. A line passes through the points $(5, 17)$ and $(-10, -7)$. Find this line's slope.

23. A line passes through the points $(-2, 0)$ and $(4, -9)$. Find this line's slope.

24. A line passes through the points $(-16, 10)$ and $(8, 1)$. Find this line's slope.

25. A line passes through the points $(2, -4)$ and $(-5, -4)$. Find this line's slope.

26. A line passes through the points $(5, -2)$ and $(-3, -2)$. Find this line's slope.

27. A line passes through the points $(1, -2)$ and $(1, 1)$. Find this line's slope.

28. A line passes through the points $(3, -4)$ and $(3, 3)$. Find this line's slope.

Slope and Graphs

29. Below is a line's graph.

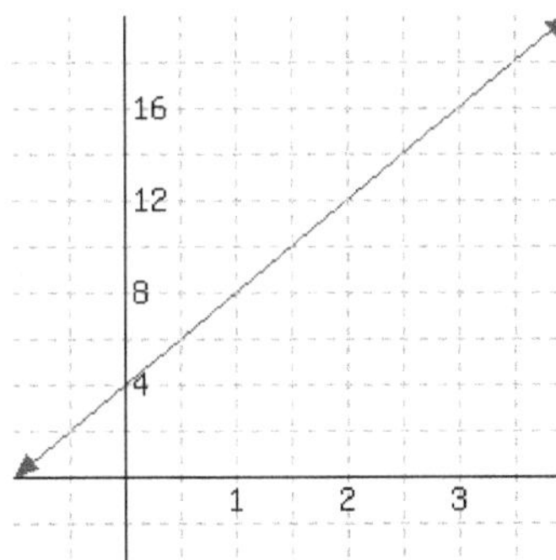

The slope of this line is ☐.

30. Below is a line's graph.

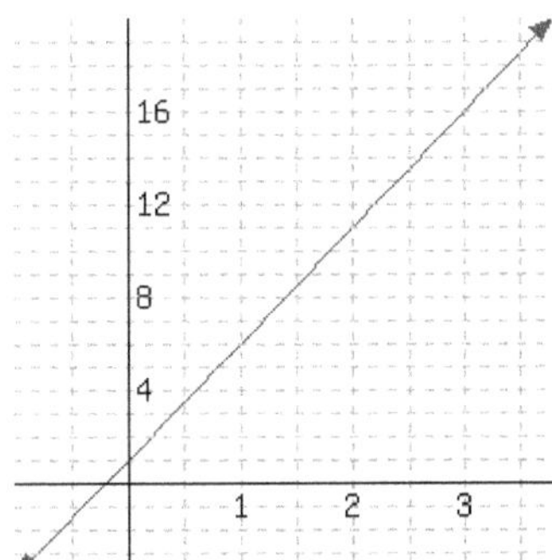

The slope of this line is ☐.

31. Below is a line's graph.

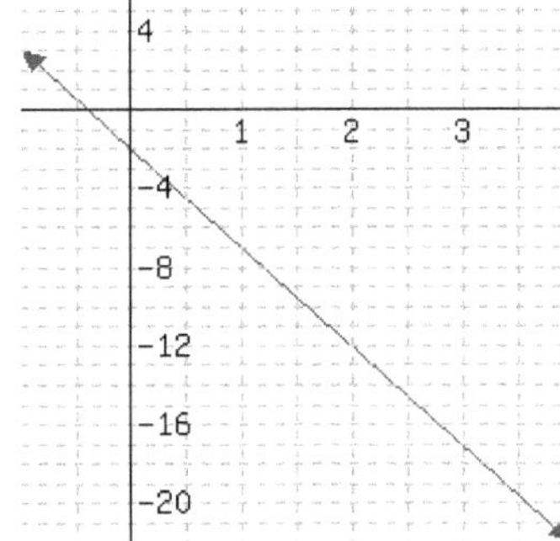

The slope of this line is ☐.

32. Below is a line's graph.

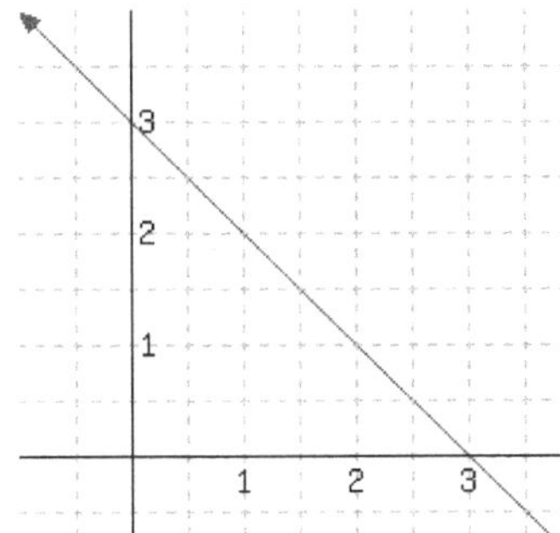

The slope of this line is ☐.

33. Below is a line's graph.

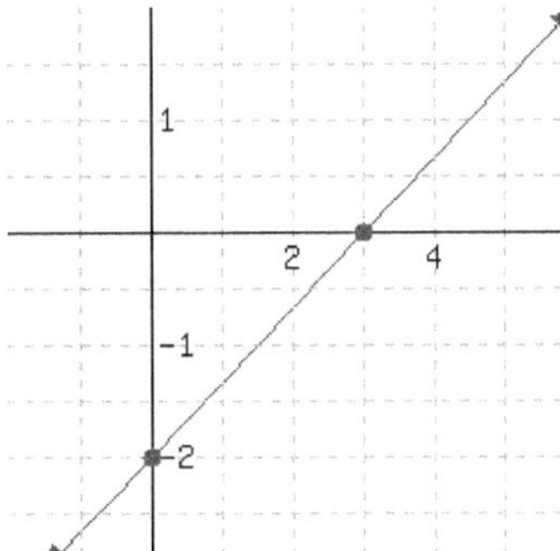

The slope of this line is ☐.

34. Below is a line's graph.

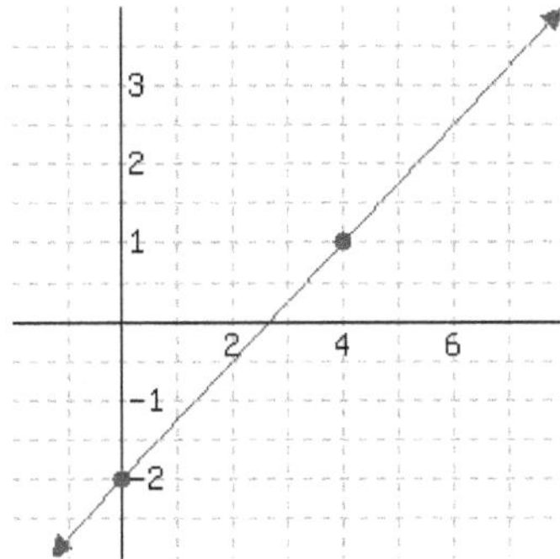

The slope of this line is ☐.

35. Below is a line's graph.

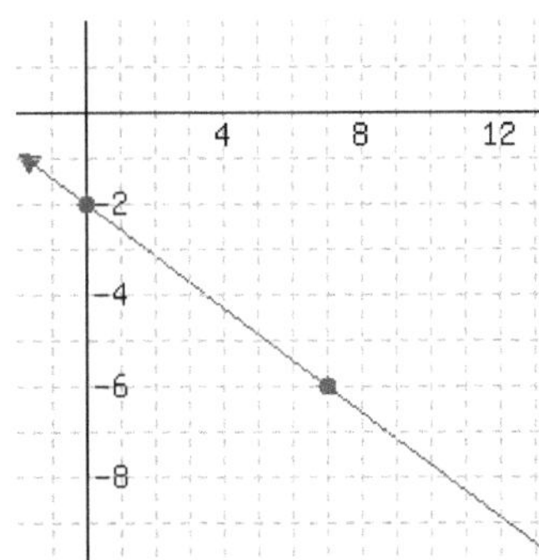

The slope of this line is ☐.

36. Below is a line's graph.

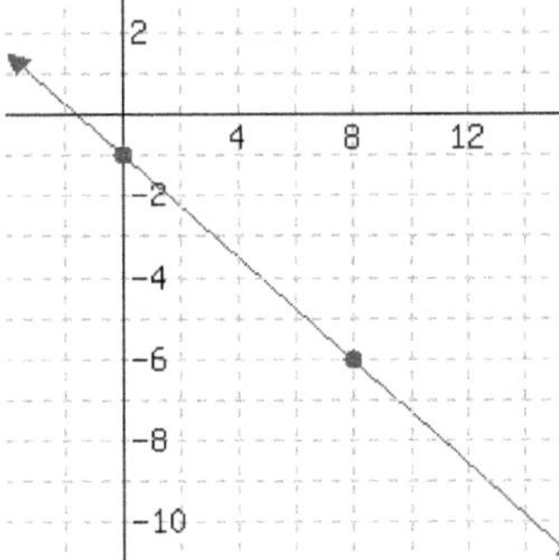

The slope of this line is ☐.

37. Below is a line's graph.

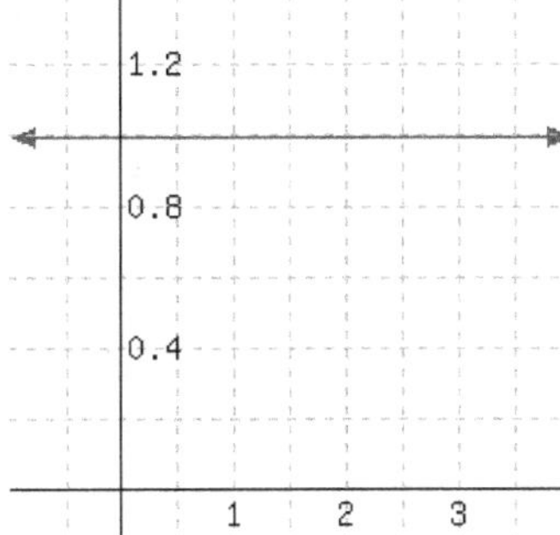

The slope of this line is ☐.

38. Below is a line's graph.

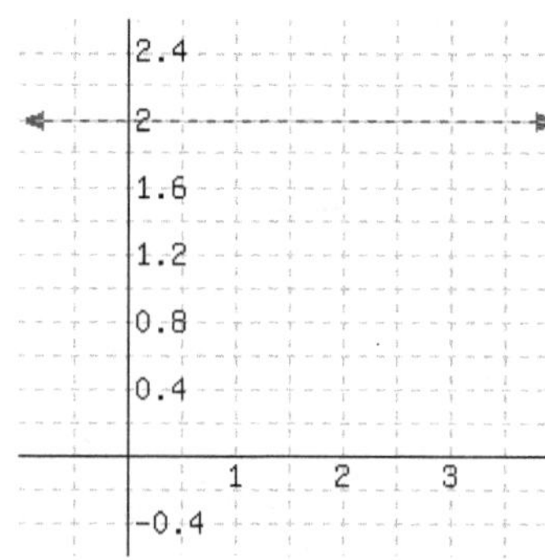

The slope of this line is ⬚.

39. Below is a line's graph.

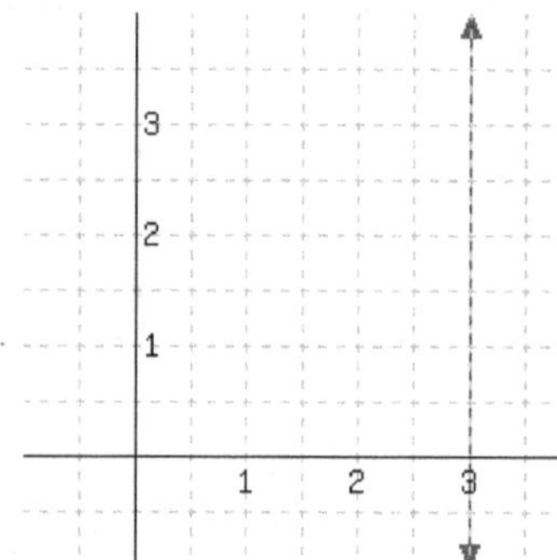

The slope of this line is ⬚.

40. Below is a line's graph.

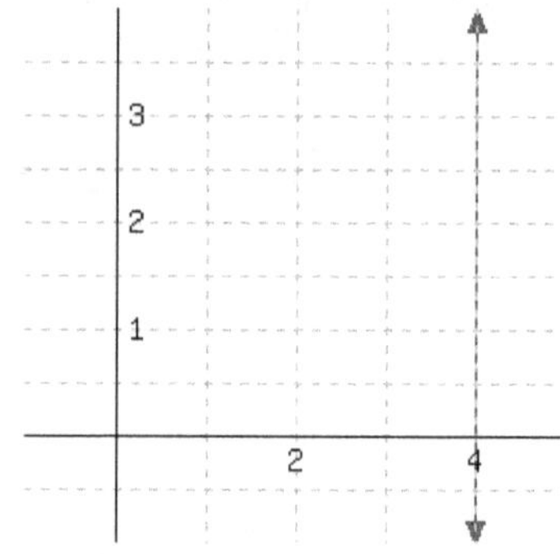

The slope of this line is ⬚.

41. A line's graph is shown below.

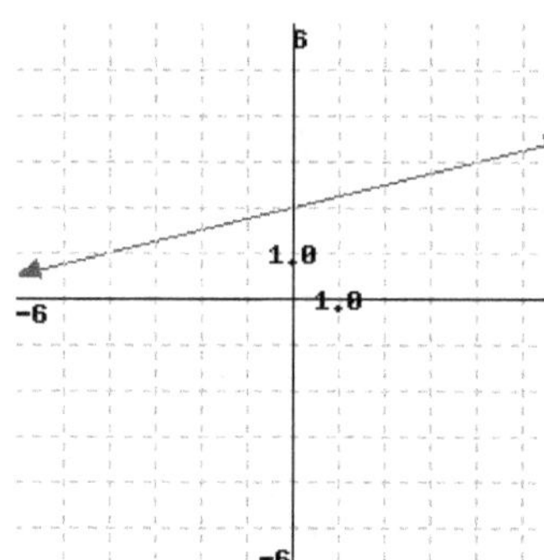

The slope is ⬚.

42. A line's graph is shown below.

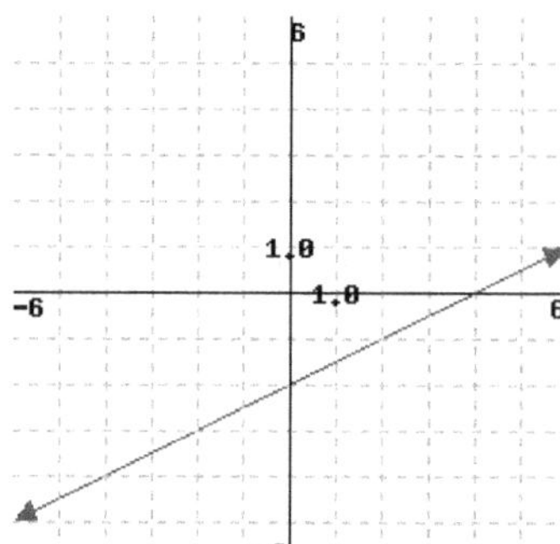

The slope is ⬚.

43. A line's graph is shown below.

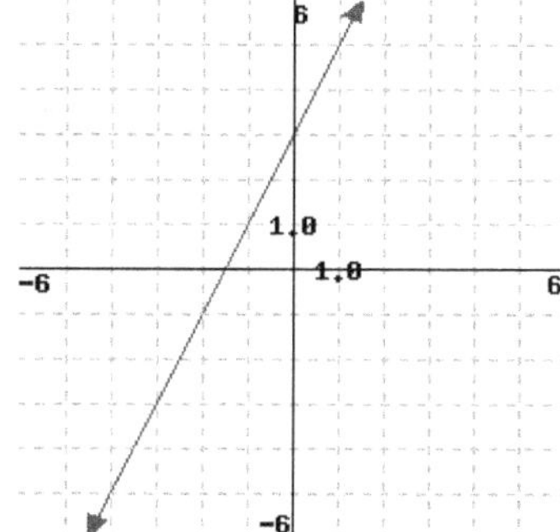

The slope is ⬚.

44. A line's graph is shown below.

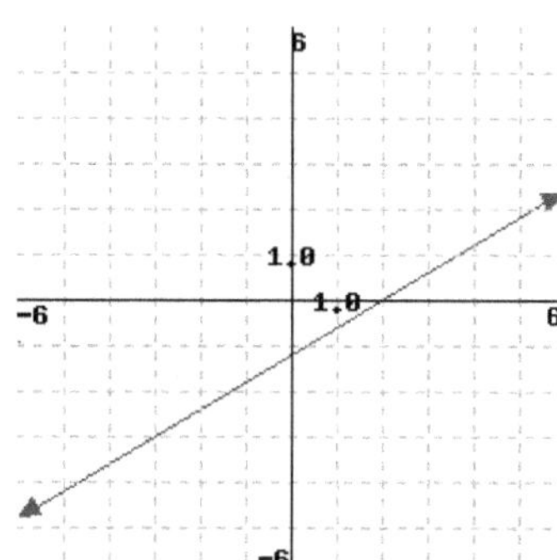

The slope is ⬚.

Slope in Context

45. By your cell phone contract, you pay a monthly fee plus some money for each minute you use the phone during the month. In one month, you spent 260 minutes on the phone, and paid $25.70. In another month, you spent 390 minutes on the phone, and paid $31.55. What is the rate (in dollars per minute) that the phone company is charging you? That is, what is the slope of the line if you plotted the bill versus the number of minutes spent on the phone?

    The rate is ____________ per minute.

46. By your cell phone contract, you pay a monthly fee plus some money for each minute you use the phone during the month. In one month, you spent 300 minutes on the phone, and paid $24.50. In another month, you spent 360 minutes on the phone, and paid $27.20. What is the rate (in dollars per minute) that the phone company is charging you? That is, what is the slope of the line if you plotted the bill versus the number of minutes spent on the phone?

    The rate is ____________ per minute.

47. A company set aside a certain amount of money in the year 2000. The company spent exactly the same amount from that fund each year on perks for its employees. In 2003, there was still $743,000 left in the fund. In 2005, there was $659,000 left. What is the rate (in dollars per year) at which this company is spending from this fund?

    The company is spending ____________ per year on perks for its employees.

48. A company set aside a certain amount of money in the year 2000. The company spent exactly the same amount from that fund each year on perks for its employees. In 2004, there was still $546,000 left in the fund. In 2005, there was $500,000 left. What is the rate (in dollars per year) at which this company is spending from this fund?

    The company is spending ____________ per year on perks for its employees.

49. A biologist has been observing a tree's height. Eleven months into the observation, the tree was 22.2 feet tall. Eleven months into the observation, the tree was 24.6 feet tall. What is the rate at which the tree is growing? In other words, what is the slope if you plotted heigth versus time?

50. A biologist has been observing a tree's height. Thirteen months into the observation, the tree was 17.53 feet tall. Thirteen months into the observation, the tree was 18.08 feet tall. What is the rate at which the tree is growing? In other words, what is the slope if you plotted heigth versus time?

51. Scientists are conducting an experiment with a gas in a sealed container. The mass of the gas is measured, and the scientists realize that the gas is leaking over time in a linear way. Five minutes since the experiment started, the gas had a mass of 133.2 grams. Twelve minutes since the experiment started, the gas had a mass of 108 grams. At what rate is the gas leaking?

52. Scientists are conducting an experiment with a gas in a sealed container. The mass of the gas is measured, and the scientists realize that the gas is leaking over time in a linear way. Seven minutes since the experiment started, the gas had a mass of 338.4 grams. Nineteen minutes since the experiment started, the gas had a mass of 225.6 grams. At what rate is the gas leaking?

53. A liquid solution is slowly leaking from a container. This graph shows the milliters of solution y remaining in the container after x minutes.

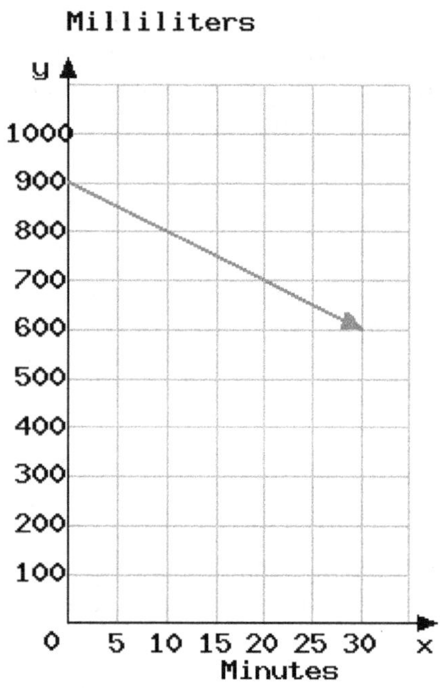

54. The graph plots the number of invasive cancer diagnoses in Oregon over time, and a trend-line has been drawn.

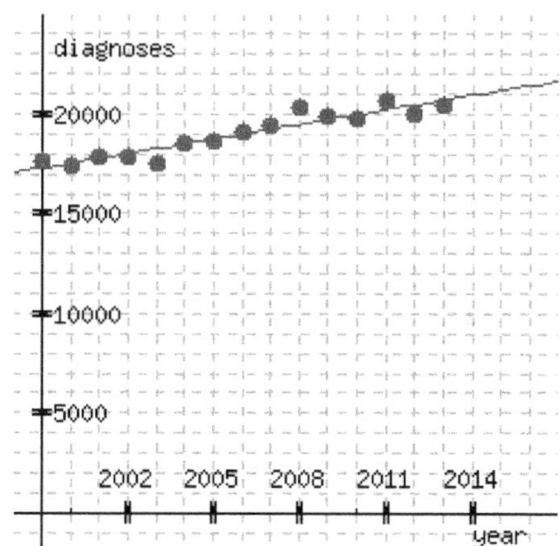

Estimate the slope of the trend-line. Just give the numerical value; the units are provided.

$$\boxed{}\ \dfrac{\text{diagnoses}}{\text{year}}$$

a. The y coordinate of the line is $\boxed{}$.

b. The slope of the line is $\boxed{}$.

c. Use the graph and your answer to part b to predict the number of minutes it will take for the container to empty if the solution continues leaking at the same rate. That time is $\boxed{}$ minutes.

Challenge

55. True or False: A slope of $\frac{2}{5}$ is steeper than a slope of $\frac{1}{4}$.   (□ true   □ false)

56. True or False: A slope of $\frac{1}{8}$ is steeper than a slope of $\frac{2}{5}$.   (□ true   □ false)

4.5 Slope-Intercept Form

In this section, we will explore one of the "standard" ways to write the equation of a line. It's known as **slope-intercept form**.

4.5.1 Slope-Intercept Definition

Recall Example 4.4.5, where Yara had $50 in her savings account when the year began, and decided to deposit $20 each week without withdrawing any money. In that example, we model using x to represent how many weeks have passed. After x weeks, Yara has added $20x$ dollars. And since she started with $50, she has

$$y = 20x + 50$$

in her account after x weeks. In this example, there is a constant rate of change of 20 dollars per week, so we call that the **slope** as discussed in Section 4.4. We also saw in Figure 4.4.7 that plotting Yara's balance over time gives us a straight-line graph.

The graph of Yara's savings has some things in common with almost every straight-line graph. There is a **slope**, and there is a place where the line crosses the y-axis. Figure 4.5.2 illustrates this in the abstract.

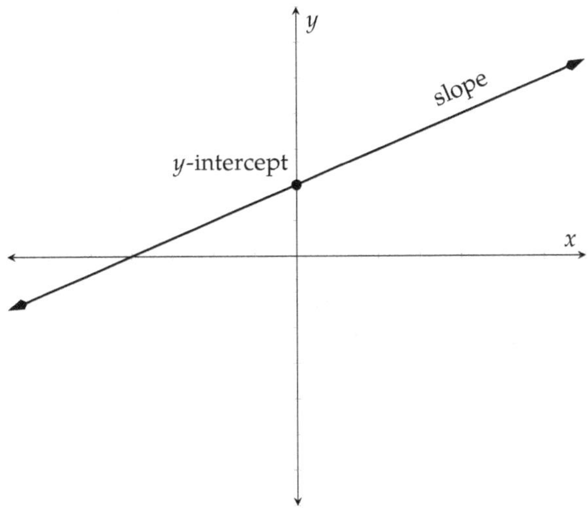

Figure 4.5.2: Generic line with slope and y-intercept

We already have an accepted symbol, m, for the slope of a line. The y-**intercept** is a *point* on the y-axis where the line crosses. Since it's on the y-axis, the x-coordinate of this point is 0. It is standard to call the y-intercept $(0, b)$ where b represents the position of the y-intercept on the y-axis.

What else is there?. Can you think of a type of straight line that does not have a notion of slope? Or that does not cross the y-axis somewhere?

 Checkpoint 4.5.3. Use Figure 4.4.7 to answer this question.

What was the value of b in the plot of Yara's savings?

What is the y-intercept?

Explanation. The line crosses the y-axis at $(0, 50)$, so the value of b is 50. And the y-intercept is $(0, 50)$

One way to write the equation for Yara's savings was

$$y = 20x + 50,$$

where both $m = 20$ and $b = 50$ are immediately visible in the equation. Now we are ready to generalize this.

Definition 4.5.4 Slope-Intercept Form. When x and y have a linear relationship where m is the slope and $(0, b)$ is the y-intercept, one equation for this relationship is

$$y = mx + b \tag{4.5.1}$$

and this equation is called the **slope-intercept form** of the line. It is called this because the slope and y-intercept are immediately discernible from the numbers in the equation.

Checkpoint 4.5.5. What are the slope and y-intercept for each of the following line equations?

Equation	Slope	y-intercept
$y = 3.1x + 1.78$	______	______
$y = -17x + 112$	______	______
$y = \frac{3}{7}x - \frac{2}{3}$	______	______
$y = 13 - 8x$	______	______
$y = 1 - \frac{2x}{3}$	______	______
$y = 2x$	______	______
$y = 3$	______	______

Explanation. In the first three equations, simply read the slope m according to slope-intercept form. The slopes are 3.1, -17, and $\frac{3}{7}$.

The fourth equation was written with the terms not in the slope-intercept form order. It could be written $y = -8x + 13$, and then it is clear that its slope is -8. In any case, the slope is the coefficient of x.

The fifth equation is also written with the terms not in the slope-intercept form order. Changing the order of the terms, it could be written $y = -\frac{2x}{3} + 1$, but this still does not match the pattern of slope-intercept form. Considering how fraction multiplication works, $\frac{2x}{3} = \frac{2}{3} \cdot \frac{x}{1} = \frac{2}{3}x$. So we can write this equation as $y = -\frac{2}{3}x + 1$, and we see the slope is $-\frac{2}{3}$.

The last two equations could be written $y = 2x + 0$ and $y = 0x + 3$, allowing us to read their slopes as 2 and 0.

For the y-intercepts, remember that we are expected to answer using an ordered pair $(0, b)$, not just a single number b. We can simply read that the first two y-intercepts are $(0, 1.78)$ and $(0, 112)$.

The third equation does not exactly match the slope-intercept form, until you view it as $y = \frac{3}{7}x + \left(-\frac{2}{3}\right)$, and then you can see that its y-intercept is $-\frac{2}{3}$.

With the fourth equation, after rewriting it as $y = -8x + 13$, we can see that its y-intercept is $(0, 13)$.

We already explored rewriting the fifth equation as $y = -\frac{2}{3}x + 1$, where we can see that its y-intercept is $(0, 1)$.

The last two equations could be written $y = 2x + 0$ and $y = 0x + 3$, allowing us to read their y-intercepts as $(0, 0)$ and $(0, 3)$.

Alternatively, we know that y-intercepts happen where $x = 0$, and substituting $x = 0$ into each equation gives you the y-value of the y-intercept.

Remark 4.5.6. The number b is the y-value when $x = 0$. Therefore it is common to refer to b as the **initial value** or **starting value** of a linear relationship.

Example 4.5.7 With a simple equation like $y = 2x + 3$, we can see that this is a line whose slope is 2 and which has initial value 3. So starting at $y = 3$ when $x = 0$ (that is, on the y-axis), each time we increase the x-value by 1, the y-value increases by 2. With these basic observations, we can quickly produce a table and/or a graph.

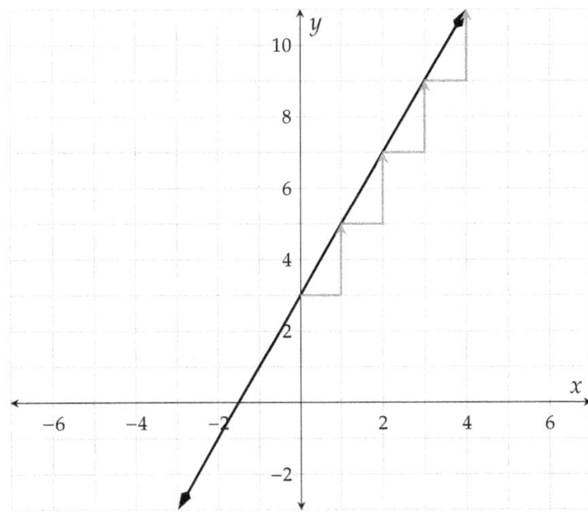

	x	y	
start on			initial
y-axis $\longrightarrow$	0	3	$\longleftarrow$ value
increase			increase
by 1 $\longrightarrow$	1	5	$\longleftarrow$ by 2
increase			increase
by 1 $\longrightarrow$	2	7	$\longleftarrow$ by 2
increase			increase
by 1 $\longrightarrow$	3	9	$\longleftarrow$ by 2
increase			increase
by 1 $\longrightarrow$	4	11	$\longleftarrow$ by 2

Example 4.5.8 Decide whether data in the table has a linear relationship. If so, write the linear equation in slope-intercept form (4.5.1).

x-values	y-values
0	−4
2	2
5	11
9	23

Explanation. To assess whether the relationship is linear, we have to recall from Section 4.3 that we should examine rates of change between data points. Note that the changes in y-values are not consistent. However, the rates of change are calculated as follows:

- When x increases by 2, y increases by 6. The first rate of change is $\frac{6}{2} = 3$.

- When x increases by 3, y increases by 9. The second rate of change is $\frac{9}{3} = 3$.

- When x increases by 4, y increases by 12. The third rate of change is $\frac{12}{4} = 3$.

Since the rates of change are all the same, 3, the relationship is linear and the slope m is 3.

According to the table, when $x = 0$, $y = -4$. So the starting value, b, is −4.

So in slope-intercept form, the line's equation is $y = 3x - 4$.

Checkpoint 4.5.9. Decide whether data in the table has a linear relationship. If so, write the linear equation in slope-intercept form. This may not be as easy as the previous example. Read the solution for a full explanation.

x-values	y-values
3	−2
6	−8
8	−12
11	−18

The data   (□ does   □ does not)   have a linear relationship, because: (□ changes in x are not constant   □ rates of change between data points are constant   □ rates of change between data points are not constant)

The slope-intercept form of the equation for this line is $\boxed{}$.

Explanation.   To assess whether the relationship is linear, we examine rates of change between data points.

- The first rate of change is $\frac{-6}{3} = -2$.

- The second rate of change is $\frac{-4}{2} = -2$.

- The third rate of change is $\frac{-6}{3} = -2$.

Since the rates of change are all the same, −2, the relationship is linear and the slope m is −2.

So we know that the slope-intercept equation is $y = -2x + b$, but what number is b? The table does not directly tell us what the initial y-value is.

One approach is to use any point that we know the line passes through, and use algebra to solve for b. We know the line passes through $(3, -2)$, so

$$y = -2x + b$$
$$-2 = -2(3) + b$$
$$-2 = -6 + b$$
$$4 = b$$

So the equation is $y = -2x + 4$.

4.5.2 Graphing Slope-Intercept Equations

Example 4.5.10 The conversion formula for a Celsius temperature into Fahrenheit is $F = \frac{9}{5}C + 32$. This appears to be in slope-intercept form, except that x and y are replaced with C and F. Suppose you are asked to graph this equation. How will you proceed? You *could* make a table of values as we do in Section 4.2 but that takes time and effort. Since the equation here is in slope-intercept form, there is a nicer way.

Since this equation is for starting with a Celsius temperature and obtaining a Fahrenheit temperature, it makes sense to let C be the horizontal axis variable and F be the vertical axis variable. Note the slope is $\frac{9}{5}$ and the vertical intercept (here, the F-intercept) is $(0, 32)$.

1. Set up the axes using an appropriate window and labels. Considering the freezing and boiling temperatures of water, it's reasonable to let C run through at least 0 to 100. Similarly it's reasonable to let F run through at least 32 to 212.

2. Plot the F-intercept, which is at $(0, 32)$.

3. Starting at the F-intercept, use slope triangles to reach the next point. Since our slope is $\frac{9}{5}$, that suggests a "run" of 5 and a "rise" of 9 might work. But as Figure 4.5.11 indicates, such slope triangles are too tiny. Since $\frac{9}{5} = \frac{90}{50}$, we can try a "run" of 50 and a rise of 90.

4. Connect your points with a straight line, use arrowheads, and label the equation.

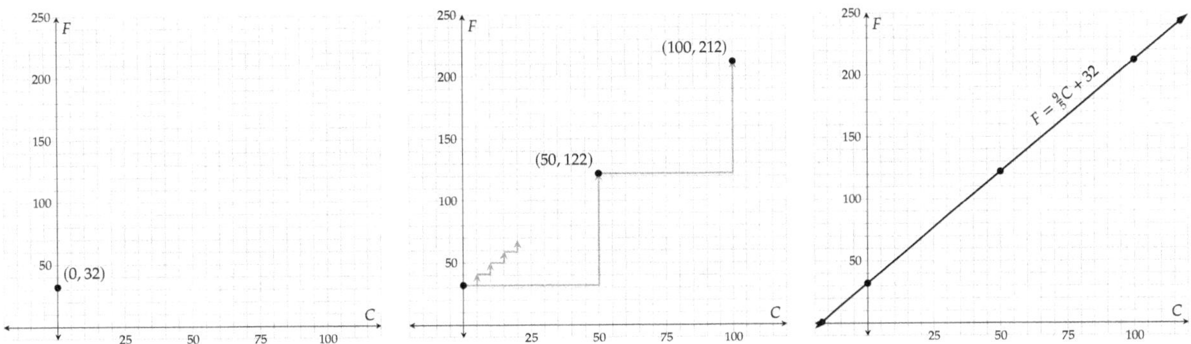

Figure 4.5.11: Graphing $F = \frac{9}{5}C + 32$

Example 4.5.12 Graph $y = -\frac{2}{3}x + 10$.

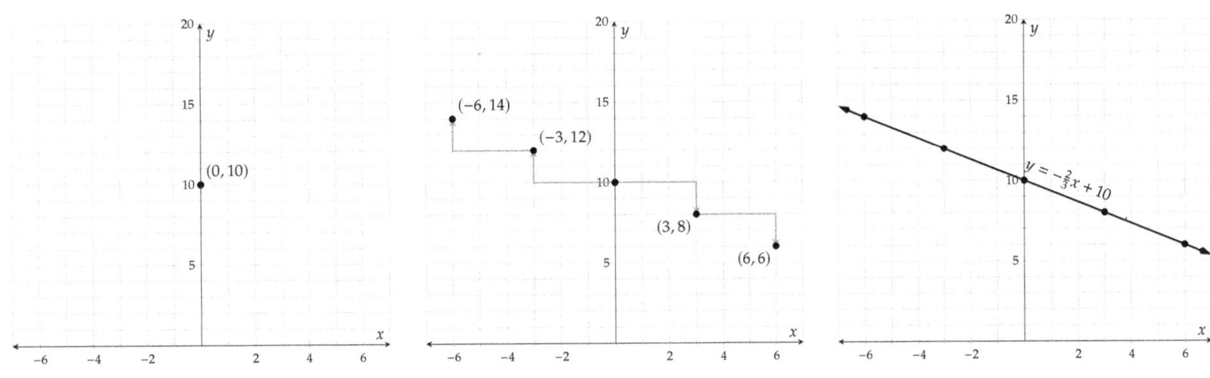

(a) Setting up the axes in an appropriate window and making sure that the y-intercept will be visible, and that any "run" and "rise" amounts we wish to use will not make triangles that are too big or too small.

(b) The slope is $-\frac{2}{3} = \frac{-2}{3} = \frac{2}{-3}$. So we can try using a "run" of 3 and a "rise" of -2 or a "run" of -3 and a "rise" of 2.

(c) Connecting the points with a straight line and adding labels.

Figure 4.5.12: Graphing $y = -\frac{2}{3}x + 10$

Example 4.5.13 Graph $y = 3x + 5$.

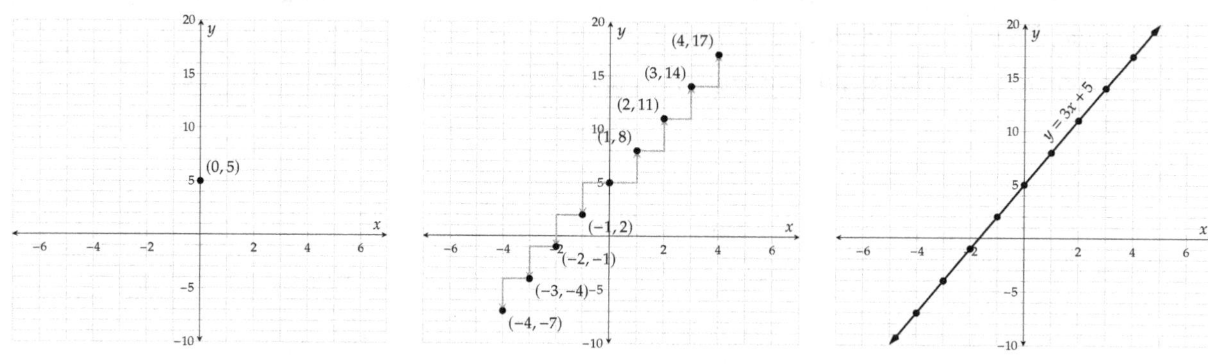

(a) Setting up the axes to make sure that the y-intercept will be visible, and that any "run" and "rise" amounts we wish to use will not make triangles that are too big or too small.

(b) The slope is a whole number 3. Every 1 unit forward causes a change of positive 3 in the y-values.

(c) Connecting the points with a straight line and adding labels.

Figure 4.5.13: Graphing $y = 3x + 5$

4.5.3 Writing a Slope-Intercept Equation Given a Graph

We can write a linear equation in slope-intercept form based on its graph. We need to be able to calculate the line's slope and see its y-intercept.

Checkpoint 4.5.14. Use the graph to write an equation of the line in slope-intercept form.

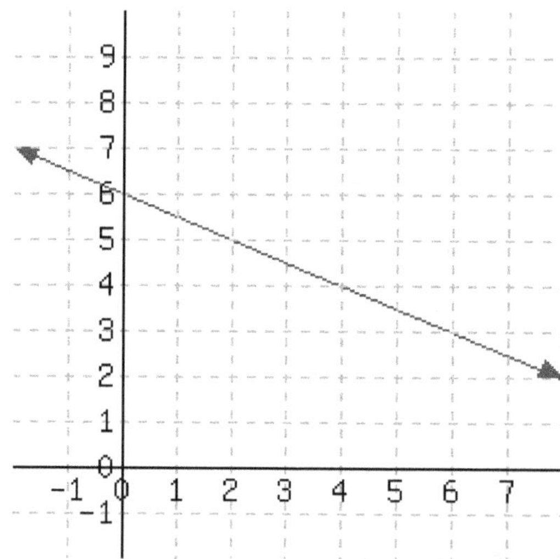

Explanation. On the line, pick two points with easy-to-read integer coordinates so that we can calculate slope. It doesn't matter which two points we use; the slope will be the same.

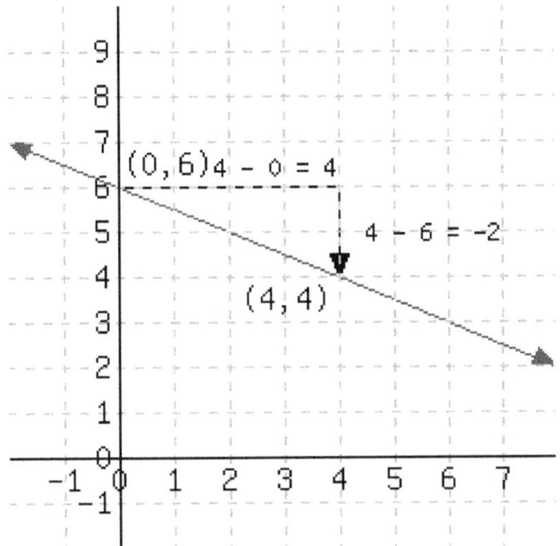

Using the slope triangle, we can calculate the line's slope:

$$\text{slope} = \frac{\Delta y}{\Delta x} = \frac{-2}{4} = -\frac{1}{2}.$$

From the graph, we can see the y-intercept is $(0, 6)$.

With the slope and y-intercept found, we can write the line's equation:

$$y = -\frac{1}{2}x + 6.$$

Checkpoint 4.5.15. There are seven public four-year colleges in Oregon. The graph plots the annual in-state tuition for each school on the x-axis, and the median income of former students ten years after first enrolling on the y-axis.

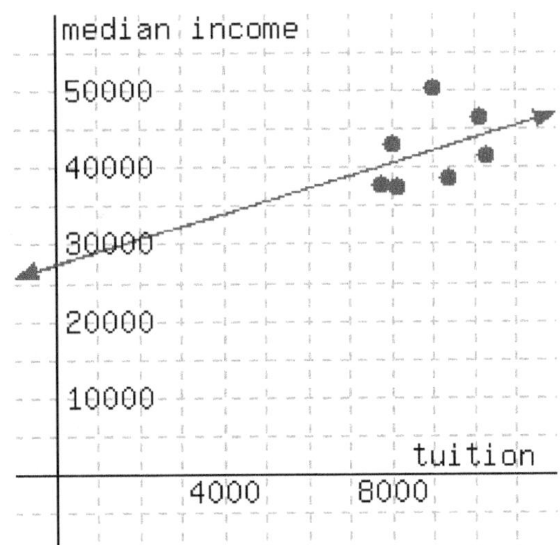

Write an equation for this line in slope-intercept form.

Explanation.   Do your best to identify two points on the line. We go with $(0, 27500)$ and $(8000, 41000)$.

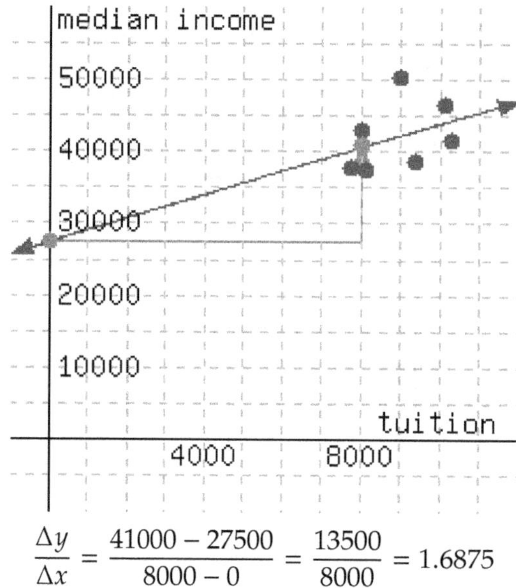

$$\frac{\Delta y}{\Delta x} = \frac{41000 - 27500}{8000 - 0} = \frac{13500}{8000} = 1.6875$$

So the slope is about 1.6875 dollars of median income per dollar of tuition. This is only an estimate since we are not all certain the two points we chose are actually on the line.

Estimating the y-intercept to be at $(0, 27500)$, we have $y = 1.6875x + 27500$.

4.5.4 Writing a Slope-Intercept Equation Given Two Points

The idea that any two points uniquely determine a line has been understood for thousands of years in many cultures around the world. Once you have two specific points, there is a straightforward process to find the slope-intercept form of the equation of the line that connects them.

Example 4.5.16 Find the slope-intercept form of the equation of the line that passes through the points $(0, 5)$ and $(8, -5)$.

Explanation.   We are trying to write down $y = mx + b$, but with specific numbers for m and b. So the first step is to find the slope, m. To do this, recall the slope formula (4.4.3) from Section 4.4. It says that if a line passes through the points (x_1, y_1) and (x_2, y_2), then the slope is found by the formula $m = \frac{y_2 - y_1}{x_2 - x_1}$.

Applying this to our two points $(\overset{x_1\ \ y_1}{0, 5})$ and $(\overset{x_2\ \ y_2}{8, -5})$, we see that the slope is:

$$\begin{aligned} m &= \frac{y_2 - y_1}{x_2 - x_1} \\ &= \frac{-5 - 5}{8 - 0} \\ &= \frac{-10}{8} \\ &= -\frac{5}{4} \end{aligned}$$

We are trying to write $y = mx + b$. Since we already found the slope, we know that we want to write $y = -\frac{5}{4}x + b$ but we need a specific number for b. We *happen* to know that one point on this line is $(0, 5)$, which is on the y-axis because its x-value is 0. So $(0, 5)$ is this line's y-intercept, and therefore $b = 5$. (We're only able to make this conclusion because this point has 0 for its x-coordinate.) So, our equation is

$$y = -\frac{5}{4}x + 5.$$

Example 4.5.17 Find the slope-intercept form of the equation of the line that passes through the points $(3, -8)$ and $(-6, 1)$.

Explanation. The first step is always to find the slope between our two points: $(\overset{x_1}{3}, \overset{y_1}{-8})$ and $(\overset{x_2}{-6}, \overset{y_2}{1})$. Using the slope formula (4.4.3) again, we have:

$$\begin{aligned}
m &= \frac{y_2 - y_1}{x_2 - x_1} \\
&= \frac{1 - (-8)}{-6 - 3} \\
&= \frac{9}{-9} \\
&= -1
\end{aligned}$$

Now that we have the slope, we can write $y = -1x + b$, which simplifies to $y = -x + b$. Unlike in Example 4.5.16, we are not given the value of b because neither of our two given points have an x-value of 0. The trick to finding b is to remember that we have two points that we know make the equation true! This means all we have to do is substitute *either* point into the equation for x and y and solve for b. Let's arbitrarily choose $(3, -8)$ to plug in.

$$\begin{aligned}
y &= -x + b \\
-8 &= -(3) + b &&\text{(Now solve for } b.) \\
-8 &= -3 + b \\
-8 + 3 &= -3 + b + 3 \\
-5 &= b
\end{aligned}$$

In conclusion, the equation for which we were searching is $y = -x - 5$.

Don't be tempted to plug in values for x and y at this point. The general equation of a line in any form should have (at least one, and in this case two) variables in the final answer.

Checkpoint 4.5.18. Find the slope-intercept form of the equation of the line that passes through the points $(-3, 150)$ and $(0, 30)$.

Explanation. The first step is always to find the slope between our points: $(\overset{x_1}{-3}, \overset{y_1}{150})$ and $(\overset{x_2}{0}, \overset{y_2}{30})$. Using the slope formula, we have:

$$\begin{aligned}
m &= \frac{y_2 - y_1}{x_2 - x_1} \\
&= \frac{30 - 150}{0 - (-3)} \\
&= \frac{-120}{3} \\
&= -40
\end{aligned}$$

Now we can write $y = -40x + b$ and to find b we need look no further than one of the given points: $(0, 30)$. Since the x-value is 0, the value of b must be 30. So, the slope-intercept form of the line is

$$y = -40x + 30$$

Checkpoint 4.5.19. Find the slope-intercept form of the equation of the line that passes through the points $\left(-3, \frac{3}{4}\right)$ and $\left(-6, -\frac{17}{4}\right)$.

Explanation. First find the slope through our points: $\left(-3, \frac{3}{4}\right)$ and $\left(-6, -\frac{17}{4}\right)$. For this problem, we choose to do all of our algebra with improper fractions as it often simplifies the process.

$$
\begin{aligned}
m &= \frac{y_2 - y_1}{x_2 - x_1} \\
&= \frac{-\frac{17}{4} - \frac{3}{4}}{-6 - (-3)} \\
&= \frac{\frac{-20}{4}}{-3} \\
&= \frac{-5}{-3} \\
&= \frac{5}{3}
\end{aligned}
$$

So far we have $y = \frac{5}{3}x + b$. Now we need to solve for b since neither of the points given were the vertical intercept. Recall that to do this, we will choose one of the two points and plug it into our equation. We choose $\left(-3, \frac{3}{4}\right)$.

$$
\begin{aligned}
y &= \frac{5}{3}x + b \\
\frac{3}{4} &= \frac{5}{3}(-3) + b \\
\frac{3}{4} &= -5 + b \\
\frac{3}{4} + 5 &= -5 + b + 5 \\
\frac{3}{4} + \frac{20}{4} &= b \\
\frac{23}{4} &= b
\end{aligned}
$$

Lastly, we write our equation.

$$
y = \frac{5}{3}x + \frac{23}{4}
$$

4.5.5 Modeling with Slope-Intercept Form

We can model many relatively simple relationships using slope-intercept form, and then solve related questions using algebra. Here are a few examples.

Example 4.5.20 Uber is a ride-sharing company. Its pricing in Portland factors in how much time and how many miles a trip takes. But if you assume that rides average out at a speed of 30 mph, then their pricing scheme boils down to a base of \$7.35 for the trip, plus \$3.85 per mile. Use a slope-intercept equation and algebra to answer these questions.

a. How much is the fare if a trip is 5.3 miles long?

b. With $100 available to you, how long of a trip can you afford?

Explanation. The rate of change (slope) is $3.85 per mile, and the starting value is $7.35. So the slope-intercept equation is

$$y = 3.85x + 7.35.$$

In this equation, x stands for the number of miles in a trip, and y stands for the amount of money to be charged.

If a trip is 5 miles long, we substitute $x = 5$ into the equation and we have:

$$\begin{aligned} y &= 3.85x + 7.35 \\ &= 3.85(5) + 7.35 \\ &= 19.25 + 7.35 \\ &= 26.60 \end{aligned}$$

And the 5-mile ride will cost you about $26.60. (We say "about," because this was all assuming you average 30 mph.)

Next, to find how long of a trip would cost $100, we substitute $y = 100$ into the equation and solve for x:

$$\begin{aligned} y &= 3.85x + 7.35 \\ 100 &= 3.85x + 7.35 \\ 100 - 7.35 &= 3.85x \\ 92.65 &= 3.85x \\ \frac{92.65}{3.85} &= x \\ 24.06 &\approx x \end{aligned}$$

So with $100 you could afford a little more than a 24-mile trip.

Checkpoint 4.5.21. In a certain wildlife reservation in Africa, there are approximately 2400 elephants. Unfortunately, the population has been decreasing by 30 elephants per year. Use a slope-intercept equation and algebra to answer these questions.

a. If the trend continues, what would the elephant population be 15 years from now?

$\boxed{}$ elephants

b. If the trend continues, how many years will it be until the elephant population dwindles to 1200?

$\boxed{}$ years

Explanation. The rate of change (slope) is -30 elephants per year. Notice that since we are losing elephants, the slope is a negative number. The starting value is 2400 elephants. So the slope-intercept equation is

$$y = -30x + 2400.$$

In this equation, x stands for a number of years into the future, and y stands for the elephant population.

To estimate the elephant population 15 years later, we substitute x in the equation with 15, and we have:

$$\begin{aligned} y &= -30x + 2400 \\ &= -30(15) + 2400 \\ &= -450 + 2400 \\ &= 1950 \end{aligned}$$

So if the trend continues, there would be 1950 elephants on this reservation 15 years later.

Next, to find when the elephant population would decrease to 1200, we substitute y in the equation with 1200, and solve for x:

$$\begin{aligned} y &= -30x + 2400 \\ 1200 &= -30x + 2400 \\ 1200 - 2400 &= -30x \\ -1200 &= -30x \\ \frac{-1200}{-30} &= x \\ 40 &= x \end{aligned}$$

So if the trend continues, 40 years later, the elephant population would dwindle to 1,200.

Exercises

Review and Warmup

1. Evaluate $10B + 2c$ for $B = 7$ and $c = -4$.

2. Evaluate $-9C - a$ for $C = -5$ and $a = -10$.

3. Evaluate

$$\frac{y_2 - y_1}{x_2 - x_1}$$

for $x_1 = 19$, $x_2 = 9$, $y_1 = 8$, and $y_2 = 11$:

4. Evaluate

$$\frac{y_2 - y_1}{x_2 - x_1}$$

for $x_1 = -18$, $x_2 = -5$, $y_1 = -16$, and $y_2 = -1$:

Identifying Slope and y-Intercept   Find the line's slope and y-intercept.

5. A line has equation $y = 3x + 1$.

This line's slope is ☐.

This line's y-intercept is ☐.

6. A line has equation $y = 4x + 7$.

This line's slope is ☐.

This line's y-intercept is ☐.

7. A line has equation $y = -7x - 7$.

This line's slope is ☐.

This line's y-intercept is ☐.

8. A line has equation $y = -6x - 1$.

This line's slope is ☐.

This line's y-intercept is ☐.

9. A line has equation $y = x + 3$.

This line's slope is ☐.

This line's y-intercept is ☐.

10. A line has equation $y = x + 5$.

This line's slope is ☐.

This line's y-intercept is ☐.

11. A line has equation $y = -x + 7$.

This line's slope is ☐.

This line's y-intercept is ☐.

12. A line has equation $y = -x + 9$.

This line's slope is ☐.

This line's y-intercept is ☐.

13. A line has equation $y = -\dfrac{2}{3}x + 8$.

This line's slope is ☐.

This line's y-intercept is ☐.

14. A line has equation $y = -\dfrac{2}{9}x - 5$.

This line's slope is ☐.

This line's y-intercept is ☐.

15. A line has equation $y = \dfrac{1}{2}x + 8$.

This line's slope is ☐.

This line's y-intercept is ☐.

16. A line has equation $y = \dfrac{1}{4}x - 7$.

This line's slope is ☐.

This line's y-intercept is ☐.

17. A line has equation $y = 7 + 6x$.

This line's slope is ☐.

This line's y-intercept is ☐.

18. A line has equation $y = 9 + 7x$.

This line's slope is ☐.

This line's y-intercept is ☐.

19. A line has equation $y = 8 - x$.

This line's slope is ☐.

This line's y-intercept is ☐.

20. A line has equation $y = 9 - x$.

This line's slope is ☐.

This line's y-intercept is ☐.

Graphs and Slope-Intercept Form

21. Graph the equation $y = 4x$.   **22.** Graph the equation $y = 5x$.   **23.** Graph the equation $y = -3x$.

24. Graph the equation $y = -2x$.   **25.** Graph the equation $y = \frac{5}{2}x$.   **26.** Graph the equation $y = \frac{1}{4}x$.

27. Graph the equation $y = -\frac{1}{3}x$.   **28.** Graph the equation $y = -\frac{5}{4}x$.

29. Graph the equation $y = 5x + 2$.   **30.** Graph the equation $y = 3x + 6$.

31. Graph the equation $y = -4x + 3$.   **32.** Graph the equation $y = -2x + 5$.

33. Graph the equation $y = x - 4$.   **34.** Graph the equation $y = x + 2$.

35. Graph the equation $y = -x + 3$.   **36.** Graph the equation $y = -x - 5$.

37. Graph the equation $y = \frac{2}{3}x + 4$.   **38.** Graph the equation $y = \frac{3}{2}x - 5$.

39. Graph the equation $y = -\frac{3}{5}x - 1$.   **40.** Graph the equation $y = -\frac{1}{5}x + 1$.

A line's graph is given.

41.
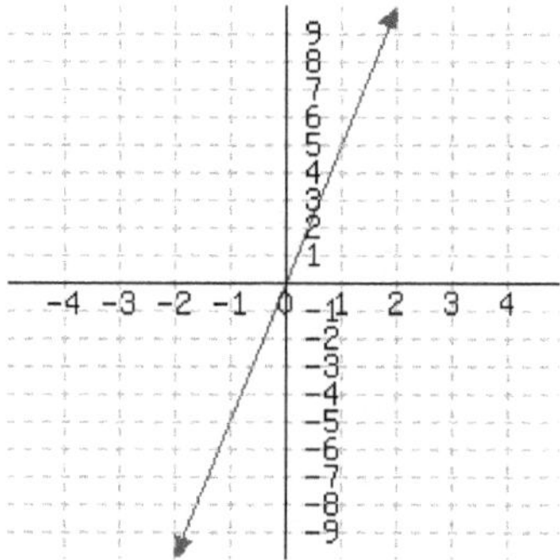

42.
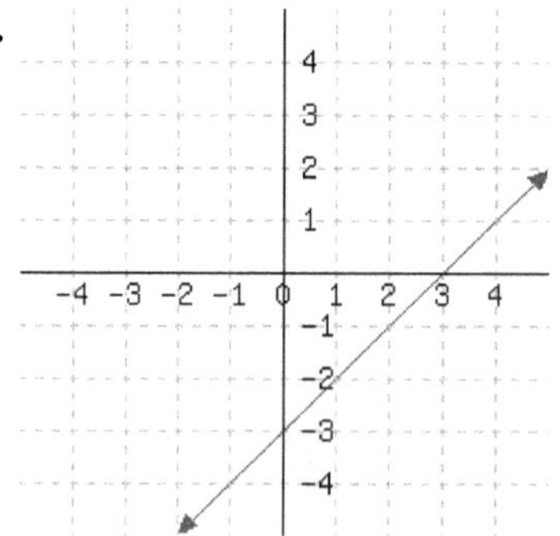

43.
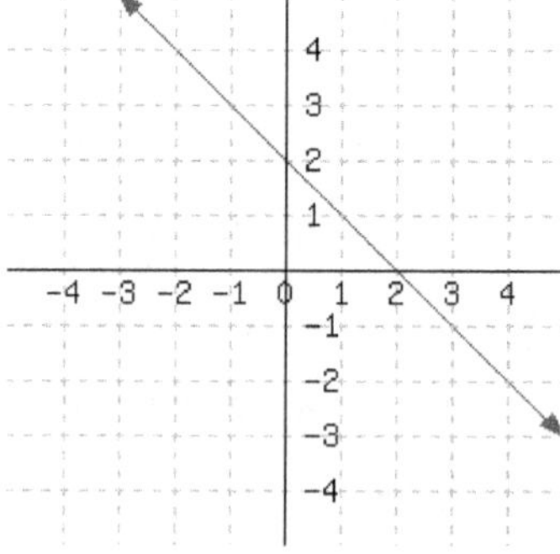

This line's slope-intercept equation is [ ]

This line's slope-intercept equation is [ ]

This line's slope-intercept equation is [ ]

44.

This line's slope-intercept equation is

45.

This line's slope-intercept equation is

46.

This line's slope-intercept equation is

47.

This line's slope-intercept equation is

48.

This line's slope-intercept equation is

Writing a Slope-Intercept Equation Given Two Points

49. A line passes through the points $(3, 22)$ and $(4, 27)$. Find this line's equation in slope-intercept form.

This line's slope-intercept equation is ___________.

50. A line passes through the points $(5, 29)$ and $(2, 14)$. Find this line's equation in slope-intercept form.

This line's slope-intercept equation is ___________.

51. A line passes through the points $(-2, 20)$ and $(-5, 35)$. Find this line's equation in slope-intercept form.

This line's slope-intercept equation is ___________.

52. A line passes through the points $(3, -12)$ and $(2, -7)$. Find this line's equation in slope-intercept form.

This line's slope-intercept equation is ___________.

53. A line passes through the points $(-2, -3)$ and $(-3, -2)$. Find this line's equation in slope-intercept form.

This line's slope-intercept equation is ___________.

54. A line passes through the points $(5, -7)$ and $(1, -3)$. Find this line's equation in slope-intercept form.

This line's slope-intercept equation is ___________.

55. A line passes through the points $(18, 16)$ and $(0, 1)$. Find this line's equation in slope-intercept form.

This line's slope-intercept equation is [____].

56. A line passes through the points $(0, 7)$ and $(-15, -11)$. Find this line's equation in slope-intercept form.

This line's slope-intercept equation is [____].

57. A line passes through the points $(-9, 16)$ and $(0, 9)$. Find this line's equation in slope-intercept form.

This line's slope-intercept equation is [____].

58. A line passes through the points $(-5, 9)$ and $(-15, 25)$. Find this line's equation in slope-intercept form.

This line's slope-intercept equation is [____].

Applications

59. A gym charges members $40 for a registration fee, and then $24 per month. You became a member some time ago, and now you have paid a total of $448 to the gym. How many months have passed since you joined the gym?

[____] months have passed since you joined the gym.

60. Your cell phone company charges a $11 monthly fee, plus $0.18 per minute of talk time. One month your cell phone bill was $68.60. How many minutes did you spend talking on the phone that month?

You spent [____] talking on the phone that month.

61. A school purchased a batch of T-shirts from a company. The company charged $4 per T-shirt, and gave the school a $75 rebate. If the school had a net expense of $1,565 from the purchase, how many T-shirts did the school buy?

The school purchased [____] T-shirts.

62. Izabelle hired a face-painter for a birthday party. The painter charged a flat fee of $65, and then charged $2.50 per person. In the end, Izabelle paid a total of $137.50. How many people used the face-painter's service?

[____] people used the face-painter's service.

63. A certain country has 406.56 million acres of forest. Every year, the country loses 4.84 million acres of forest mainly due to deforestation for farming purposes. If this situation continues at this pace, how many years later will the country have only 227.48 million acres of forest left? (Use an equation to solve this problem.)

After [____] years, this country would have 227.48 million acres of forest left.

64. Anthony has $80 in his piggy bank. He plans to purchase some Pokemon cards, which costs $1.75 each. He plans to save $60.75 to purchase another toy. At most how many Pokemon cards can he purchase?

Write an equation to solve this problem.

Anthony can purchase at most [____] Pokemon cards.

65. By your cell phone contract, you pay a monthly fee plus \$0.06 for each minute you spend on the phone. In one month, you spent 220 minutes over the phone, and had a bill totaling \$25.20.

Let x be the number of minutes you spend on the phone in a month, and let y be your total cell phone bill for that month, in dollars. Use a linear equation to model your monthly bill based on the number of minutes you spend on the phone.

   a. This line's slope-intercept equation is ☐.

   b. If you spend 140 minutes on the phone in a month, you would be billed ☐.

   c. If your bill was \$39.60 one month, you must have spent ☐ minutes on the phone in that month.

66. A company set aside a certain amount of money in the year 2000. The company spent exactly \$42,000 from that fund each year on perks for its employees. In 2003, there was still \$782,000 left in the fund.

Let x be the number of years since 2000, and let y be the amount of money, in dollars, left in the fund that year. Use a linear equation to model the amount of money left in the fund after so many years.

   a. The linear model's slope-intercept equation is ☐.

   b. In the year 2009, there was ☐ left in the fund.

   c. In the year ☐, the fund will be empty.

67. A biologist has been observing a tree's height. This type of tree typically grows by 0.27 feet each month. Ten months into the observation, the tree was 17.1 feet tall.

Let x be the number of months passed since the observations started, and let y be the tree's height at that time, in feet. Use a linear equation to model the tree's height as the number of months pass.

   a. This line's slope-intercept equation is ☐.

   b. 26 months after the observations started, the tree would be ☐ feet in height.

   c. ☐ months after the observation started, the tree would be 29.52 feet tall.

68. Scientists are conducting an experiment with a gas in a sealed container. The mass of the gas is measured, and the scientists realize that the gas is leaking over time in a linear way. Each minute, they lose 1.7 grams. Seven minutes since the experiment started, the remaining gas had a mass of 73.1 grams.

Let x be the number of minutes that have passed since the experiment started, and let y be the mass of the gas in grams at that moment. Use a linear equation to model the weight of the gas over time.

   a. This line's slope-intercept equation is ☐.

   b. 33 minutes after the experiment started, there would be ☐ grams of gas left.

   c. If a linear model continues to be accurate, ☐ minutes since the experiment started, all gas in the container will be gone.

69. A company set aside a certain amount of money in the year 2000. The company spent exactly the same amount from that fund each year on perks for its employees. In 2004, there was still $807,000 left in the fund. In 2005, there was $786,000 left.

Let x be the number of years since 2000, and let y be the amount of money, in dollars, left in the fund that year. Use a linear equation to model the amount of money left in the fund after so many years.

   a. The linear model's slope-intercept equation is [＿＿＿＿].

   b. In the year 2009, there was [＿＿＿＿] left in the fund.

   c. In the year [＿＿＿＿], the fund will be empty.

70. By your cell phone contract, you pay a monthly fee plus some money for each minute you use the phone during the month. In one month, you spent 230 minutes on the phone, and paid $17.45. In another month, you spent 380 minutes on the phone, and paid $19.70.

Let x be the number of minutes you talk over the phone in a month, and let y be your cell phone bill, in dollars, for that month. Use a linear equation to model your monthly bill based on the number of minutes you talk over the phone.

   a. This linear model's slope-intercept equation is [＿＿＿＿].

   b. If you spent 130 minutes over the phone in a month, you would pay [＿＿＿＿].

   c. If in a month, you paid $20.15 of cell phone bill, you must have spent [＿＿＿＿] minutes on the phone in that month.

71. Scientists are conducting an experiment with a gas in a sealed container. The mass of the gas is measured, and the scientists realize that the gas is leaking over time in a linear way.

Nine minutes since the experiment started, the gas had a mass of 42.9 grams.

Fifteen minutes since the experiment started, the gas had a mass of 35.1 grams.

Let x be the number of minutes that have passed since the experiment started, and let y be the mass of the gas in grams at that moment. Use a linear equation to model the weight of the gas over time.

   a. This line's slope-intercept equation is [＿].

   b. 32 minutes after the experiment started, there would be [＿＿＿＿] grams of gas left.

   c. If a linear model continues to be accurate, [＿] minutes since the experiment started, all gas in the container will be gone.

72. A biologist has been observing a tree's height. 10 months into the observation, the tree was 18.2 feet tall. 16 months into the observation, the tree was 19.22 feet tall.

Let x be the number of months passed since the observations started, and let y be the tree's height at that time, in feet. Use a linear equation to model the tree's height as the number of months pass.

   a. This line's slope-intercept equation is [＿].

   b. 26 months after the observations started, the tree would be [＿＿＿＿] feet in height.

   c. [＿＿＿＿] months after the observation started, the tree would be 25 feet tall.

Challenge

73. Line *S* has the equation $y = ax + b$ and Line *T* has the equation $y = cx + d$. Suppose $a > b > c > d > 0$.

    a. What can you say about Line *S* and Line *T*, given that $a > c$? Give as much information about Line *S* and Line *T* as possible.

    b. What can you say about Line *S* and Line *T*, given that $b > d$? Give as much information about Line *S* and Line *T* as possible.

4.6 Point-Slope Form

In Section 4.5, we learned that a linear equation can be written in slope-intercept form, $y = mx + b$. This section covers an alternative that can often be more useful depending on the application: **point-slope form**.

4.6.1 Point-Slope Motivation and Definition

Starting in 1990, the population of the United States has been growing by about 2.865 million per year. Also, back in 1990, the population was 253 million. Since the rate of growth has been roughly constant, a linear model is appropriate. Let's try to write an equation to model this.

We consider using slope-intercept form (4.5.1), but we would need to know the y-intercept, and nothing in the background tells us that. We'd need to know the population of the United States in the year 0, before there even was a United States.

We could do some side work to calculate the y-intercept, but let's try something else. Here are some things we know:

1. The slope equation is $m = \frac{y_2 - y_1}{x_2 - x_1}$.

2. The slope is $m = 2.865$ (million per year).

3. One point on the line is $(1990, 253)$, because in 1990, the population was 253 million.

If we use the generic (x, y) to represent a point *somewhere* on this line, then the rate of change between $(1990, 253)$ and (x, y) has to be 2.865. So

$$\frac{y - 253}{x - 1990} = 2.865.$$

There is good reason[1] to want to isolate y in this equation:

$$\frac{y - 253}{x - 1990} = 2.865$$
$$y - 253 = 2.865 \cdot (x - 1990) \qquad \text{(could distribute, but not going to)}$$
$$y = 2.865(x - 1990) + 253$$

This is a good place to stop. We have isolated y, and three *meaningful* numbers appear in the population: the rate of growth, a certain year, and the population in that year. This is a specific example of **point-slope form**. Before we look deeper at point-slope form, let's continue reducing the line equation into slope-intercept form.

$$y = 2.865(x - 1990) + 253$$
$$y = 2.865x - 5701.35 + 253$$
$$y = 2.865x - 5448.35$$

One concern with slope-intercept form (4.5.1) is that it uses the y-intercept, which might be somewhat meaningless in the context of an application. For example, here we have found that the y-intercept is at $(0, -5448.35)$, but what practical use is that? It's nonsense to say that in the year 0, the population of the United States was -5448.35 million. It doesn't make sense to have a negative population. It doesn't make sense to talk about the United States population before there even was a United States. And it doesn't make

---

[1] It will help us to see that y (population) *depends* on x (whatever year it is).

sense to use this model for years earlier than 1990 because the background information says clearly that the rate of change we have applies to years 1990 and later.

For all these reasons, we prefer the equation when it was in the form

$$y = 2.865(x - 1990) + 253$$

Definition 4.6.2 Point-Slope Form. When x and y have a linear relationship where m is the slope and (x_0, y_0) is some specific point that the line passes through, one equation for this relationship is

$$y = m\,(x - x_0) + y_0 \qquad (4.6.1)$$

and this equation is called the **point-slope form** of the line. It is called this because the slope and one point on the line are immediately discernible from the numbers in the equation.

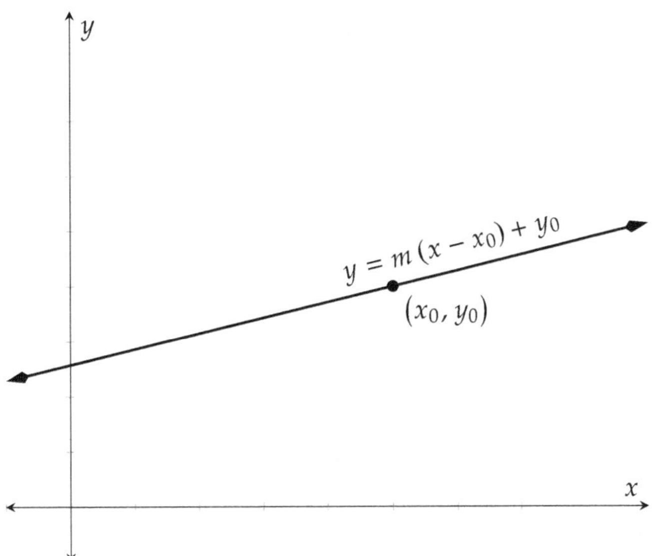

Figure 4.6.3

Remark 4.6.4 Alternative Point-Slope Form. It is also common to define point-slope form as

$$y - y_0 = m\,(x - x_0) \qquad (4.6.2)$$

by subtracting y_0 from each side. Some exercises may appear using this form.

Checkpoint 4.6.5. Consider the line in this graph:

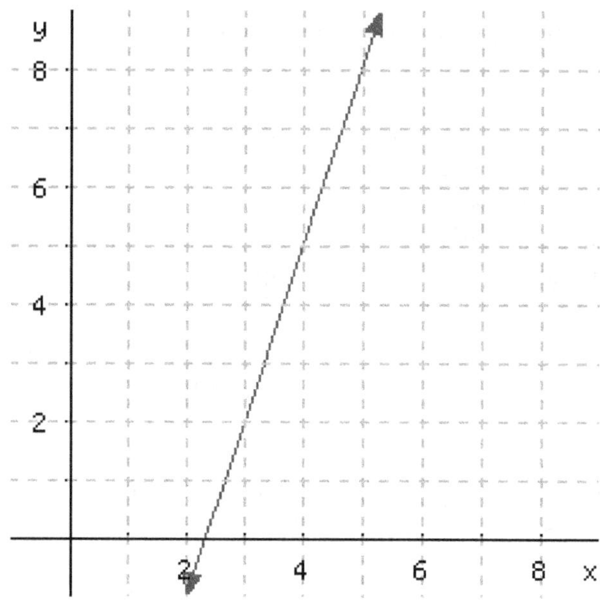

a. Identify a point visible on this line that has integer coordinates.

b. What is the slope of the line?

c. Use point-slope form to write an equation for this line, making use of a point with integer coordinates.

Explanation.

a. The visible points with integer coordinates are $(2, -1)$, $(3, 2)$, $(4, 5)$, and $(5, 8)$.

b. Several slope triangles are visible where the "run" is 1 and the "rise" is 3. So the slope is $\frac{3}{1} = 3$.

c. Using $(3, 2)$, the point-slope equation is $y = 3(x - 3) + 2$. (You could use other points, like $(2, -1)$, and get a different-looking equation like $y = 3(x - 2) + (-1)$ which simplifies to $y = 3(x - 2) - 1$.)

In Checkpoint 4.6.5, the solution explains that each of the following are acceptable equations for the same line:

$$y = 3(x - 3) + 2 \qquad\qquad y = 3(x - 2) - 1$$

The first uses $(3, 2)$ as a point on the line, and the second uses $(2, -1)$. Are those two equations really equivalent? Let's distribute and simplify each of them to get slope-intercept form (4.5.1).

$$
\begin{aligned}
y &= 3(x - 3) + 2 & \qquad\qquad y &= 3(x - 2) - 1 \\
y &= 3x - 9 + 2 & y &= 3x - 6 - 1 \\
y &= 3x - 7 & y &= 3x - 7
\end{aligned}
$$

So, yes. It didn't matter which point we used to write a point-slope equation. We get different-looking equations that still represent the same line.

Point-slope form is preferable when we know a line's slope and a point on it, but we don't know the y-intercept.

Example 4.6.6 A spa chain has been losing customers at a roughly constant rate since the year 2010. In 2013, it had 2,975 customers; in 2016, it had 2,585 customers. Management estimated that the company will go out of business once its customer base decreases to 1,800. If this trend continues, when will the company close?

The given information tells us two points on the line: $(2013, 2975)$ and $(2016, 2585)$. The slope formula (4.4.3) will give us the slope. After labeling those two points as $(\overset{x_1}{2013}, \overset{y_1}{2975})$ and $(\overset{x_2}{2016}, \overset{y_2}{2585})$, we have:

$$\begin{aligned}
\text{slope} &= \frac{y_2 - y_1}{x_2 - x_1} \\
&= \frac{2585 - 2975}{2016 - 2013} \\
&= \frac{-390}{3} \\
&= -130
\end{aligned}$$

And considering units, this means they are losing 130 customers per year.

Let's note that we could try to make an equation for this line in slope-intercept form, but then we would need to calculate the y-intercept, which in context would correspond to the number of customers in year 0. We could do it, but we'd be working with numbers that have no real-world meaning in this context.

For point-slope form, since we calculated the slope, we know at least this much:

$$y = -130(x - x_0) + y_0.$$

Now we can pick one of those two given points, say $(2013, 2975)$, and get the equation

$$y = -130(x - 2013) + 2975.$$

Note that all three numbers in this equation have meaning in the context of the spa chain.

We're ready to answer the question about when the chain might go out of business. Substitute y in the equation with 1800 and solve for x, and we will get the answer we seek.

$$\begin{aligned}
y &= -130(x - 2013) + 2975 \\
1800 &= -130(x - 2013) + 2975 \\
1800 - 2975 &= -130(x - 2013) \\
-1175 &- 130(x - 2013) \\
\frac{-1175}{-130} &= \frac{-130(x - 2013)}{-130} \\
9.038 &\approx x - 2013 \\
9.038 + 2013 &\approx x \\
2022 &\approx x
\end{aligned}$$

And so we find that at this rate, the company is headed toward a collapse in 2022.

Shown is a graph that represents the scenario. Note that to make a graph of $y = -130(x-2013)+2975$, we must first find the point $(2013, 2975)$ and from there use the slope of -130 to draw the line.

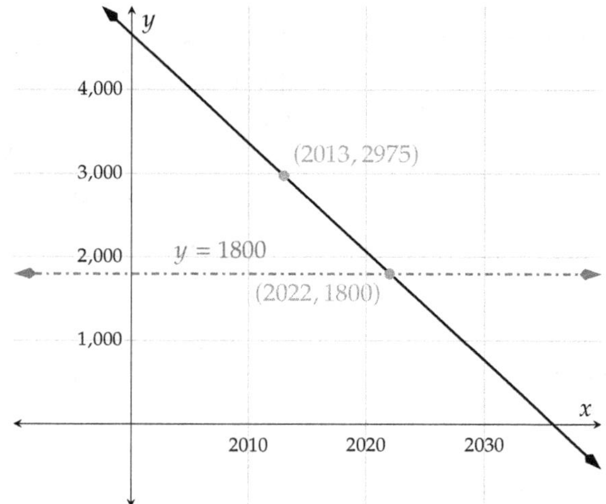

Figure 4.6.7: A Graph of $y = -130(x - 2013) + 2975$

Checkpoint 4.6.8. If we go state by state and compare the Republican candidate's 2012 vote share (x) to the Republican candidate's 2016 vote share (y), we get the following graph where a trendline has been superimposed.

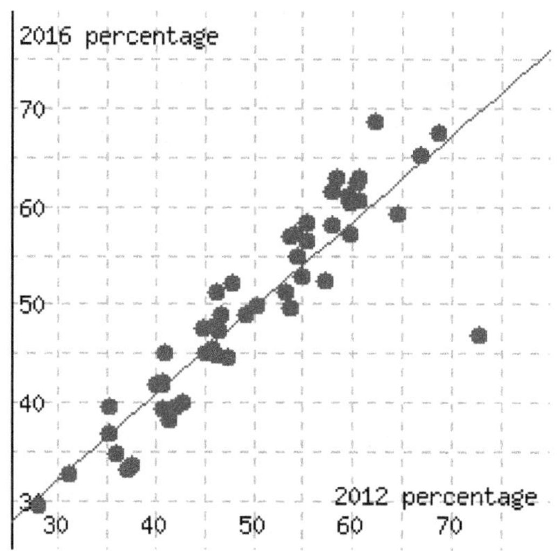

Find a point-slope equation for this line. (Note that a slope-intercept equation would use the y-intercept cooridnate b, and that would not be meaningful in context, since no state had anywhere near zero percent Republican vote.)

Explanation. We need to calculate slope first. And for that, we need to identify two points on the line. conveniently, the line appears to pass right through $(50, 50)$. We have to take a second point somewhere, and $(75, 72)$ seems like a reasonable roughly accurate choice. The slope equation gives us that

$$m = \frac{72 - 50}{75 - 50} = \frac{22}{25} = 0.88.$$

Using $(50, 50)$ as the point, the point-slope equation would then be

$$y = 0.88(x - 50) + 50.$$

4.6.2 Using Two Points to Build a Linear Equation

Since two points can determine a line's location, we can calculate a line's equation using just the coordinates from any two points it passes through.

Example 4.6.9 A line passes through $(-6, 0)$ and $(9, -10)$. Find this line's equation in both point-slope and slope-intercept form.

Explanation. We will use the slope formula (4.4.3) to find the slope first. After labeling those two points as $(\overset{x_1}{-6}, \overset{y_1}{0})$ and $(\overset{x_2}{9}, \overset{y_2}{-10})$, we have:

$$\begin{aligned}
\text{slope} &= \frac{y_2 - y_1}{x_2 - x_1} \\
&= \frac{-10 - 0}{9 - (-6)} \\
&= \frac{-10}{15} \\
&= -\frac{2}{3}
\end{aligned}$$

The point-slope equation is $y = -\frac{2}{3}(x - x_0) + y_0$. Next, we will use $(9, -10)$ and substitute x_0 with 9 and y_0 with -10, and we have:

$$\begin{aligned}
y &= -\frac{2}{3}(x - x_0) + y_0 \\
y &= -\frac{2}{3}(x - 9) + (-10) \\
y &= -\frac{2}{3}(x - 9) - 10
\end{aligned}$$

Next, we will change the point-slope equation into slope-intercept form:

$$\begin{aligned}
y &= -\frac{2}{3}(x - 9) - 10 \\
y &= -\frac{2}{3}x + 6 - 10 \\
y &= -\frac{2}{3}x - 4
\end{aligned}$$

Remark 4.6.10. Note that many other resources use the alternate point-slope form (4.6.2) to write their equations. Those equations will always be equivalent to those created using our point-slope form. In Example 4.6.9, we found the point-slope form $y = -\frac{2}{3}(x - 9) - 10$. The alternate point-slope form equation[2] would have given us $y + 10 = -\frac{2}{3}(x - 9)$. If you solve this equation for y and simplify, you should still get $y = -\frac{2}{3}x - 4$, as we did earlier.

Checkpoint 4.6.11. A line passes through $(13, -108)$ and $(-42, 23)$. Find equations for this line using both point-slope and slope-intercept form.

A point-slope equation: ⬚

A slope-intercept equation: ⬚

---

[2]khanacademy.org/math/algebra/two-var-linear-equations/point-slope/a/point-slope-form-review

Explanation.   First, use the slope formula to find the slope of this line:

$$m = \frac{y_2 - y_1}{x_2 - x_1} = \frac{23 - (-108)}{-42 - 13}$$
$$= \frac{131}{-55}$$
$$= -\frac{131}{55}.$$

The generic point-slope equation is $y = m(x - x_0) + y_0$. We have found the slope, m, and we may use $(13, -108)$ for (x_0, y_0). So an equation in point-slope form is $y = \frac{-131}{55}(x - 13) - 108$.

To find a slope-intercept form equation, we can take the generic $y = mx + b$ and substitute in the value of m we found. Also, we know that $(x, y) = (13, -108)$ should make the equation true. So we have

$$y = mx + b$$
$$-108 = -\frac{131}{55}(13) + b \qquad \text{Now we may solve for } b.$$
$$-108 \cdot 55 = \left(-\frac{131}{55}(13) + b\right) \cdot 55$$
$$-5940 = -131(13) + 55b$$
$$-5940 = -1703 + 55b$$
$$-5940 + 1703 = -1703 + 55b + 1703$$
$$-4237 = 55b$$
$$\frac{-4237}{55} = \frac{55b}{55}$$
$$b = -\frac{4237}{55}.$$

So the slope-intercept equation is $y = \frac{-131}{55}x - \frac{4237}{55}$.

4.6.3 More on Point-Slope Form

We can tell a lot about a linear equation now that we have learned both slope-intercept form (4.5.1) and point-slope form (4.6.1). For example, we can know that $y = 4x + 2$ is in slope-intercept form because it looks like $y = mx + b$. It will graph as a line with slope 4 and vertical intercept $(0, 2)$. Likewise, we know that the equation $y = -5(x - 3) + 2$ is in point-slope form because it looks like $y = m(x - x_0) + y_0$. It will graph as a line that has slope -5 and will pass through the point $(3, 2)$.

Example 4.6.12 For the equations below, state whether they are in slope-intercept form or point-slope form. Then identify the slope of the line and at least one point that the line will pass through.

a. $y = -3x + 2$

b. $y = 9(x + 1) - 6$

c. $y = 5 - x$

d. $y = -\frac{12}{5}(x - 9) + 1$

Explanation.

a. The equation $y = -3x + 2$ is in slope-intercept form. The slope is -3 and the vertical intercept is $(0, 2)$.

b. The equation $y = 9(x + 1) - 6$ is in point-slope form. The slope is 9 and the line passes through the point $(-1, 6)$.

c. The equation $y = 5 - x$ is almost in slope-intercept form. If we rearrange the right hand side to be $y = -x + 5$, we can see that the slope is -1 and the vertical intercept is $(0, 5)$.

d. The equation $y = -\frac{12}{5}(x - 9) + 1$ is in point-slope form. The slope is $-\frac{12}{5}$ and the line passes through the point $(9, 1)$.

Remark 4.6.13. Again, we should note that the alternate point-slope form (4.6.2) can be used to identify equations. For example, the equation $y + 10 = -\frac{2}{3}(x - 9)$ matches the alternate point-slope form equation[3] with slope $-\frac{2}{3}$ and the line passes through the point $(9, -10)$. Note that both coordinates are the opposite of what they appear to be in the equation with this form.

Consider the graph in Figure 4.6.15.

Example 4.6.14   a. Find three equations that describe the line shown written in point-slope form. Three integer-valued points are shown for convenience.

b. Determine the slope-intercept form of the equation of this line.

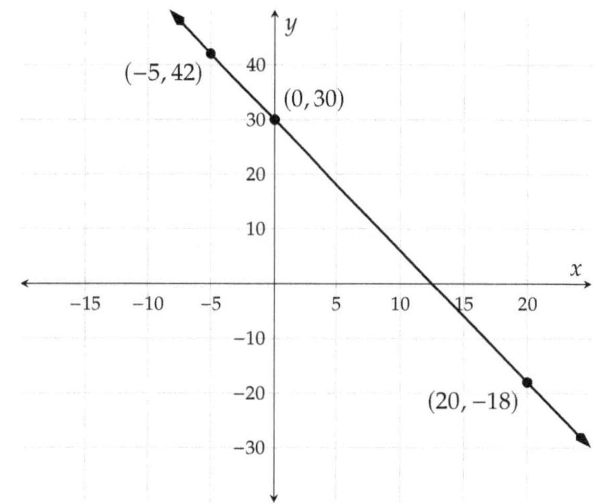

Figure 4.6.15

Explanation.

a. To write *any* of the equations representing this line in point-slope form, we must first find the slope of the line and we can use the slope formula (4.4.3) to do so. We will arbitrarily choose $(0, 30)$ and $(-5, 42)$ as the two points. Inputting these points into the slope formula yields:

$$
\begin{aligned}
m &= \frac{y_2 - y_1}{x_2 - x_1} \\
&= \frac{42 - 30}{-5 - 0} \\
&= \frac{12}{-5}
\end{aligned}
$$

---

[3]en.wikipedia.org/wiki/Linear_equation#Point–slope_form

$$= -\frac{12}{5}$$

Thus the slope of the line is $-\frac{12}{5}$.

Next, we need to write an equation in point-slope form based on each point shown. Using the point $(0, 30)$, we have:

$$y = -\frac{12}{5}(x - 0) + 30$$

(This simplifies to $y = -\frac{12}{5}x + 30$.)

The next point is $(20, -18)$. Using this point, we can write an equation for this line as:

$$y = -\frac{12}{5}(x - 20) - 18$$

Finally, we can also use the point $(-5, 42)$ to write an equation for this line:

$$y = -\frac{12}{5}(x - (-5)) + 42$$

which can also be written as:

$$y = -\frac{12}{5}(x + 5) + 42$$

b. As $(0, 30)$ is the vertical intercept, we can write the equation of this line in slope-intercept form as $y = -\frac{12}{5}x + 30$. It's important to note that each of the equations that were written in point-slope form simplify to this, making all four equations equivalent.

Exercises

Review and Warmup

1. Evaluate $-5C - 7b$ for $C = 6$ and $b = -7$.

2. Evaluate $-a + 3A$ for $a = 2$ and $A = -7$.

3. Evaluate

$$\frac{y_2 - y_1}{x_2 - x_1}$$

for $x_1 = -14$, $x_2 = -10$, $y_1 = -19$, and $y_2 = -10$:

4. Evaluate

$$\frac{y_2 - y_1}{x_2 - x_1}$$

for $x_1 = -10$, $x_2 = 17$, $y_1 = -2$, and $y_2 = 19$:

Point-Slope Form

5. A line's equation is given in point-slope form:

$y = 5(x - 5) + 28$

This line's slope is [ ].

A point on this line that is apparent from the given equation is [ ].

6. A line's equation is given in point-slope form:

$y = 2(x - 1) + 5$

This line's slope is [ ].

A point on this line that is apparent from the given equation is [ ].

7. A line's equation is given in point-slope form:

$y = -2(x + 2) + 5$

This line's slope is [ ].

A point on this line that is apparent from the given equation is [ ].

8. A line's equation is given in point-slope form:

$y = -3(x + 4) + 7$

This line's slope is [ ].

A point on this line that is apparent from the given equation is [ ].

9. A line's equation is given in point-slope form:

$y = \dfrac{8}{3}(x + 9) - 23$

This line's slope is [ ].

A point on this line that is apparent from the given equation is [ ].

10. A line's equation is given in point-slope form:

$y = \dfrac{9}{8}(x + 24) - 29$

This line's slope is [ ].

A point on this line that is apparent from the given equation is [ ].

11. A line passes through the points $(2, 9)$ and $(1, 7)$. Find this line's equation in point-slope form.

Using the point $(2, 9)$, this line's point-slope form equation is [ ].

Using the point $(1, 7)$, this line's point-slope form equation is [ ].

12. A line passes through the points $(4, 10)$ and $(1, 4)$. Find this line's equation in point-slope form.

Using the point $(4, 10)$, this line's point-slope form equation is [ ].

Using the point $(1, 4)$, this line's point-slope form equation is [ ].

13. A line passes through the points $(-3, 17)$ and $(0, 8)$. Find this line's equation in point-slope form.

Using the point $(-3, 17)$, this line's point-slope form equation is [ ].

Using the point $(0, 8)$, this line's point-slope form equation is [ ].

14. A line passes through the points $(1, 2)$ and $(-3, 14)$. Find this line's equation in point-slope form.

Using the point $(1, 2)$, this line's point-slope form equation is [ ].

Using the point $(-3, 14)$, this line's point-slope form equation is [ ].

15. A line passes through the points $(6,5)$ and $(-6,-25)$. Find this line's equation in point-slope form.

Using the point $(6,5)$, this line's point-slope form equation is [   ].

Using the point $(-6,-25)$, this line's point-slope form equation is [   ].

16. A line passes through the points $(7,5)$ and $(21,17)$. Find this line's equation in point-slope form.

Using the point $(7,5)$, this line's point-slope form equation is [   ].

Using the point $(21,17)$, this line's point-slope form equation is [   ].

17. A line's slope is 4. The line passes through the point $(5,22)$. Find an equation for this line in both point-slope and slope-intercept form.

An equation for this line in point-slope form is: [   ].

An equation for this line in slope-intercept form is: [   ].

18. A line's slope is 5. The line passes through the point $(2,11)$. Find an equation for this line in both point-slope and slope-intercept form.

An equation for this line in point-slope form is: [   ].

An equation for this line in slope-intercept form is: [   ].

19. A line's slope is -2. The line passes through the point $(2,-2)$. Find an equation for this line in both point-slope and slope-intercept form.

An equation for this line in point-slope form is: [   ].

An equation for this line in slope-intercept form is: [   ].

20. A line's slope is -5. The line passes through the point $(-4,18)$. Find an equation for this line in both point-slope and slope-intercept form.

An equation for this line in point-slope form is: [   ].

An equation for this line in slope-intercept form is: [   ].

21. A line's slope is 1. The line passes through the point $(5,1)$. Find an equation for this line in both point-slope and slope-intercept form.

An equation for this line in point-slope form is: [   ].

An equation for this line in slope-intercept form is: [   ].

22. A line's slope is 1. The line passes through the point $(2,-1)$. Find an equation for this line in both point-slope and slope-intercept form.

An equation for this line in point-slope form is: [   ].

An equation for this line in slope-intercept form is: [   ].

23. A line's slope is -1. The line passes through the point $(-2, 1)$. Find an equation for this line in both point-slope and slope-intercept form.

An equation for this line in point-slope form is: [________].

An equation for this line in slope-intercept form is: [________].

24. A line's slope is -1. The line passes through the point $(3, 2)$. Find an equation for this line in both point-slope and slope-intercept form.

An equation for this line in point-slope form is: [________].

An equation for this line in slope-intercept form is: [________].

25. A line's slope is $\frac{6}{5}$. The line passes through the point $(10, 8)$. Find an equation for this line in both point-slope and slope-intercept form.

An equation for this line in point-slope form is: [________].

An equation for this line in slope-intercept form is: [________].

26. A line's slope is $\frac{7}{4}$. The line passes through the point $(12, 22)$. Find an equation for this line in both point-slope and slope-intercept form.

An equation for this line in point-slope form is: [________].

An equation for this line in slope-intercept form is: [________].

27. A line's slope is $-\frac{8}{9}$. The line passes through the point $(9, -13)$. Find an equation for this line in both point-slope and slope-intercept form.

An equation for this line in point-slope form is: [________].

An equation for this line in slope-intercept form is: [________].

28. A line's slope is $-\frac{9}{5}$. The line passes through the point $(10, -20)$. Find an equation for this line in both point-slope and slope-intercept form.

An equation for this line in point-slope form is: [________].

An equation for this line in slope-intercept form is: [________].

Point-Slope and Slope-Intercept Change this equation from point-slope form to slope-intercept form.

29. $y = 2(x - 4) + 5$

In slope-intercept form: [________]

30. $y = 2(x + 2)$

In slope-intercept form: [________]

31. $y = -4(x - 3) - 10$

In slope-intercept form: [________]

32. $y = -4(x + 4) + 12$

In slope-intercept form: [________]

33. $y = \frac{5}{8}(x - 16) + 15$

In slope-intercept form: [________]

34. $y = \frac{6}{5}(x - 5) + 11$

In slope-intercept form: [________]

35. $y = -\dfrac{7}{3}(x + 3) + 6$

In slope-intercept form:

36. $y = -\dfrac{8}{7}(x + 21) + 29$

In slope-intercept form:

Point-Slope Form and Graphs Determine the point-slope form of the linear equation from its graph.

37. 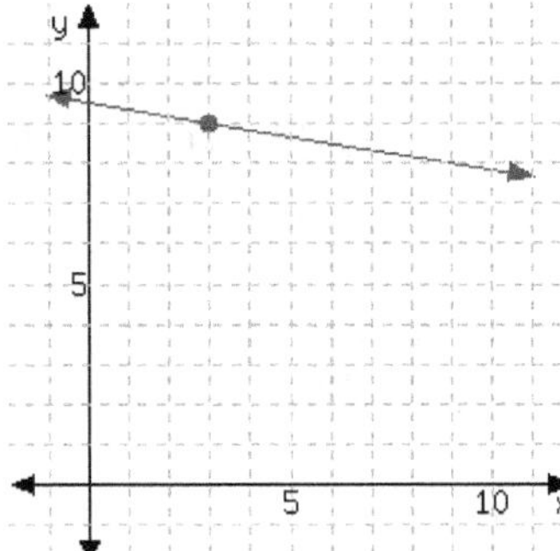

38. 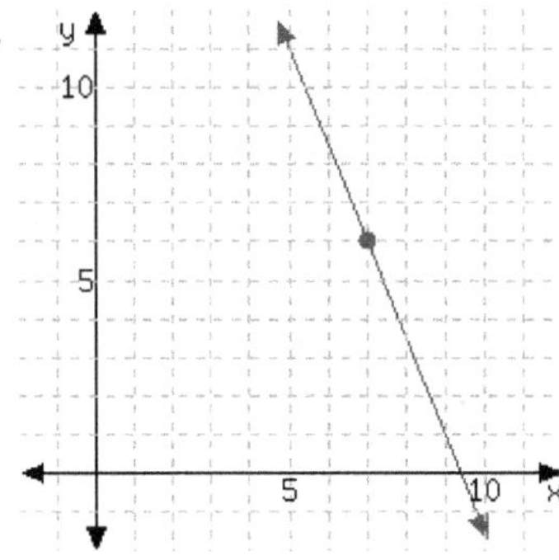

39. 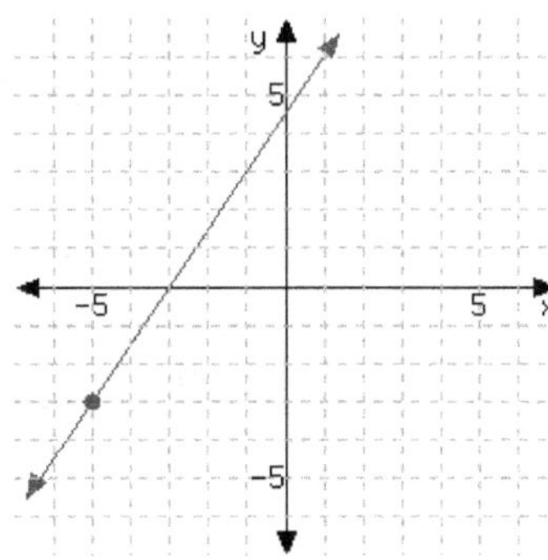

40. 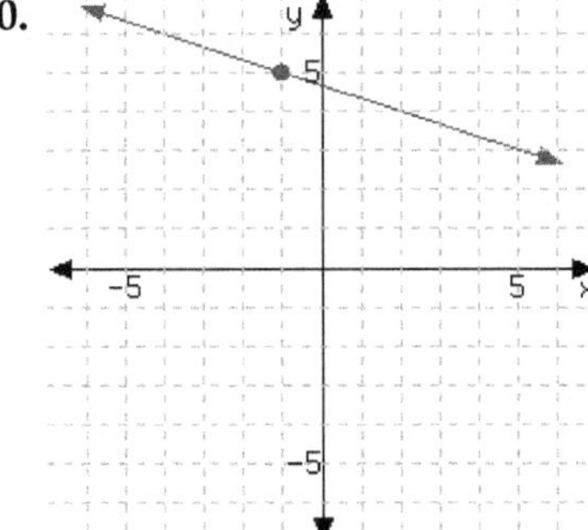

41. 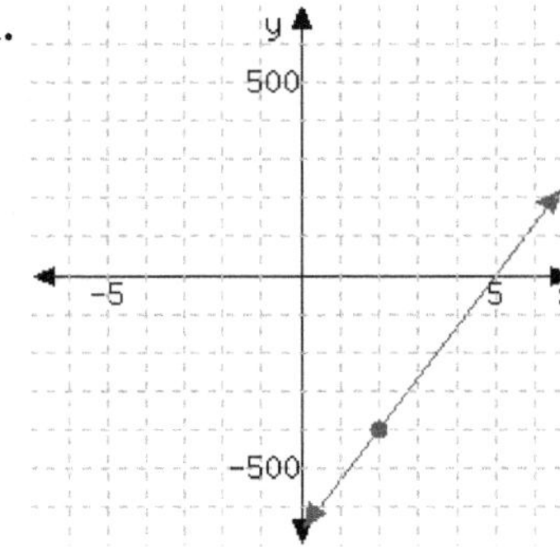

42. 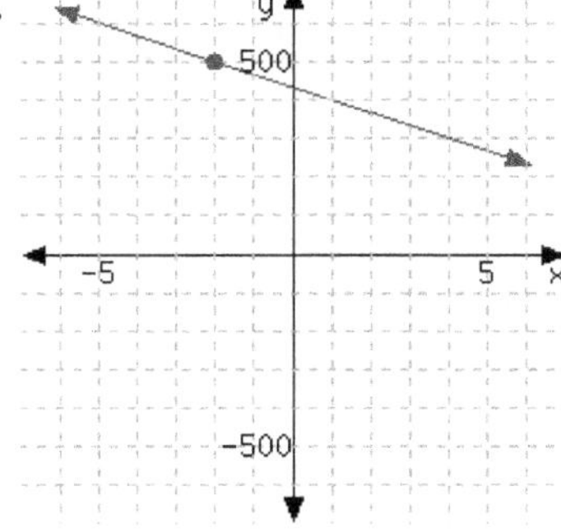

43. 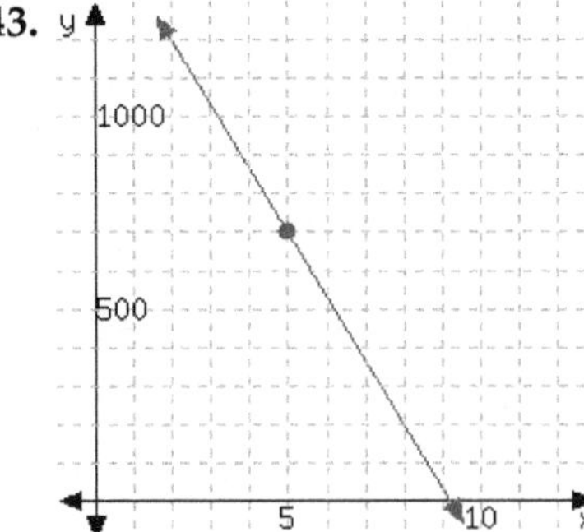

44. 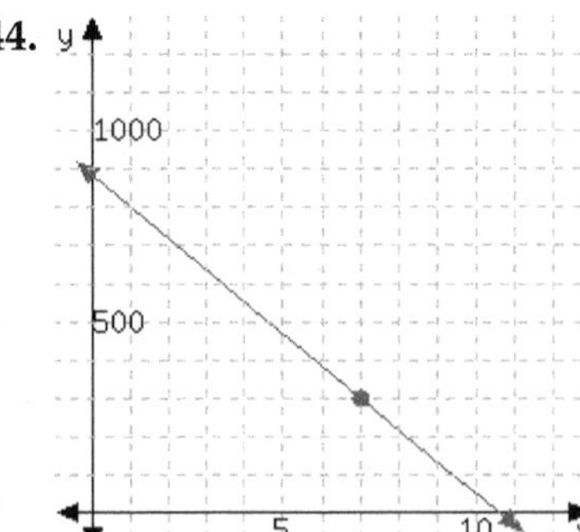

45. 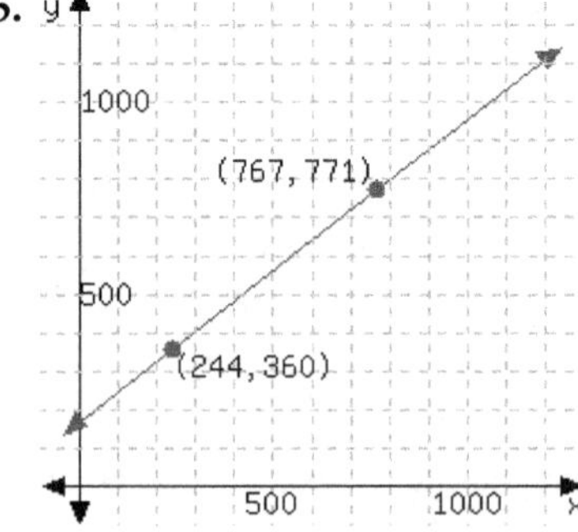

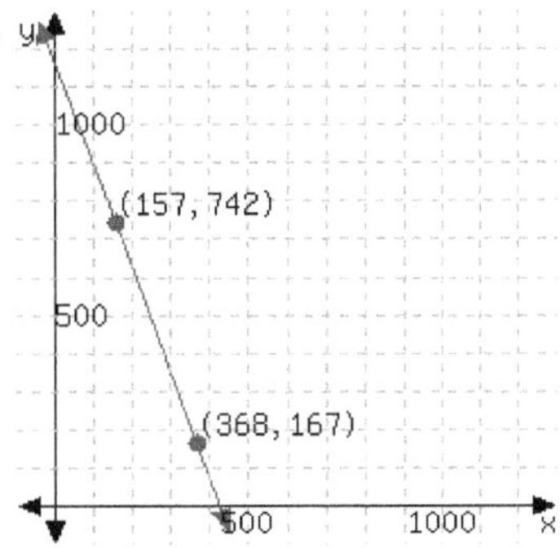

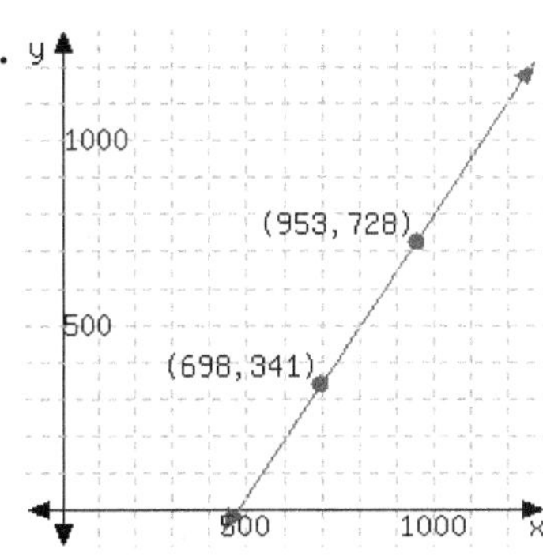

46. y (1029, 978) (462, 756) 1000 500 500 1000 x

47. y (157, 742) (368, 167) 1000 500 500 1000 x

48. y (953, 728) (698, 341) 1000 500 500 1000 x

49. Graph the linear equation $y = -\frac{8}{3}(x - 4) - 5$ by identifying the slope and one point on this line.

50. Graph the linear equation $y = \frac{5}{7}(x + 3) + 2$ by identifying the slope and one point on this line.

51. Graph the linear equation $y = \frac{3}{4}(x + 2) + 1$ by identifying the slope and one point on this line.

52. Graph the linear equation $y = -\frac{5}{2}(x - 1) - 5$ by identifying the slope and one point on this line.

53. Graph the linear equation $y = -3(x - 9) + 4$ by identifying the slope and one point on this line.

54. Graph the linear equation $y = 7(x + 3) - 10$ by identifying the slope and one point on this line.

55. Graph the linear equation $y = 8(x + 12) - 20$ by identifying the slope and one point on this line.

56. Graph the linear equation $y = -5(x - 20) - 70$ by identifying the slope and one point on this line.

Applications

57. By your cell phone contract, you pay a monthly fee plus \$0.04 for each minute you spend on the phone. In one month, you spent 230 minutes over the phone, and had a bill totaling \$22.20.

Let x be the number of minutes you spend on the phone in a month, and let y be your total cell phone bill for that month, in dollars. Use a linear equation to model your monthly bill based on the number of minutes you spend on the phone.

a. A point-slope equation to model this is ____________ .

b. If you spend 160 minutes on the phone in a month, you would be billed ____________ .

c. If your bill was \$30.60 one month, you must have spent ____________ minutes on the phone in that month.

58. A company set aside a certain amount of money in the year 2000. The company spent exactly $31,000 from that fund each year on perks for its employees. In 2003, there was still $872,000 left in the fund.

Let x be the number of years since 2000, and let y be the amount of money, in dollars, left in the fund that year. Use a linear equation to model the amount of money left in the fund after so many years.

    a. A point-slope equation to model this is ⬚.

    b. In the year 2010, there was ⬚ left in the fund.

    c. In the year ⬚, the fund will be empty.

59. A biologist has been observing a tree's height. This type of tree typically grows by 0.19 feet each month. Ten months into the observation, the tree was 17.4 feet tall.

Let x be the number of months passed since the observations started, and let y be the tree's height at that time, in feet. Use a linear equation to model the tree's height as the number of months pass.

    a. A point-slope equation to model this is ⬚.

    b. 28 months after the observations started, the tree would be ⬚ feet in height.

    c. ⬚ months after the observation started, the tree would be 25.76 feet tall.

60. Scientists are conducting an experiment with a gas in a sealed container. The mass of the gas is measured, and the scientists realize that the gas is leaking over time in a linear way. Each minute, they lose 3 grams. Seven minutes since the experiment started, the remaining gas had a mass of 117 grams.

Let x be the number of minutes that have passed since the experiment started, and let y be the mass of the gas in grams at that moment. Use a linear equation to model the weight of the gas over time.

    a. A point-slope equation to model this is ⬚.

    b. 35 minutes after the experiment started, there would be ⬚ grams of gas left.

    c. If a linear model continues to be accurate, ⬚ minutes since the experiment started, all gas in the container will be gone.

61. A company set aside a certain amount of money in the year 2000. The company spent exactly the same amount from that fund each year on perks for its employees. In 2004, there was still $783,000 left in the fund. In 2006, there was $701,000 left.

    Let x be the number of years since 2000, and let y be the amount of money, in dollars, left in the fund that year. Use a linear equation to model the amount of money left in the fund after so many years.

    a. A point-slope equation to model this is ______________.

    b. In the year 2010, there was ______________ left in the fund.

    c. In the year ______________, the fund will be empty.

62. By your cell phone contract, you pay a monthly fee plus some money for each minute you use the phone during the month. In one month, you spent 230 minutes on the phone, and paid $27.65. In another month, you spent 320 minutes on the phone, and paid $32.60.

    Let x be the number of minutes you talk over the phone in a month, and let y be your cell phone bill, in dollars, for that month. Use a linear equation to model your monthly bill based on the number of minutes you talk over the phone.

    a. A point-slope equation to model this is ______________.

    b. If you spent 150 minutes over the phone in a month, you would pay ______________.

    c. If in a month, you paid $41.95 of cell phone bill, you must have spent ______________ minutes on the phone in that month.

63. Scientists are conducting an experiment with a gas in a sealed container. The mass of the gas is measured, and the scientists realize that the gas is leaking over time in a linear way.

    Nine minutes since the experiment started, the gas had a mass of 106.6 grams.

    Eighteen minutes since the experiment started, the gas had a mass of 83.2 grams.

    Let x be the number of minutes that have passed since the experiment started, and let y be the mass of the gas in grams at that moment. Use a linear equation to model the weight of the gas over time.

    a. A point-slope equation to model this is ______________.

    b. 35 minutes after the experiment started, there would be ______________ grams of gas left.

    c. If a linear model continues to be accurate, ______________ minutes since the experiment started, all gas in the container will be gone.

64. A biologist has been observing a tree's height. 10 months into the observation, the tree was 18.6 feet tall. 18 months into the observation, the tree was 19.4 feet tall.

Let x be the number of months passed since the observations started, and let y be the tree's height at that time, in feet. Use a linear equation to model the tree's height as the number of months pass.

    a. A point-slope equation to model this is ⬚.

    b. 27 months after the observations started, the tree would be ⬚ feet in height.

    c. ⬚ months after the observation started, the tree would be 23.5 feet tall.

4.7 Standard Form

We've seen that a linear relationship can be expressed with an equation in slope-intercept form (4.5.1) or with an equation in point-slope-form (4.6.1). There is a third standard form that you can use to write line equations. It's so "standard" that it's actually known as **standard form**.

4.7.1 Standard Form Definition

Imagine trying to gather donations to pay for a $10,000 medical procedure you cannot afford. Oversimplifying the mathematics a bit, suppose that there were only two types of donors in the world: those who will donate $20 and those who will donate $100. How many of each, or what combination, do you need to reach the funding goal? As in, if x people donate $20 and y people donate $100, what numbers could x and y be? The donors of the first type have collectively donated $20x$ dollars, and the donors of the second type have collectively donated $100y$. So altogether you'd need

$$20x + 100y = 10000.$$

This is an example of a line equation in **standard form**.

Definition 4.7.2 Standard Form. It is always possible to write an equation for a line in the form

$$Ax + By = C \tag{4.7.1}$$

where A, B, and C are three numbers (each of which might be 0, although at least one of A and B must be nonzero). This form of a line equation is called **standard form**. In the context of an application, the meaning of A, B, and C depends on that context. This equation is called **standard form** perhaps because *any* line can be written this way, even vertical lines which cannot be written using the two previous forms we've studied.

Checkpoint 4.7.3. For each of the following equations, identify what form they are in.

$2.7x + 3.4y = -82$ ($\square$ slope-intercept $\square$ point-slope $\square$ standard $\square$ other linear $\square$ not linear)

$y = \frac{2}{7}(x - 3) + \frac{1}{10}$ ($\square$ slope-intercept $\square$ point-slope $\square$ standard $\square$ other linear $\square$ not linear)

$12x - 3 = y + 2$ ($\square$ slope-intercept $\square$ point-slope $\square$ standard $\square$ other linear $\square$ not linear)

$y = x^2 + 5$ ($\square$ slope-intercept $\square$ point-slope $\square$ standard $\square$ other linear $\square$ not linear)

$x - y = 10$ ($\square$ slope-intercept $\square$ point-slope $\square$ standard $\square$ other linear $\square$ not linear)

$y = 4x + 1$ ($\square$ slope-intercept $\square$ point-slope $\square$ standard $\square$ other linear $\square$ not linear)

Explanation. $2.7x + 3.4y = -82$ is in standard form, with $A = 2.7$, $B = 3.4$, and $C = -82$.

$y = \frac{2}{7}(x - 3) + \frac{1}{10}$ is in point-slope form, with slope $\frac{2}{7}$, and passing through $\left(3, \frac{1}{10}\right)$.

$12x - 3 = y + 2$ is linear, but not in any of the forms we have studied. Using algebra, you can rearrange it to read $y = 12x - 5$.

$y = x^2 + 5$ is not linear. The exponent on x is a dead giveaway.

$x - y = 10$ is in standard form, with $A = 1$, $B = -1$, and $C = 10$.

$y = 4x + 1$ is in slope-intercept form, with slope 4 and y-intercept at $(0, 1)$.

Returning to the example with donations for the medical procedure, let's examine the equation

$$20x + 100y = 10000.$$

What units are attached to all of the parts of this equation? Both x and y are numbers of people. The 10000 is in dollars. Both the 20 and the 100 are in dollars per person. Note how both sides of the equation are in dollars. On the right, that fact is clear. On the left, $20x$ is in dollars since 20 is in dollars per person, and x is in people. The same is true for $100y$, and the two dollar amounts $20x$ and $100y$ add to a dollar amount.

What is the slope of the linear relationship? It's not immediately visible since m is not part of the standard form equation. But we can use algebra to isolate y:

$$20x + 100y = 10000$$
$$100y = -20x + 10000$$
$$y = \frac{-20x + 10000}{100}$$
$$y = \frac{-20x}{100} + \frac{10000}{100}$$
$$y = -\frac{1}{5}x + 100.$$

And we see that the slope is $-\frac{1}{5}$. OK, what units are on that slope? As always, the units on slope are $\frac{y\text{-unit}}{x\text{-unit}}$. In this case that's $\frac{\text{person}}{\text{person}}$, which sounds a little weird and seems like it should be simplified away to unitless. But this slope of $-\frac{1}{5}\frac{\text{person}}{\text{person}}$ is saying that for every one extra person who donates \$20, you only need $\frac{1}{5}$ fewer people donating \$100 to still reach your goal.

What is the y-intercept? Since we've already converted the equation into slope-intercept form, we can see that it is at $(0, 100)$. This tells us that if 0 people donate \$20, then you will need 100 people to each donate \$100.

What does a graph for this line look like? We've already converted into slope-intercept form, and we could use that to make the graph. But when given a line in standard form, there is another approach that is often used. Returning to

$$20x + 100y = 10000,$$

let's calculate the y-intercept and the x-intercept. Recall that these are *points* where the line crosses the y-axis and x-axis. To be on the y-axis means that $x = 0$, and to be on the x-axis means that $y = 0$. All these zeros make the resulting algebra easy to solve:

$$20x + 100y = 10000 \qquad\qquad 20x + 100y = 10000$$
$$20(0) + 100y = 10000 \qquad\qquad 20x + 100(0) = 10000$$
$$100y = 10000 \qquad\qquad 20x = 10000$$
$$y = \frac{10000}{100} \qquad\qquad x = \frac{10000}{20}$$
$$y = 100 \qquad\qquad x = 500$$

So we have a y-intercept at $(0, 100)$ and an x-intercept at $(500, 0)$. If we plot these, we get to mark especially relevant points given the context, and then drawing a straight line between them gives us Figure 4.7.4.

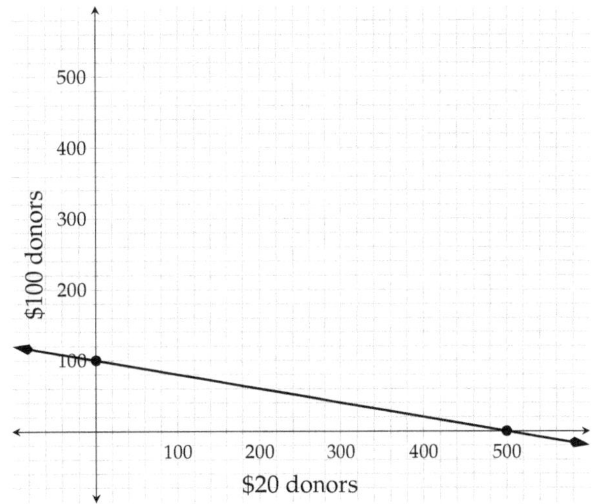

Figure 4.7.4

4.7.2 The x- and y-Intercepts

With a linear relationship (and other types of equations), we are often interested in the x-intercept and y-intercept because they are important in the context. For example, in Figure 4.7.4, the x-intercept implies that if *no one* donates \$100, you need 500 people to donate \$20 to get us to \$10,000. And the y-intercept implies if *no one* donates \$20, you need 100 people to donate \$100. Let's look at another example.

Example 4.7.5 James owns a restaurant that uses about 32 lb of flour every day. He just purchased 1200 lb of flour. Model the amount of flour that remains x days later with a linear equation, and interpret the meaning of its x-intercept and y-intercept.

Since the rate of change is constant (-32 lb every day), and we know the initial value, we can model the amount of flour at the restaurant with a slope-intercept equation (4.5.1):

$$y = -32x + 1200$$

where x represents the number of days passed since the initial purchase, and y represents the amount of flour left (in lb.)

A line's x-intercept is in the form of $(x, 0)$, since to be on the x-axis, the y-coordinate must be 0. To find this line's x-intercept, we substitute y in the equation with 0, and solve for x:

$$y = -32x + 1200$$
$$0 = -32x + 1200$$
$$0 - 1200 = -32x$$
$$-1200 = -32x$$
$$\frac{-1200}{-32} = x$$
$$37.5 = x$$

So the line's x-intercept is at $(37.5, 0)$. In context this means the flour would last for 37.5 days.

A line's y-intercept is in the form of $(0, y)$. This line equation is already in slope-intercept form, so we can just see that its y-intercept is at $(0, 1200)$. In general though, we would substitute x in the equation with 0, and we have:

$$y = -32x + 1200$$
$$y = -32(0) + 1200$$
$$y = 1200$$

So yes, the line's y-intercept is at $(0, 1200)$. This means that when the flour was purchased, there was $1200\,\text{lb}$ of it. In other words, the y-intercept tells us one of the original pieces of information: in the beginning, James purchased $1200\,\text{lb}$ of flour.

If a line is in standard form, it's often easiest to graph it using its two intercepts.

Example 4.7.6 Graph $2x - 3y = -6$ using its intercepts. And then use the intercepts to calculate the line's slope.

Explanation. To graph a line by its x-intercept and y-intercept, it might help to first set up a table like Table 4.7.7:

	x-value	y-value	Intercepts
x-intercept		0	
y-intercept	0		

Table 4.7.7: Intercepts of $2x - 3y = -6$

A table like this might help you stay focused on the fact that we are searching for *two* points. As we've noted earlier, an x-intercept is on the x-axis, and so its y-coordinate must be 0. This is worth taking special note of: to find an x-intercept, y must be 0. This is why we put 0 in the y-value cell of the x-intercept. Similarly, a line's y-intercept has $x = 0$, and we put 0 into the x-value cell of the y-intercept.

Next, we calculate the line's x-intercept by substituting $y = 0$ into the equation

$$2x - 3y = -6$$
$$2x - 3(0) = -6$$
$$2x = -6$$
$$x = -3$$

So the line's x-intercept is $(-3, 0)$.

Similarly, we substitute $x = 0$ into the equation to calculate the y-intercept:

$$2x - 3y = -6$$
$$2(0) - 3y = -6$$
$$-3y = -6$$
$$y = 2$$

So the line's y-intercept is $(0, 2)$.

Now we can complete the table:

	x-value	y-value	Intercepts
x-intercept	-3	0	$(-3,0)$
y-intercept	0	2	$(0,2)$

Table 4.7.8: Intercepts of $2x - 3y = -6$

With both intercepts' coordinates, we can graph the line:

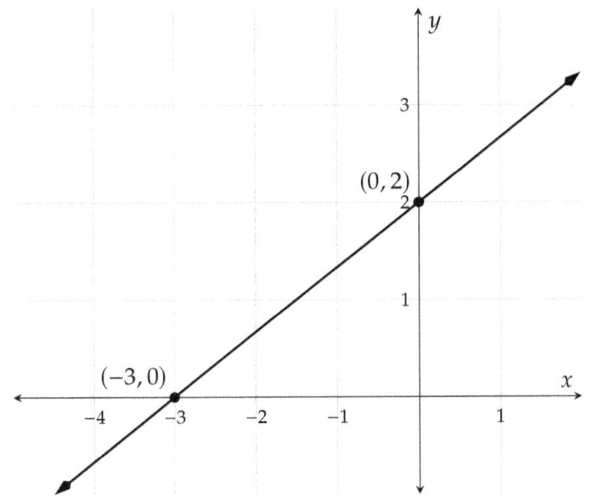

There is a slope triangle from the x-intercept to the origin up to the y-intercept. It tells us that the slope is

$$m = \frac{\Delta y}{\Delta x} = \frac{2}{3}.$$

Figure 4.7.9: Graph of $2x - 3y = -6$

This last example generalizes to a fact worth noting.

Fact 4.7.10. *If a line's x-intercept is at $(r, 0)$ and its y-intercept is at $(0, b)$, then the slope of the line is $-\frac{b}{r}$. (Unless the line passes through the origin. Then both r and b equal 0, and then this fraction is undefined. And the slope of the line could be anything.)*

Checkpoint 4.7.11. Consider the line with equation $2x + 4.3y = \frac{1000}{99}$.

a. What is its x-intercept?

b. What is its y-intercept?

c. What is its slope?

Explanation.

a. To find the x-intercept:

$$2x + 4.3y = \frac{1000}{99}$$
$$2x + 4.3(0) = \frac{1000}{99}$$
$$2x = \frac{1000}{99}$$
$$x = \frac{500}{99}$$

So the x-intercept is at $\left(\frac{500}{99}, 0\right)$.

b. To find the y-intercept:

$$2x + 4.3y = \frac{1000}{99}$$
$$2(0) + 4.3y = \frac{1000}{99}$$
$$4.3y = \frac{1000}{99}$$
$$y = \frac{1}{4.3} \cdot \frac{1000}{99}$$
$$y \approx 2.349\ldots$$

So the y-intercept is at about $(0, 2.349)$.

c. Since we have the x- and y-intercepts, we can calulate the slope:

$$m \approx -\frac{2.349}{\frac{500}{99}} = -\frac{2.349 \cdot 99}{500} \approx -0.4561.$$

4.7.3 Transforming between Standard Form and Slope-Intercept Form

Sometimes a linear equation arises in standard form (4.7.1), but it would be useful to see that equation in slope-intercept form (4.5.1). Or perhaps, vice versa.

A linear equation in slope-intercept form (4.5.1) tells us important information about the line: its slope m and y-intercept $(0, b)$. However, a line's standard form does not show those two important values. As a result, we often need to change a line's equation from standard form to slope-intercept form. Let's look at some examples.

Example 4.7.12 Change $2x - 3y = -6$ to slope-intercept form, and then graph it.

Explanation. Since a line in slope-intercept form looks like $y = \ldots$, we will solve for y in $2x - 3y = -6$:

$$2x - 3y = -6$$
$$-3y = -6 - 2x$$
$$-3y = -2x - 6$$
$$y = \frac{-2x - 6}{-3}$$
$$y = \frac{-2x}{-3} - \frac{6}{-3}$$
$$y = \frac{2}{3}x + 2$$

In the third line, we wrote $-2x - 6$ on the right side, instead of $-6 - 2x$. The only reason we did this is because we are headed to slope-intercept form, where the x-term is traditionally written first.

Now we can see that the slope is $\frac{2}{3}$ and the y-intercept is at $(0, 2)$. With these things found, we can graph the line using slope triangles.

Compare this graphing method with the Graphing by Intercepts method in Example 4.7.6. We have more points in this graph, thus we can graph the line more accurately.

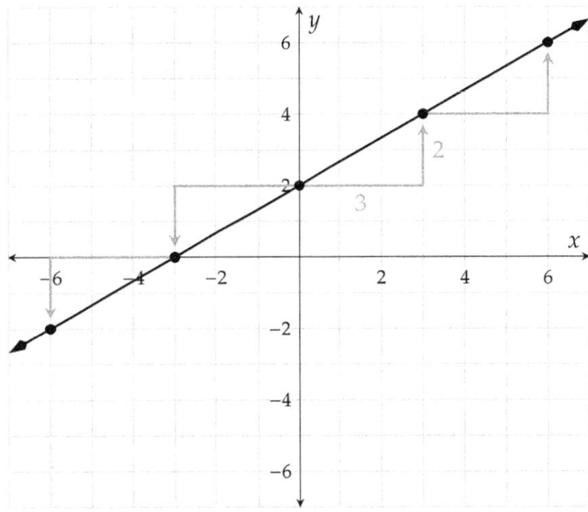

Figure 4.7.13: Graphing $2x - 3y = -6$ with Slope Triangles

Example 4.7.14 Graph $2x - 3y = 0$.

Explanation. First, we will try (and fail) to graph this line using its x- and y-intercepts.

Trying to find the x-intercept:

$$\begin{aligned}
2x - 3y &= 0 \\
2x - 3(0) &= 0 \\
2x &= 0 \\
x &= 0
\end{aligned}$$

So the line's x-intercept is at $(0, 0)$, at the origin.

Huh, that is *also* on the y-axis…

Trying to find the y-intercept:

$$\begin{aligned}
2x - 3y &= 0 \\
2(0) - 3y &= 0 \\
-3y &= 0 \\
y &= 0
\end{aligned}$$

So the line's y-intercept is also at $(0, 0)$.

Since both intercepts are the same point, there is no way to use the intercepts alone to graph this line. So what can be done?

Several approaches are out there, but one is to convert the line equation into slope-intercept form:

$$2x - 3y = 0$$
$$-3y = 0 - 2x$$
$$-3y = -2x$$
$$y = \frac{-2x}{-3}$$
$$y = \frac{2}{3}x$$

So the line's slope is $\frac{2}{3}$, and we can graph the line using slope triangles and the intercept at $(0,0)$, as in Figure 4.7.15.

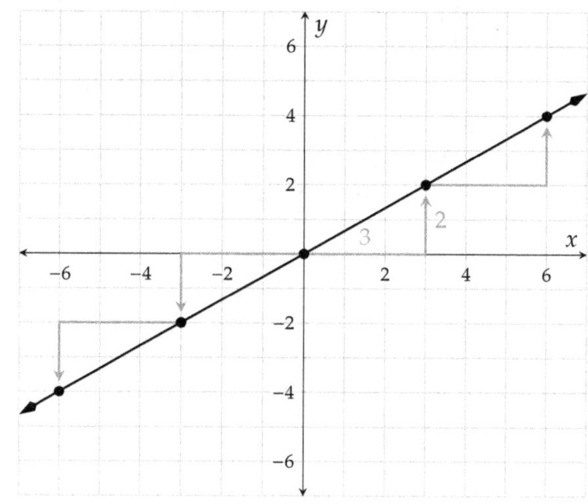

Figure 4.7.15: Graphing $2x - 3y = 0$ with Slope Triangles

In summary, if $C = 0$ in a standard form equation (4.7.1), it's convenient to graph it by first converting the equation to slope-intercept form (4.5.1).

Example 4.7.16 Write the equation $y = \frac{2}{3}x + 2$ in standard form.

Explanation. Once we subtract $\frac{2}{3}x$ on both sides of the equation, we have

$$-\frac{2}{3}x + y = 2$$

Technically, this equation is already in standard form $Ax + By = C$. However, you might like to end up with an equation that has no fractions, and so you can take some extra steps.

$$y = \frac{2}{3}x + 2$$
$$y - \frac{2}{3}x = 2$$
$$-\frac{2}{3}x + y = 2$$

This is in standard form, but we keep going to clear away the fraction.

$$3 \cdot \left(-\frac{2}{3}x + y\right) = 3 \cdot 2$$
$$-2x + 3y = 6$$

Exercises

Review and Warmup   Solve the linear equation for y.

1.
$$3y - 6x = 27$$

2.
$$-18x - 3y = 6$$

3.
$$-x - y = 16$$

4.
$$-4x - y = -9$$

5.
$$2x - 8y = -5$$

6.
$$-5x - 9y = -5$$

Slope and y-intercept   Find the line's slope and y-intercept.

7. A line has equation $-2x + y = 4$.

This line's slope is [   ].

This line's y-intercept is [   ].

8. A line has equation $-x - y = -8$.

This line's slope is [   ].

This line's y-intercept is [   ].

9. A line has equation $2x + 2y = 4$.

This line's slope is [   ].

This line's y-intercept is [   ].

10. A line has equation $12x - 3y = -9$.

This line's slope is [   ].

This line's y-intercept is [   ].

11. A line has equation $x + 3y = -6$.

This line's slope is [   ].

This line's y-intercept is [   ].

12. A line has equation $7x + 6y = -24$.

This line's slope is [   ].

This line's y-intercept is [   ].

13. A line has equation $7x - 6y = 24$.

This line's slope is [   ].

This line's y-intercept is [   ].

14. A line has equation $20x + 12y = -36$.

This line's slope is [   ].

This line's y-intercept is [   ].

15. A line has equation $12x - 10y = 0$.

This line's slope is [   ].

This line's y-intercept is [   ].

16. A line has equation $3x - 12y = 0$.

This line's slope is [   ].

This line's y-intercept is [   ].

17. A line has equation $2x + 6y = 5$.

This line's slope is [   ].

This line's y-intercept is [   ].

18. A line has equation $8x + 12y = 5$.

This line's slope is [   ].

This line's y-intercept is [   ].

Converting to Standard Form

19. Rewrite $y = 4x + 7$ in standard form.

20. Rewrite $y = 5x - 6$ in standard form.

21. Rewrite $y = \frac{6}{7}x - 6$ in standard form.

22. Rewrite $y = -\frac{7}{5}x - 7$ in standard form.

Graphs and Standard Form

23. Find the y-intercept and x-intercept of the line given by the equation. If a particular intercept does not exist, enter none into all the answer blanks for that row.

$$7x + 2y = 28$$

	x-value	y-value	Location
y-intercept	_______	_______	_______________
x-intercept	_______	_______	_______________

24. Find the y-intercept and x-intercept of the line given by the equation. If a particular intercept does not exist, enter none into all the answer blanks for that row.

$$8x + 7y = -168$$

	x-value	y-value	Location
y-intercept	_______	_______	_______________
x-intercept	_______	_______	_______________

25. Find the y-intercept and x-intercept of the line given by the equation. If a particular intercept does not exist, enter none into all the answer blanks for that row.

$$2x - 5y = -20$$

	x-value	y-value	Location
y-intercept	_______	_______	_______________
x-intercept	_______	_______	_______________

26. Find the y-intercept and x-intercept of the line given by the equation. If a particular intercept does not exist, enter none into all the answer blanks for that row.

$$x - 3y = -3$$

	x-value	y-value	Location
y-intercept	_______	_______	_______________
x-intercept	_______	_______	_______________

27. Find the x- and y-intercepts of the line with equation $4x + 6y = 24$. Then find one other point on the line. Use your results to graph the line.

28. Find the x- and y-intercepts of the line with equation $4x + 5y = -40$. Then find one other point on the line. Use your results to graph the line.

29. Find the x- and y-intercepts of the line with equation $5x - 2y = 10$. Then find one other point on the line. Use your results to graph the line.

30. Find the x- and y-intercepts of the line with equation $5x - 6y = -90$. Then find one other point on the line. Use your results to graph the line.

31. Find the x- and y-intercepts of the line with equation $x + 5y = -15$. Then find one other point on the line. Use your results to graph the line.

32. Find the x- and y-intercepts of the line with equation $6x + y = -18$. Then find one other point on the line. Use your results to graph the line.

33. Make a graph of the line $x + y = 2$.

34. Make a graph of the line $-5x - y = -3$.

35. Make a graph of the line $x + 5y = 5$.

36. Make a graph of the line $x - 2y = 2$.

37. Make a graph of the line $20x - 4y = 8$.

38. Make a graph of the line $3x + 5y = 10$.

39. Make a graph of the line $-3x + 2y = 6$.

40. Make a graph of the line $-4x - 5y = 10$.

41. Make a graph of the line $4x - 5y = 0$.

42. Make a graph of the line $5x + 7y = 0$.

Interpreting Intercepts in Context

43. Scot is buying some tea bags and some sugar bags. Each tea bag costs 6 cents, and each sugar bag costs 2 cents. He can spend a total of $1.80.

Assume Scot will purchase x tea bags and y sugar bags. Use a linear equation to model the number of tea bags and sugar bags he can purchase.

Find this line's x-intercept, and interpret its meaning in this context.

- ⊙ A. The x-intercept is $(0, 90)$. It implies Scot can purchase 90 sugar bags with no tea bags.
- ⊙ B. The x-intercept is $(30,0)$. It implies Scot can purchase 30 tea bags with no sugar bags.
- ⊙ C. The x-intercept is $(90,0)$. It implies Scot can purchase 90 tea bags with no sugar bags.
- ⊙ D. The x-intercept is $(0,30)$. It implies Scot can purchase 30 sugar bags with no tea bags.

44. Douglas is buying some tea bags and some sugar bags. Each tea bag costs 2 cents, and each sugar bag costs 9 cents. He can spend a total of $0.90.

Assume Douglas will purchase x tea bags and y sugar bags. Use a linear equation to model the number of tea bags and sugar bags he can purchase.

Find this line's y-intercept, and interpret its meaning in this context.

- ⊙ A. The y-intercept is $(0,45)$. It implies Douglas can purchase 45 sugar bags with no tea bags.
- ⊙ B. The y-intercept is $(45,0)$. It implies Douglas can purchase 45 tea bags with no sugar bags.
- ⊙ C. The y-intercept is $(10,0)$. It implies Douglas can purchase 10 tea bags with no sugar bags.
- ⊙ D. The y-intercept is $(0, 10)$. It implies Douglas can purchase 10 sugar bags with no tea bags.

45. An engine's tank can hold 70 gallons of gasoline. It was refilled with a full tank, and has been running without breaks, consuming 3.5 gallons of gas per hour.

Assume the engine has been running for x hours since its tank was refilled, and assume there are y gallons of gas left in the tank. Use a linear equation to model the amount of gas in the tank as time passes.

Find this line's x-intercept, and interpret its meaning in this context.

    ⊙ *A.* The x-intercept is (20,0). It implies the engine will run out of gas 20 hours after its tank was refilled.

    ⊙ *B.* The x-intercept is (0,70). It implies the engine started with 70 gallons of gas in its tank.

    ⊙ *C.* The x-intercept is (0,20). It implies the engine started with 20 gallons of gas in its tank.

    ⊙ *D.* The x-intercept is (70,0). It implies the engine will run out of gas 70 hours after its tank was refilled.

46. An engine's tank can hold 120 gallons of gasoline. It was refilled with a full tank, and has been running without breaks, consuming 3 gallons of gas per hour.

Assume the engine has been running for x hours since its tank was refilled, and assume there are y gallons of gas left in the tank. Use a linear equation to model the amount of gas in the tank as time passes.

Find this line's y-intercept, and interpret its meaning in this context.

    ⊙ *A.* The y-intercept is (40,0). It implies the engine will run out of gas 40 hours after its tank was refilled.

    ⊙ *B.* The y-intercept is (120,0). It implies the engine will run out of gas 120 hours after its tank was refilled.

    ⊙ *C.* The y-intercept is (0,120). It implies the engine started with 120 gallons of gas in its tank.

    ⊙ *D.* The y-intercept is (0,40). It implies the engine started with 40 gallons of gas in its tank.

47. A new car of a certain model costs \$43,200.00. According to Blue Book, its value decreases by \$2,400.00 every year.

Assume x years since its purchase, the car's value is y dollars. Use a linear equation to model the car's value.

Find this line's x-intercept, and interpret its meaning in this context.

    ⊙ *A.* The x-intercept is (0,43200). It implies the car's initial value was 43200.

    ⊙ *B.* The x-intercept is (0,18). It implies the car would have no more value 18 years since its purchase.

    ⊙ *C.* The x-intercept is (18,0). It implies the car would have no more value 18 years since its purchase.

    ⊙ *D.* The x-intercept is (43200,0). It implies the car's initial value was 43200.

48. A new car of a certain model costs \$39,000.00. According to Blue Book, its value decreases by \$2,600.00 every year.

Assume x years since its purchase, the car's value is y dollars. Use a linear equation to model the car's value.

Find this line's y-intercept, and interpret its meaning in this context.

⊙ A. The y-intercept is (15,0). It implies the car would have no more value 15 years since its purchase.

⊙ B. The y-intercept is (0,39000). It implies the car's initial value was 39000.

⊙ C. The y-intercept is (0,15). It implies the car would have no more value 15 years since its purchase.

⊙ D. The y-intercept is (39000,0). It implies the car's initial value was 39000.

Challenge

49. Fill in the variables A, B, and C in $Ax + By = C$ with the numbers $10, 11$ and 14. You may only use each number once.

a. To make a line with the steepest slope possible, A must equal [______], B must equal [______], and C must equal [______].

b. To make a line with the shallowest slope possible, A must equal [______], B must equal [______], and C must equal [______].

4.8 Horizontal, Vertical, Parallel, and Perpendicular Lines

Horizontal and vertical lines have some special features worth our attention. Also if a pair of lines are parallel or perpendicular to each other, we have some interesting things to say about them. This section looks at these geometric features that lines may have.

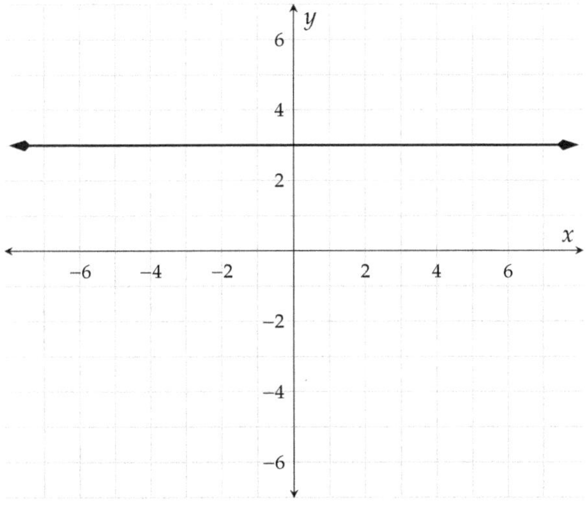

Figure 4.8.2: Horizontal Line

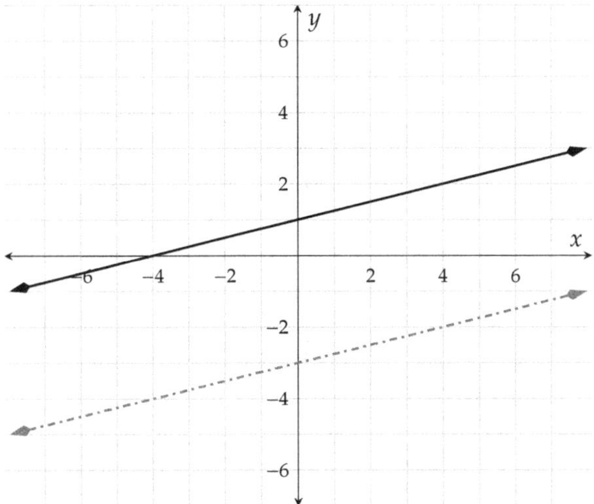

Figure 4.8.4: Parallel Lines

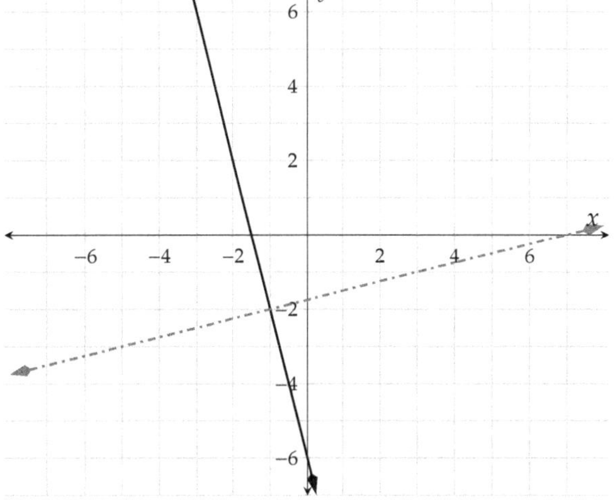

Figure 4.8.5: Perpendicular Lines

4.8.1 Horizontal Lines and Vertical Lines

We learned in Section 4.7 that all lines can be written in standard form (4.7.1). When either A or B equal 0, we end up with a horizontal or vertical line, as we will soon see. Let's take the standard form (4.7.1) line equation, let $A = 0$ and $B = 0$ one at a time and simplify each equation.

$$\begin{aligned} Ax + By &= C \\ 0x + By &= C \\ By &= C \\ y &= \frac{C}{B} \\ y &= k \end{aligned} \qquad\qquad \begin{aligned} Ax + By &= C \\ Ax + 0y &= C \\ Ax &= C \\ x &= \frac{C}{A} \\ x &= h \end{aligned}$$

At the end we just renamed the constant numbers $\frac{C}{B}$ and $\frac{C}{A}$ to k and h because of tradition. What is important, is that you view h and k (as well as A, B, and C) as constants: numbers that have some specific value and don't change in the context of one problem.

Think about just one of these last equations: $y = k$. It says that the y-value is the same no matter where you are on the line. If you wanted to plot points on this line, you are free to move far to the left or far to the right on the x-axis, but then you always move up (or down) to make the y-value equal k. What does such a line look like?

Example 4.8.6 Let's plot the line with equation $y = 3$. (Note that this is the same as $0x + 1y = 3$.)

To plot some points, it doesn't matter what x-values we use. All that matters is that y is *always* 3.

A line like this is **horizontal**, parallel to the horizontal axis. All lines with an equation in the form

$$y = k$$

(or, in standard form, $0x + By = C$) are **horizontal**.

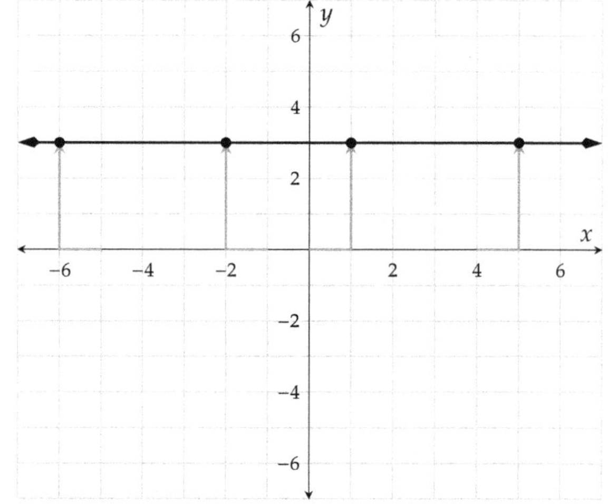

Figure 4.8.7: $y = 3$

Example 4.8.8 Let's plot the line with equation $x = 5$. (Note that this is the same as $1x + 0y = 5$.)

The line has $x = 5$, so to plot points, we are *required* to move over to $x = 5$. From there, we have complete freedom to move however far we like up or down.

A line like this is **vertical**, parallel to the vertical axis. All lines with an equation in the form

$$x = h$$

(or, in standard form, $Ax + 0y = C$) are vertical.

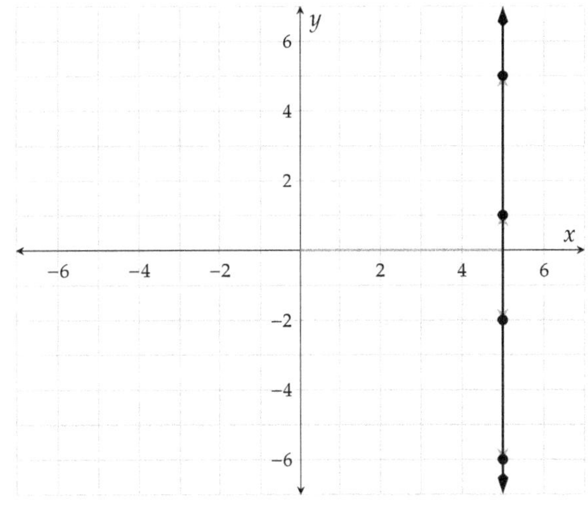

Figure 4.8.9: $x = 5$

Example 4.8.10 Zero Slope. In Checkpoint 4.4.17, we learned that a horizontal line's slope is 0, because the distance doesn't change as time moves on. So the numerator in the slope formula (4.4.3) is 0. Now, if we know a line's slope and its y-intercept, we can use slope-intercept form (4.5.1) to write its equation:

$$y = mx + b$$
$$y = 0x + b$$
$$y = b$$

This provides us with an alternative way to think about equations of horizontal lines. They have a certain y-intercept b, and they have slope 0.

We use horizontal lines to model scenarios where there is no change in y-values, like when Kato stopped for 12 hours (he deserved a rest)!

Checkpoint 4.8.11 Plotting Points. Suppose you need to plot the equation $y = -4.25$. Since the equation is in "$y =$" form, you decide to make a table of points. Fill out some points for this table.

x	y

Explanation. We can use whatever values for x that we like, as long as they are all different. The equation tells us the y-value has to be -4.25 each time.

$$
\begin{array}{cc}
x & y \\
-2 & -4.25 \\
-1 & -4.25 \\
0 & -4.25 \\
1 & -4.25 \\
2 & -4.25
\end{array}
$$

The reason we made a table was to help with plotting the line.

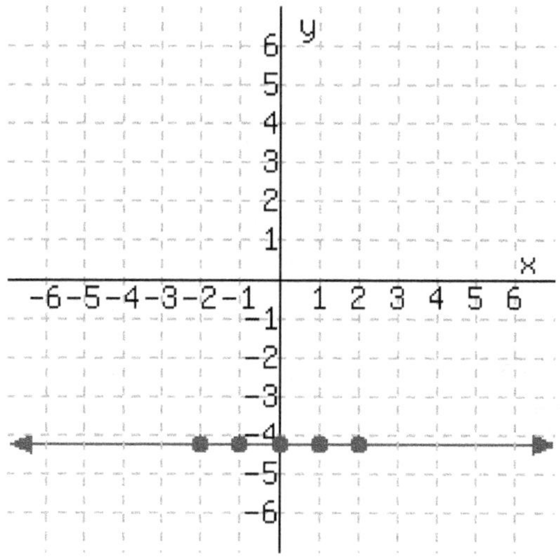

Example 4.8.12 Undefined Slope. What is the slope of a vertical line? Figure 4.8.13 shows three lines passing through the origin, each steeper than the last. In each graph, you can see a slope triangle that uses a "rise" of 4 each time.

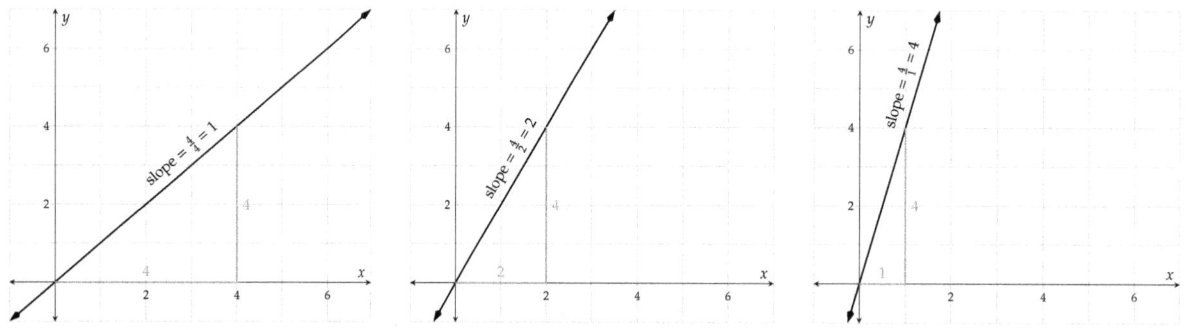

Figure 4.8.13

If we continued making the line steeper and steeper until it was vertical, the slope triangle would still have a "rise" of 4, but the "run" would become smaller and smaller, closer to 0. And then the slope would be $m = \frac{4}{\text{very small}} = $ very large. So the slope of a vertical line can be thought of as "infinitely large."

If we actually try to compute the slope using the slope triangle when the run is 0, we would have $\frac{4}{0}$, which is undefined. So we also say that the slope of a vertical line is *undefined*. Some people say that a vertical line *has no slope*.

Remark 4.8.14. Be careful not to mix up "no slope" (which means "its slope is undefined") with "has slope 0." If a line has slope 0, it *does* have a slope.

Checkpoint 4.8.15 Plotting Points. Suppose you need to plot the equation $x = 3.14$. You decide to try making a table of points. Fill out some points for this table.

x	y

Explanation. Since the equation says x is always the number 3.14, we have to use this for the x value in all the points. This is different from how we would plot a "$y =$" equation, where we would use several different x-values. We can use whatever values for y that we like, as long as they are all different.

x	y
3.14	-2
3.14	-1
3.14	0
3.14	1
3.14	2

The reason we made a table was to help with plotting the line.

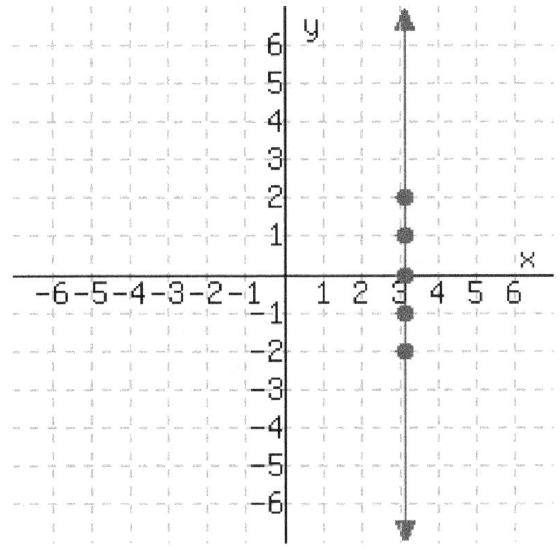

Example 4.8.16 Let x represent the price of a new 60-inch television at Target on Black Friday (which was \$650), and let y be the number of hours you will watch something on this TV over its lifetime. What is the relationship between x and y?

Well, there is no getting around the fact that $x = 650$. As for y, without any extra information about your viewing habits, it could theoretically be as low as 0 or it could be anything larger than that. If we graph this scenario, we have to graph the equation $x = 650$ which we now know to give a vertical line, and we get Figure 4.8.17.

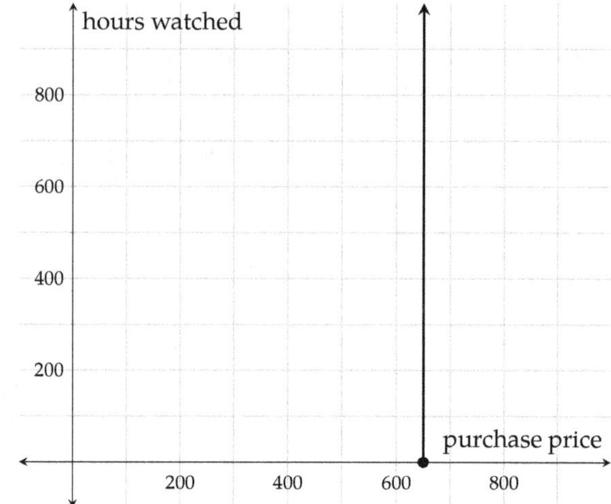

Figure 4.8.17: New TV: hours watched versus purchase price; negative y-values omitted since they make no sense in context

Summary of Horizontal and Vertical Line Equations

Horizontal Lines	Vertical Lines
A line is **horizontal** if and only if its equation can be written $$y = k$$ for some constant k.	A line is **vertical** if and only if its equation can be written $$x = h$$ for some constant h.
In standard form (4.7.1), any line with equation $$0x + By = C$$ is horizontal.	In standard form (4.7.1), any line with equation $$Ax + 0y = C$$ is vertical.
If the line with equation $y = k$ is horizontal, it has a y-intercept at $(0, k)$ and has slope 0.	If the line with equation $x = h$ is vertical, it has an x-intercept at $(h, 0)$ and its slope is *undefined*. Some say it has *no* slope, and some say the slope is *infinitely large*.
In slope-intercept form (4.5.1), any line with equation $$y = 0x + b$$ is horizontal.	It's impossible to write the equation of a vertical line in slope-intercept form (4.5.1), because vertical lines do not have a defined slope.

350

4.8.2 Parallel Lines

Two trees were planted in the same year, and their growth over time is modeled by the two lines in Figure 4.8.19. Use linear equations to model each tree's growth, and interpret their meanings in this context.

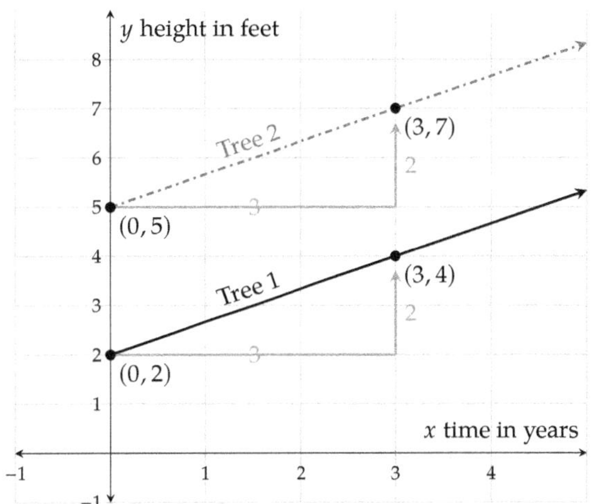

Figure 4.8.19: Two Trees' Growth Chart

Example 4.8.18 We can see Tree 1's equation is $y = \frac{2}{3}x + 2$, and Tree 2's equation is $y = \frac{2}{3}x + 5$. Tree 1 was 2 feet tall when it was planted, and Tree 2 was 5 feet tall when it was planted. Both trees have been growing at the same rate, $\frac{2}{3}$ feet per year, or 2 feet every 3 years.

An important observation right now is that those two lines are parallel. Why? For lines with positive slopes, the bigger a line's slope, the steeper the line is slanted. As a result, if two lines have the same slope, they are slanted at the same angle, thus they are parallel.

Fact 4.8.20. *Any two vertical lines are parallel to each other. For two non-vertical lines, they are parallel if and only if they have the same slope.*

Checkpoint 4.8.21. A line ℓ is parallel to the line with equation $y = 17.2x - 340.9$, but ℓ has y-intercept at $(0, 128.2)$. What is an equation for ℓ?

Explanation. Parallel lines have the same slope, and the slope of $y = 17.2x - 340.9$ is 17.2. So ℓ has slope 17.2. And we have been given that ℓ's y-intercept is at $(0, 128.2)$. So we can use slope-intercept form to write its equation as

$$y = 17.2x + 128.2.$$

Checkpoint 4.8.22. A line κ is parallel to the line with equation $y = -3.5x + 17$, but κ passes through the point $(-12, 23)$. What is an equation for κ?

Explanation. Parallel lines have the same slope, and the slope of $y = -3.5x + 17$ is -3.5. So κ has slope -3.5. And we know a point that κ passes through, so we can use point-slope form to write its equation as

$$y = -3.5(x + 12) + 23.$$

4.8.3 Perpendicular Lines

The slopes of two perpendicular lines have a special relationship too.

Figure 4.8.22 walks you through an explanation of this realationship.

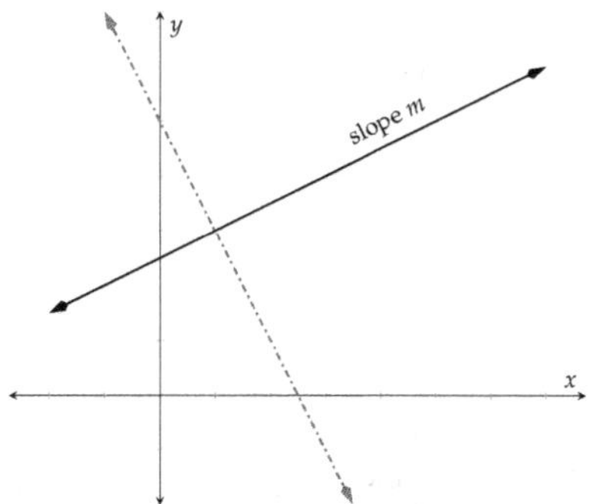

(a) Two generic perpendicular lines, where one has slope m.

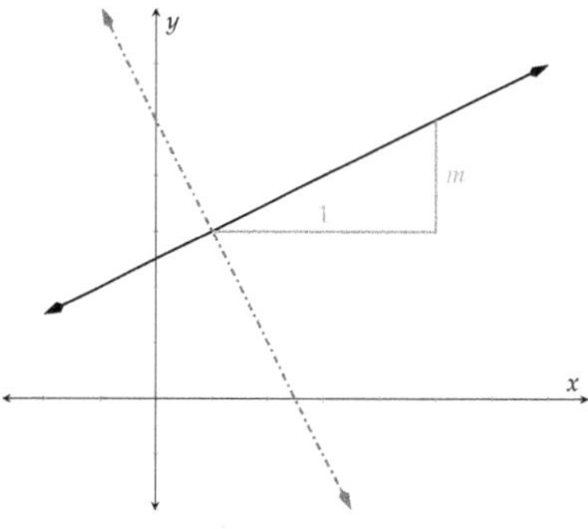

(b) Since the one slope is m, we can draw a slope triangle with "run" 1 and "rise" m.

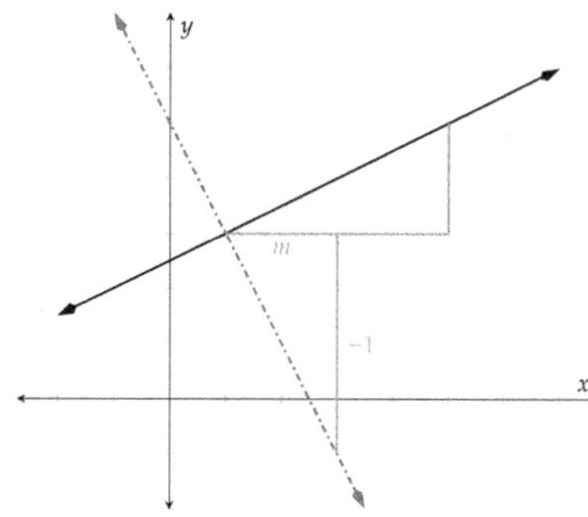

(c) A *congruent* slope triangle can be drawn for the perpendicular line. It's legs have the same lengths, but in different positions, and one is negative.

Figure 4.8.22: The relationship between slopes of perpendicular lines

The second line in Figure 4.8.22 has slope

$$\frac{\Delta y}{\Delta x} = \frac{-1}{m} = -\frac{1}{m}.$$

Fact 4.8.23. *A vertical line and a horizontal line are perpendicular. For lines that are neither vertical nor horizontal, they are perpendicular if and only if the slope of one is the negative reciprocal of the slope of the other. That is, if one has slope m, the other has slope $-\frac{1}{m}$.*

Another way to say this is that the product of the slopes of two perpendicular lines is -1 (assuming both of the lines have a slope in the first place).

Not convinced? Here are three pairs of perpendicular lines where we can see if the pattern holds.

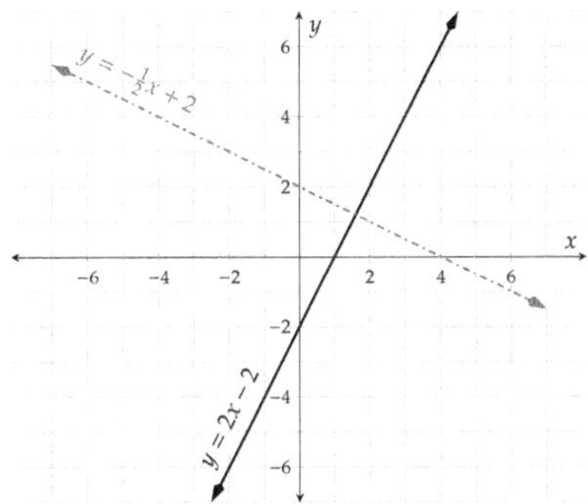

Figure 4.8.24: Graphing $y = 2x - 2$ and $y = -\frac{1}{2}x + 2$. Note the relationship between their slopes: $2 = -\frac{1}{-1/2}$

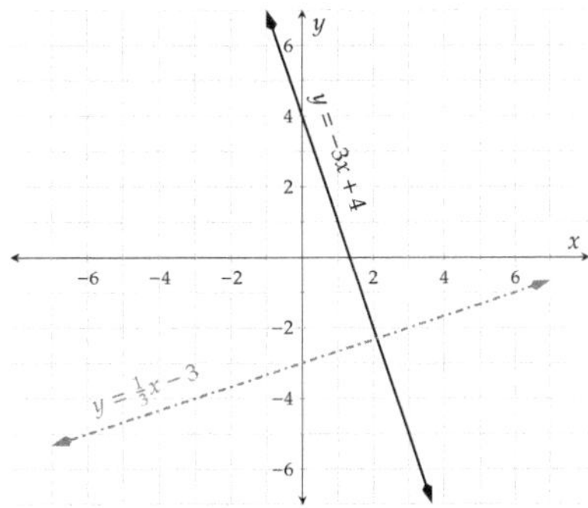

Figure 4.8.25: Graphing $y = -3x + 4$ and $y = \frac{1}{3}x - 3$. Note the relationship between their slopes: $-3 = -\frac{1}{1/3}$

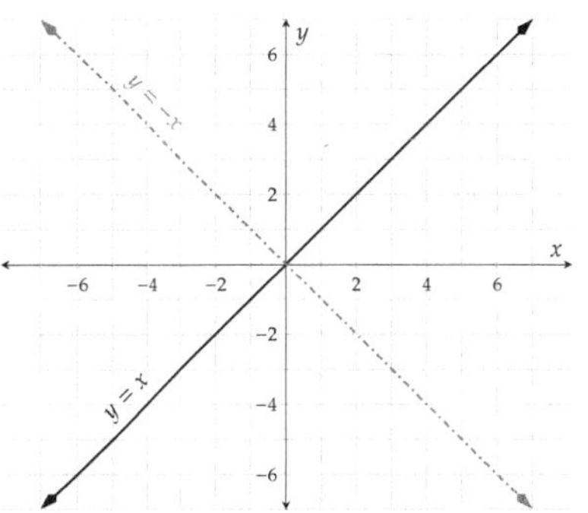

Figure 4.8.26: Graphing $y = x$ and $y = -x$. Note the relationship between their slopes: $1 = -\frac{1}{-1}$

Example 4.8.27 Line A passes through $(-2, 10)$ and $(3, -10)$. Line B passes through $(-4, -4)$ and $(8, -1)$. Determine whether these two lines are parallel, perpendicular or neither.

Explanation. We will use the slope formula to find both lines' slopes:

$$\text{Line } A\text{'s slope} = \frac{y_2 - y_1}{x_2 - x_1} \qquad\qquad \text{Line } B\text{'s slope} = \frac{y_2 - y_1}{x_2 - x_1}$$
$$= \frac{-10 - 10}{3 - (-2)} \qquad\qquad = \frac{-1 - (-4)}{8 - (-4)}$$
$$= \frac{-20}{5} \qquad\qquad = \frac{3}{12}$$
$$= -4 \qquad\qquad = \frac{1}{4}$$

Their slopes are not the same, so those two lines are not parallel.

The product of their slopes is $(-4) \cdot \frac{1}{4} = -1$, which means the two lines are perpendicular.

Checkpoint 4.8.28. Line A and Line B are perpendicular. Line A's equation is $2x + 3y = 12$. Line B passes through the point $(4, -3)$. Find an equation for Line B.

Explanation. First, we will find Line A's slope by rewriting its equation from standard form to slope-intercept form:

$$2x + 3y = 12$$
$$3y = 12 - 2x$$
$$3y = -2x + 12$$
$$y = \frac{-2x + 12}{3}$$
$$y = -\frac{2}{3}x + 4$$

So Line A's slope is $-\frac{2}{3}$. Since Line B is perpendicular to Line A, its slope is $-\frac{1}{-\frac{2}{3}} = \frac{3}{2}$. It's also given that

Line B passes through $(4, -3)$, so we can write Line B's point-slope form equation:

$$y = m(x - x_0) + y_0$$
$$y = \frac{3}{2}(x - 4) - 3$$

Exercises

Review and Warmup

1. Evaluate the following expressions. If the answer is undefined, you may answer with DNE (meaning "does not exist").

   a. $\dfrac{6}{0} = \boxed{}$

   b. $\dfrac{0}{6} = \boxed{}$

2. Evaluate the following expressions. If the answer is undefined, you may answer with DNE (meaning "does not exist").

   a. $\dfrac{0}{7} = \boxed{}$

   b. $\dfrac{7}{0} = \boxed{}$

3. A line passes through the points $(5, 6)$ and $(-5, 6)$. Find this line's slope.

4. A line passes through the points $(3, 8)$ and $(-3, 8)$. Find this line's slope.

5. A line passes through the points $(-10, -5)$ and $(-10, 5)$. Find this line's slope.

6. A line passes through the points $(-8, -1)$ and $(-8, 2)$. Find this line's slope.

7. Consider the equation:

$$y = 1$$

Which of the following ordered pairs are solutions to the given equation? There may be more than one correct answer.

☐ $(0, 9)$ ☐ $(-6, 1)$ ☐ $(1, 4)$
☐ $(4, 1)$

8. Consider the equation:

$$y = 1$$

Which of the following ordered pairs are solutions to the given equation? There may be more than one correct answer.

☐ $(-8, 1)$ ☐ $(0, 7)$ ☐ $(5, 1)$
☐ $(1, 2)$

9. Consider the equation:

$$x + 1 = 0$$

Which of the following ordered pairs are solutions to the given equation? There may be more than one correct answer.

☐ $(-1, 3)$ ☐ $(-1, 0)$
☐ $(1, -1)$ ☐ $(0, -6)$

10. Consider the equation:

$$x + 1 = 0$$

Which of the following ordered pairs are solutions to the given equation? There may be more than one correct answer.

☐ $(-1, 3)$ ☐ $(0, -8)$
☐ $(1, -1)$ ☐ $(-1, 0)$

Tables for Horizontal and Vertical Lines

11. Fill out this table for the equation $y = 7$. The first row is an example.

x	y	Points
−3	7	(−3, 7)
−2		
−1		
0		
1		
2		

12. Fill out this table for the equation $y = 8$. The first row is an example.

x	y	Points
−3	8	(−3, 8)
−2		
−1		
0		
1		
2		

13. Fill out this table for the equation $x = -2$. The first row is an example.

x	y	Points
−2	−3	(−2, −3)
	−2	
	−1	
	0	
	1	
	2	

14. Fill out this table for the equation $x = -10$. The first row is an example.

x	y	Points
−10	−3	(−10, −3)
	−2	
	−1	
	0	
	1	
	2	

Line Equations

15. A line's graph is given.

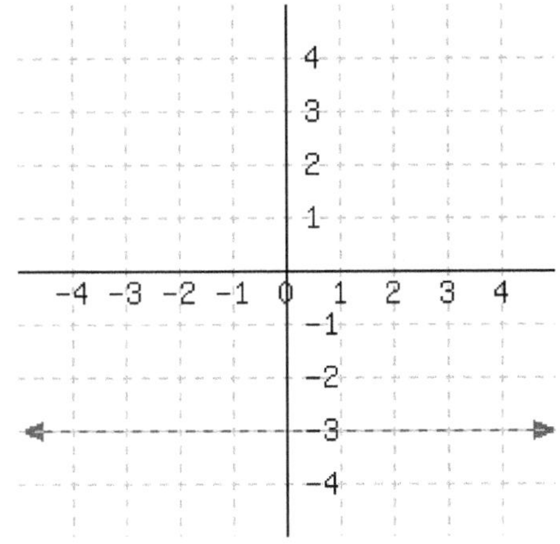

This line's equation is [______]

16. A line's graph is given.

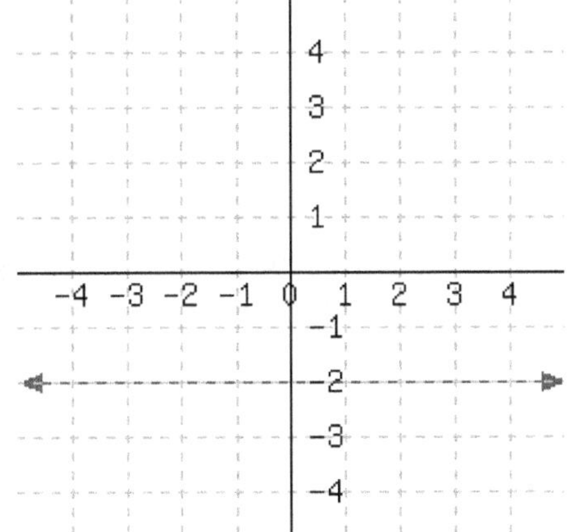

This line's equation is [______]

17. A line's graph is given.

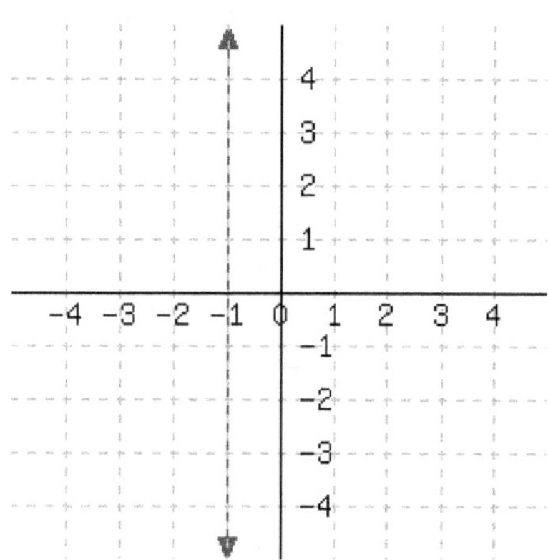

This line's equation is [____________]

18. A line's graph is given.

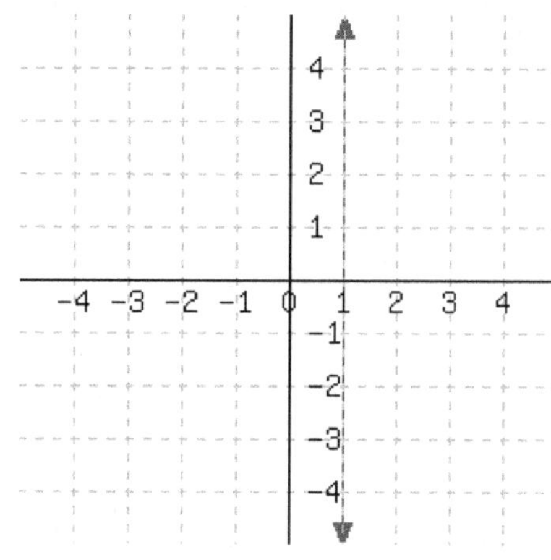

This line's equation is [____________]

19. A line passes through the points $(-2, 1)$ and $(3, 1)$. Find an equation for this line.

An equation for this line is [__________].

20. A line passes through the points $(5, 4)$ and $(-4, 4)$. Find an equation for this line.

An equation for this line is [__________].

21. A line passes through the points $(6, 1)$ and $(6, 2)$. Find an equation for this line.

An equation for this line is [__________].

22. A line passes through the points $(8, -3)$ and $(8, 5)$. Find an equation for this line.

An equation for this line is [__________].

Intercepts

23. Find the y-intercept and x-intercept of the line given by the equation. If a particular intercept does not exist, enter none into all the answer blanks for that row.

$$x = 10$$

	x-value	y-value	Location
y-intercept	_______	_______	_______
x-intercept	_______	_______	_______

24. Find the y-intercept and x-intercept of the line given by the equation. If a particular intercept does not exist, enter none into all the answer blanks for that row.

$$x = -8$$

	x-value	y-value	Location
y-intercept	_______	_______	_______
x-intercept	_______	_______	_______

25. Find the y-intercept and x-intercept of the line given by the equation. If a particular intercept does not exist, enter none into all the answer blanks for that row.

$$y = -6$$

	x-value	y-value	Location
y-intercept	_______	_______	_______
x-intercept	_______	_______	_______

26. Find the y-intercept and x-intercept of the line given by the equation. If a particular intercept does not exist, enter none into all the answer blanks for that row.

$$y = -4$$

	x-value	y-value	Location
y-intercept	_______	_______	_______
x-intercept	_______	_______	_______

Graphs of Horizontal and Vertical Lines

27. Graph the line $y = 1$.

28. Graph the line $y + 5 = 0$.

29. Graph the line $x = 2$.

30. Graph the line $x - 3 = 0$.

Parallel or Perpendicular?

31. Line *m* passes points $(-5, -14)$ and $(5, 16)$.

Line *n* passes points $(-4, -10)$ and $(4, 14)$.

Determine how the two lines are related.

These two lines are

    ⊙ parallel

    ⊙ perpendicular

    ⊙ neither parallel nor perpendicular

32. Line *m* passes points $(5, -1)$ and $(15, -13)$.

Line *n* passes points $(20, -20)$ and $(-5, 10)$.

Determine how the two lines are related.

These two lines are

    ⊙ parallel

    ⊙ perpendicular

    ⊙ neither parallel nor perpendicular

33. Line *m* passes points $(12, 4)$ and $(-8, 9)$.

Line *n* passes points $(-4, -26)$ and $(-2, -18)$.

Determine how the two lines are related.

These two lines are

    ⊙ parallel

    ⊙ perpendicular

    ⊙ neither parallel nor perpendicular

34. Line *m* passes points $(-10, 12)$ and $(5, -12)$.

Line *n* passes points $(16, 2)$ and $(8, -3)$.

Determine how the two lines are related.

These two lines are

    ⊙ parallel

    ⊙ perpendicular

    ⊙ neither parallel nor perpendicular

35. Line *m* passes points $(3, -11)$ and $(4, -12)$.

Line *n* passes points $(-3, -16)$ and $(5, 8)$.

Determine how the two lines are related.

These two lines are

    ⊙ parallel

    ⊙ perpendicular

    ⊙ neither parallel nor perpendicular

36. Line *m* passes points $(6, 10)$ and $(-7, 10)$.

Line *n* passes points $(1, 2)$ and $(3, 2)$.

Determine how the two lines are related.

These two lines are

    ⊙ parallel

    ⊙ perpendicular

    ⊙ neither parallel nor perpendicular

37. Line *m* passes points $(-8, -1)$ and $(-8, 1)$.

Line *n* passes points $(-4, 0)$ and $(-4, -7)$.

Determine how the two lines are related.

These two lines are

    ⊙ parallel

    ⊙ perpendicular

    ⊙ neither parallel nor perpendicular

38. Line *m* passes points $(-6, -8)$ and $(-6, 10)$.

Line *n* passes points $(-9, 0)$ and $(-9, 3)$.

Determine how the two lines are related.

These two lines are

    ⊙ parallel

    ⊙ perpendicular

    ⊙ neither parallel nor perpendicular

Parallel and Perpendicular Line Equations

39. A line passes through the point $(-2, 5)$, and it's parallel to the line $y = -4$. Find an equation for this line.

An equation for this line is ____________.

40. A line passes through the point $(7, -2)$, and it's parallel to the line $y = -1$. Find an equation for this line.

An equation for this line is ____________.

41. A line passes through the point $(-9, -6)$, and it's parallel to the line $x = 1$. Find an equation for this line.

An equation for this line is ____________.

42. A line passes through the point $(4, 3)$, and it's parallel to the line $x = 3$. Find an equation for this line.

An equation for this line is ____________.

43. Line k has the equation $y = 3x + 4$.

Line ℓ is parallel to line k, but passes through the point $(-5, -17)$.

Find an equation for line ℓ in both slope-intercept form and point-slope form.

An equation for ℓ in slope-intercept form is:

____________.

An equation for ℓ in point-slope form is: ____ .

44. Line k has the equation $y = 4x - 2$.

Line ℓ is parallel to line k, but passes through the point $(-1, -9)$.

Find an equation for line ℓ in both slope-intercept form and point-slope form.

An equation for ℓ in slope-intercept form is:

____________.

An equation for ℓ in point-slope form is: ____ .

45. Line k has the equation $y = -\frac{9}{7}x - 2$.

Line ℓ is parallel to line k, but passes through the point $(-21, 31)$.

Find an equation for line ℓ in both slope-intercept form and point-slope form.

An equation for ℓ in slope-intercept form is:

____________.

An equation for ℓ in point-slope form is: ____ .

46. Line k has the equation $y = -\frac{1}{4}x + 3$.

Line ℓ is parallel to line k, but passes through the point $(8, -4)$.

Find an equation for line ℓ in both slope-intercept form and point-slope form.

An equation for ℓ in slope-intercept form is:

____________.

An equation for ℓ in point-slope form is: ____ .

47. Line k has the equation $y = -x + 10$.

Line ℓ is perpendicular to line k, and passes through the point $(1, 4)$.

Find an equation for line ℓ in both slope-intercept form and point-slope form.

An equation for ℓ in slope-intercept form is:

.

An equation for ℓ in point-slope form is: ___ .

48. Line k has the equation $y = 3x - 4$.

Line ℓ is perpendicular to line k and passes through the point $(-3, -1)$.

Find an equation for ℓ in both slope-intercept form and point-slope forms.

An equation for ℓ in slope-intercept form is:

.

An equation for ℓ in point-slope form is: ___ .

49. Line k's equation is $y = \frac{5}{4}x - 5$.

Line ℓ is perpendicular to line k and passes through the point $(-15, 10)$.

Find an equation for line ℓ in both slope-intercept form and point-slope forms.

An equation for ℓ in slope-intercept form is:

.

An equation for ℓ in point-slope form is: ___ .

50. Line k has the equation $x - 6y = -30$.

Line ℓ is perpendicular to line k and passes through the point $(-2, 15)$.

Find line ℓ's equation in both slope-intercept form and point-slope form.

An equation for ℓ in slope-intercept form is:

.

An equation for ℓ in point-slope form is: ___ .

Challenge

51. Prove that a triangle with vertices at the points $(1, 1)$, $(-4, 4)$, and $(-3, 0)$ is a right triangle.

4.9 Summary of Graphing Lines

The previous several sections have demonstrated several methods for plotting a graph of a linear equation. In this section, we review these methods.

We have learned three forms to write a linear equation:

- slope-intercept form

  $y = mx + b$

- standard form

  $Ax + By = C$

- point-slope form

  $y = m(x - x_0) + y_0$

We have studied two special types of line:

- horizontal line: $y = k$

- vertical line: $x = h$

We have practiced three ways to graph a line:

- building a table of x- and y-values

- plotting *one* point (often the y-intercept) and drawing slope triangles

- plotting its x-intercept and y-intercept

4.9.1 Graphing Lines in Slope-Intercept Form

In the following examples we will graph $y = -2x + 1$, which is in slope-intercept form (4.5.1), with different methods and compare them.

Example 4.9.2 Building a Table of x- and y-values. First, we will graph $y = -2x + 1$ by building a table of values. In theory this method can be used for any type of equation, linear or not. Every student must feel comfortable with building a table of values based on an equation.

x-value	y-value	Point
-2	$y = -2(-2) + 1 = 5$	$(-2, 5)$
-1	$y = -2(-1) + 1 = 3$	$(-1, 3)$
0	$y = -2(0) + 1 = 1$	$(0, 1)$
1	$y = -2(1) + 1 = -1$	$(1, -1)$
2	$y = -2(2) + 1 = -3$	$(2, -3)$

Table 4.9.3: Table for $y = -2x + 1$

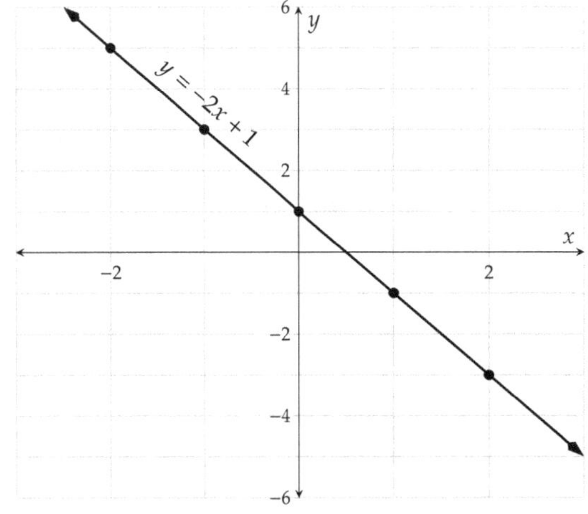

Figure 4.9.4: Graphing $y = -2x + 1$ by Building a Table of Values

Example 4.9.5 Using Slope Triangles. Although making a table is straightforward, the slope triangle method is both faster and reinforces the true meaning of slope. In the slope triangle method, we first identify some point on the line. With a line in slope-intercept form (4.5.1), we know the y-intercept, which is $(0, 1)$. Then, we can draw slope triangles in both directions to find more points.

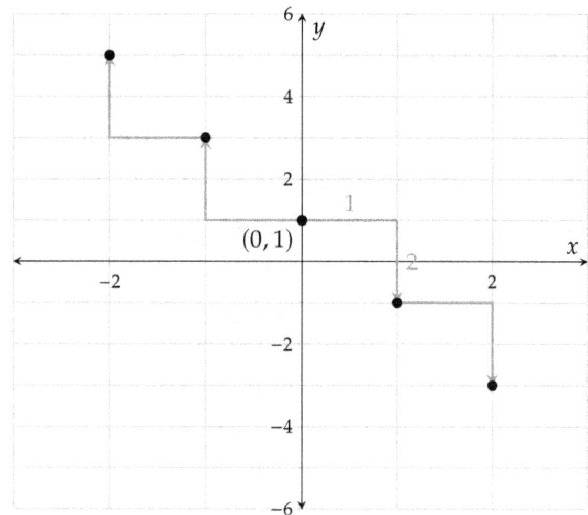

Figure 4.9.6: Marking a point and some slope triangles

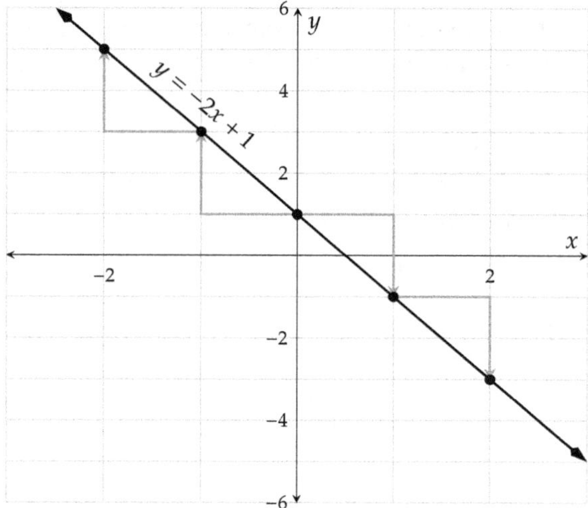

Figure 4.9.7: Graphing $y = -2x + 1$ by slope triangles

Compared to the table method, the slope triangle method:

- is less straightforward

- doesn't take the time and space to make a table

- doesn't involve lots of calculations where you might make a human error

- shows slope triangles, which reinforces the meaning of slope

Example 4.9.8 Using intercepts. If we use the x- and y-intercepts to plot $y = -2x + 1$, we have some calculation to do. While it is apparent that the y-intercept is at $(0, 1)$, where is the x-intercept? Here are two methods to find it.

Set $y = 0$.

$$y = -2x + 1$$
$$0 = -2x + 1$$
$$0 - 1 = -2x$$
$$-1 = -2x$$
$$\frac{-1}{-2} = x$$
$$\frac{1}{2} = x$$

Factor out the coefficient of x.

$$y = -2x + 1$$
$$y = -2x + (-2)\left(-\frac{1}{2}\right)1$$
$$y = -2\left(x + \left(-\frac{1}{2}\right)1\right)$$
$$y = -2\left(x - \frac{1}{2}\right)$$

And now it is easy to see that substituting $x = \frac{1}{2}$ would make $y = 0$.

So the x-intercept is at $\left(\frac{1}{2}, 0\right)$. Plotting both intercepts:

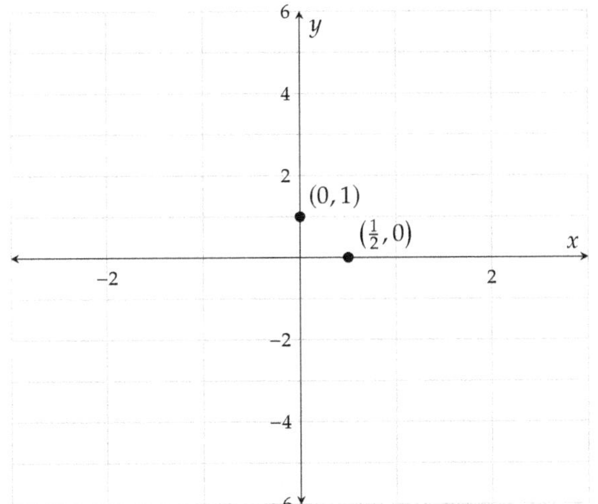

Figure 4.9.9: Marking intercepts

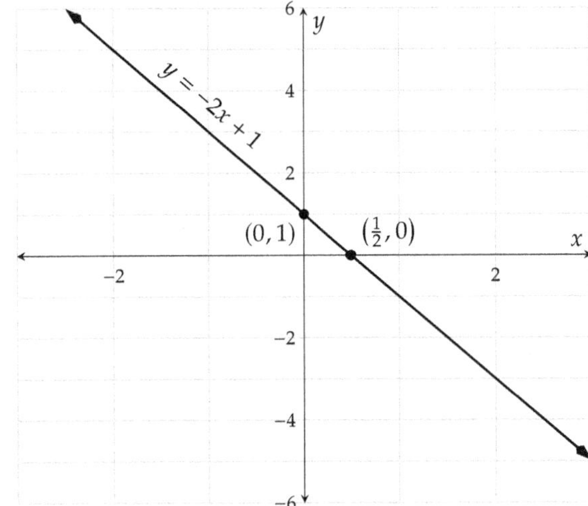

Figure 4.9.10: Using slope triangles

This worked, but here are some observations about why this method is not the greatest.

- We had to plot a point with fractional coordinates.

- We only plotted two points and they turned out very close to each other, so even the slightest inaccuracy in our drawing skills could result in a line that is way off.

When a line is presented in slope-intercept form (4.5.1), our opinion is that the slope triangle method is the best choice for making its graph.

4.9.2 Graphing Lines in Standard Form

In the following examples we will graph $3x + 4y = 12$, which is in standard form (4.7.1), with different methods and compare them.

Example 4.9.11 Building a Table of x- and y-values. To make a table, we could substitute x for various numbers and use algebra to find the corresponding y-values. Let's start with $x = -2$, planning to move on to $x = -1, 0, 1, 2$.

$$3x + 4y = 12$$
$$3(-2) + 4y = 12$$
$$-6 + 4y = 12$$
$$4y = 18$$
$$y = \frac{18}{4} = \frac{9}{2}$$

The first point we found is $\left(-2, \frac{9}{2}\right)$. This has been a lot of calculation, and we ended up with a fraction we will have to plot. *And* we have to repeat this process a few more times to get more points for the table. The table method is generally not a preferred way to graph a line in standard form (4.7.1). Let's look at other options.

Example 4.9.12 Using intercepts. Next, we will try graphing $3x + 4y = 12$ using intercepts. We set up a small table to record the two intercepts:

	x-value	y-value	Intercept
x-intercept		0	
y-intercept	0		

We have to calculate the line's x-intercept by substituting $y = 0$ into the equation:

$$3x + 4y = 12$$
$$3x + 4(0) = 12$$
$$3x = 12$$
$$x = \frac{12}{3}$$
$$x = 4$$

And similarly for the y-intercept:

$$3x + 4y = 12$$
$$3(0) + 4y = 12$$
$$4y = 12$$
$$y = \frac{12}{4}$$
$$y = 3$$

So the line's x-intercept is at $(4, 0)$ and its y-intercept is at $(0, 3)$. Now we can complete the table and then graph the line:

	x-value	y-value	Intercepts
x-intercept	4	0	$(4, 0)$
y-intercept	0	3	$(0, 3)$

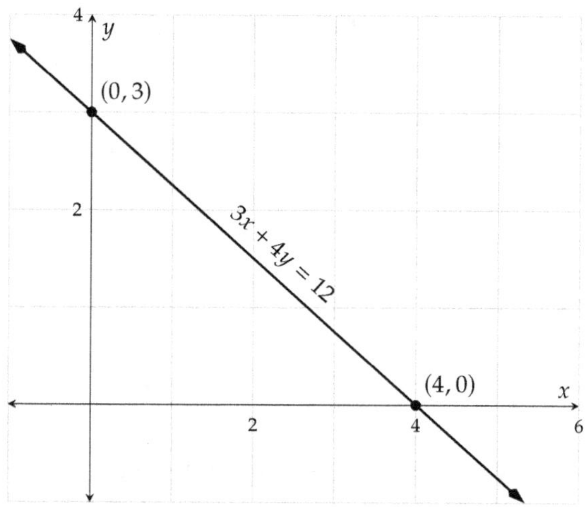

Table 4.9.13: Intercepts of $3x + 4y = 12$

Figure 4.9.14: Graph of $3x + 4y = 12$

We can always rearrange $3x + 4y = 12$ into slope-intercept form (4.5.1), and then graph it with the slope triangle method:

$$3x + 4y = 12$$
$$4y = 12 - 3x$$
$$4y = -3x + 12$$
$$y = \frac{-3x + 12}{4}$$
$$y = -\frac{3}{4}x + 3$$

Example 4.9.15 With Slope Triangles. With the y-intercept at $(0, 3)$ and slope $-\frac{3}{4}$, we can graph the line using slope triangles:

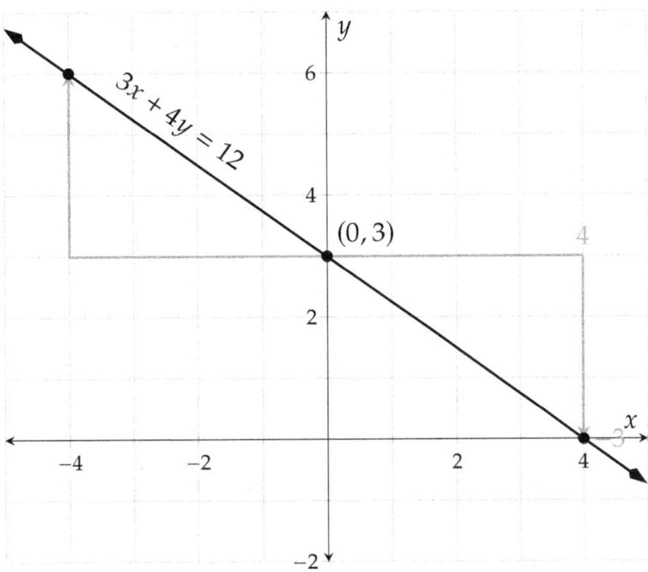

Figure 4.9.16: Graphing $3x + 4y = 12$ using slope triangles

Compared with the intercepts method, the slope triangle method takes more time, but shows more points with slope triangles, and thus a more accurate graph. Also sometimes (as with Example 4.7.14) when we graph a standard form equation like $2x - 3y = 0$, the intercepts method doesn't work because both intercepts are actually at the same point, and we have to resort to something else like slope triangles anyway.

Here are some observations about graphing a line equation that is in standard form (4.7.1):

- The intercepts method might be the quickest approach.

- The intercepts method only tells us two intercepts of the line. When we need to know more information, like the line's slope, and get a more accurate graph, we should spend more time and use the slope triangle method.

- When $C = 0$ in a standard form equation (4.7.1) we have to use something else like slope triangles anyway.

4.9.3 Graphing Lines in Point-Slope Form

When we graph a line in point-slope form (4.6.1) like $y = \frac{2}{3}(x + 1) + 3$, the slope triangle method is the obvious choice. We can see a point on the line, $(-1, 3)$, and the slope is apparent: $\frac{2}{3}$. Here is the graph:

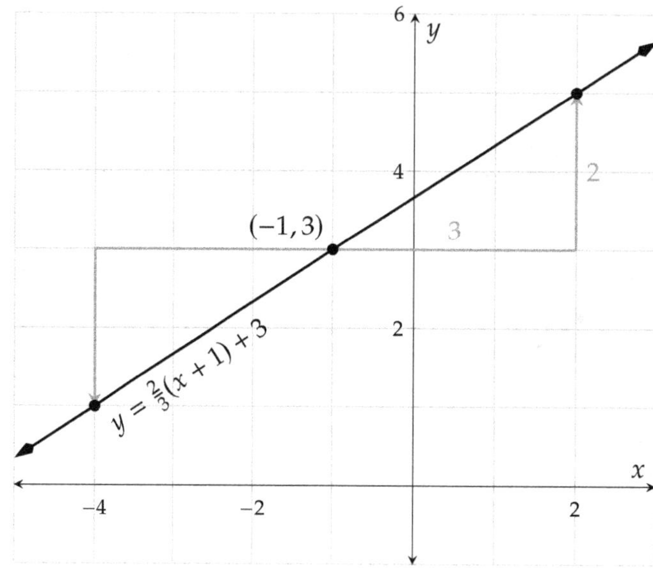

Figure 4.9.17: Graphing $y - 3 = \frac{2}{3}(x + 1)$ using slope triangles

Other graphing methods would take more work and miss the purpose of point-slope form (4.6.1). To graph a line in point-slope form (4.6.1), we recommend always using slope triangles.

4.9.4 Graphing Horizontal and Vertical Lines

We learned in Section 4.8 that equations in the form $x = h$ and $y = k$ make vertical and horizontal lines. But perhaps you will one day find yourself not remembering which is which. Making a table and plotting points can quickly remind you which type of equation makes which type of line. Let's build a table for $y = 2$ and another one for $x = -3$:

x-value	y-value	Point
0	2	$(0, 2)$
1	2	$(1, 2)$

Table 4.9.18: Table of Data for $y = 2$

x-value	y-value	Point
−3	0	$(-3, 0)$
−3	1	$(-3, 1)$

Table 4.9.19: Table of Data for $x = -3$

With two points on each line, we can graph them:

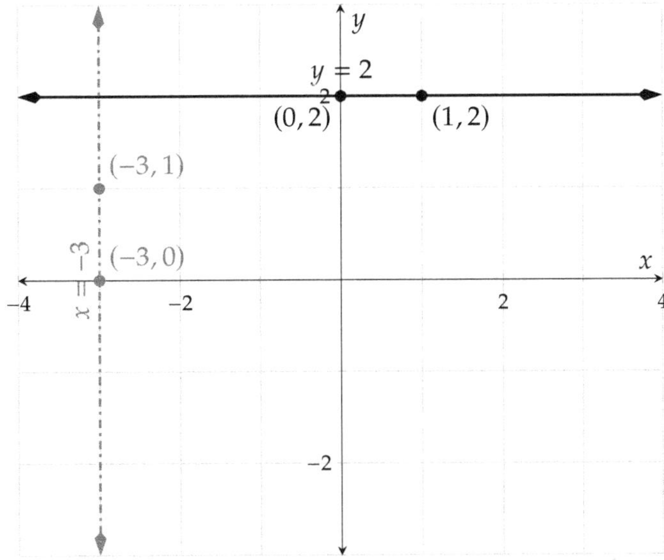

Figure 4.9.20: Graphing $y = 2$ and $x = -3$

Exercises

Graphing by Table

1. Use a table to make a plot of $y = 4x + 3$.

2. Use a table to make a plot of $y = -5x - 1$.

3. Use a table to make a plot of $y = -\frac{3}{4}x - 1$.

4. Use a table to make a plot of $y = \frac{5}{3}x + 3$.

Graphing Standard Form Equations

5. First find the x- and y-intercepts of the line with equation $6x + 5y = -90$. Then find one other point on the line. Use your results to graph the line.

6. First find the x- and y-intercepts of the line with equation $2x - 3y = -6$. Then find one other point on the line. Use your results to graph the line.

7. First find the x- and y-intercepts of the line with equation $3x + y = -9$. Then find one other point on the line. Use your results to graph the line.

8. First find the x- and y-intercepts of the line with equation $-15x + 3y = -3$. Then find one other point on the line. Use your results to graph the line.

9. First find the x- and y-intercepts of the line with equation $4x + 3y = -3$. Then find one other point on the line. Use your results to graph the line.

10. First find the x- and y-intercepts of the line with equation $-4x - 5y = 5$. Then find one other point on the line. Use your results to graph the line.

11. First find the x- and y-intercepts of the line with equation $5x-3y=0$. Then find one other point on the line. Use your results to graph the line.

12. First find the x- and y-intercepts of the line with equation $2x+9y=0$. Then find one other point on the line. Use your results to graph the line.

Graphing Slope-Intercept Equations

13. Use the slope and y-intercept from the line $y=-5x$ to plot the line. Use slope triangles.

14. Use the slope and y-intercept from the line $y=3x-6$ to plot the line. Use slope triangles.

15. Use the slope and y-intercept from the line $y=-\frac{2}{5}x+2$ to plot the line. Use slope triangles.

16. Use the slope and y-intercept from the line $y=\frac{10}{3}x-3$ to plot the line. Use slope triangles.

Graphing Horizontal and Vertical Lines

17. Plot the line $y=1$.

18. Plot the line $y=-4$.

19. Plot the line $x=-8$.

20. Plot the line $x=5$.

Choosing the Best Method to Graph Lines

21. Use whatever method you think best to plot $y=2x+2$.

22. Use whatever method you think best to plot $y=-3x+6$.

23. Use whatever method you think best to plot $y=-\frac{3}{4}x-1$.

24. Use whatever method you think best to plot $y=\frac{5}{3}x-3$.

25. Use whatever method you think best to plot $y=-\frac{3}{4}(x-5)+2$.

26. Use whatever method you think best to plot $y=\frac{2}{5}(x+1)-3$.

27. Use whatever method you think best to plot $3x+2y=6$.

28. Use whatever method you think best to plot $5x-4y=8$.

29. Use whatever method you think best to plot $3x-4y=0$.

30. Use whatever method you think best to plot $9x+6y=0$.

31. Use whatever method you think best to plot $x = -3$.

32. Use whatever method you think best to plot $x = 2$.

33. Use whatever method you think best to plot $y = -7$.

34. Use whatever method you think best to plot $y = 5$.

4.10 Linear Inequalities in Two Variables

We have learned how to graph lines like $y = 2x + 1$. In this section, we will learn how to graph linear inequalities like $y > 2x + 1$.

Example 4.10.2 Office Supplies. Isabel has a budget of \$133.00 to purchase some staplers and markers for the office supply closet. Each stapler costs \$19.00, and each marker costs \$1.75. We will define the variables so that she will purchase x staplers and y markers. Write and plot a linear inequality to model the relationship between the number of staplers and markers Isabel can purchase. Keep in mind that she might not spend all of the \$133.00.

The cost of buying x staplers would be $19x$ dollars. Similarly, the cost of buying y markers would be $1.75y$ dollars. Since whatever Isabel spends needs to be no more than 133 dollars, we have the inequality

$$19x + 1.75y \le 133.$$

This is a standard-form inequality, similar to Equation (4.7.1). Next, let's graph it.

The first method to graph the inequality is to graph the corresponding equation, $19x + 1.75y = 133$. Its x- and y-intercepts can be found this way:

$$
\begin{aligned}
19x + 1.75y &= 133 \\
19x + 1.75(0) &= 133 \\
19x &= 133 \\
\frac{19x}{19} &= \frac{133}{19} \\
x &= 7
\end{aligned}
\qquad\qquad
\begin{aligned}
19x + 1.75y &= 133 \\
19(0) + 1.75y &= 133 \\
1.75y &= 133 \\
\frac{1.75y}{1.75} &= \frac{133}{1.75} \\
y &= 76
\end{aligned}
$$

So the intercepts are $(7, 0)$ and $(0, 76)$, and we can plot the line in Figure 4.10.3.

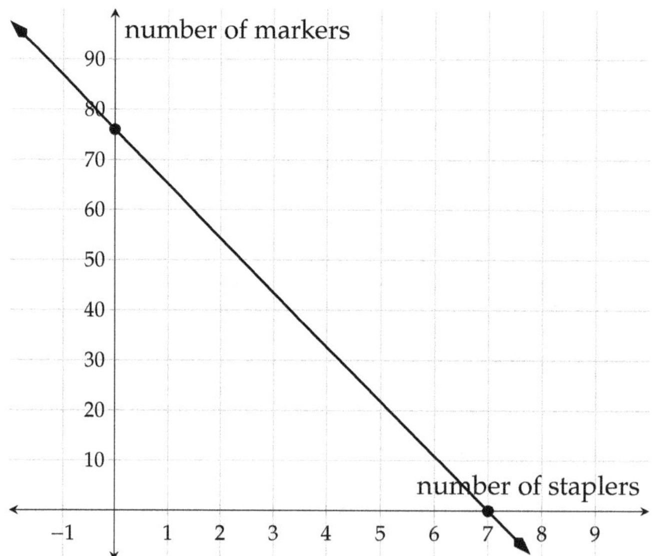

Figure 4.10.3: $19x + 1.75y = 133$

The points *on* this line represent ways in which Isabel can spend exactly all of the \$133. But what does a point like $(2, 40)$ in Figure 4.10.4, which is not on the line, mean in this context? That would mean Isabel bought 2 staplers and 40 markers, spending $19 \cdot 2 + 1.75 \cdot 40 = 108$ dollars. That is within her budget.

In fact, any point on the lower left side of this line represents a total purchase within Isabel's budget. The shading in Figure 4.10.5 captures *all* solutions to $19x + 1.75y \leq 133$. Some of those solutions have negative x- and y-values, which make no sense in context. So in Figure 4.10.6, we restrict the shading to solutions which make physical sense.

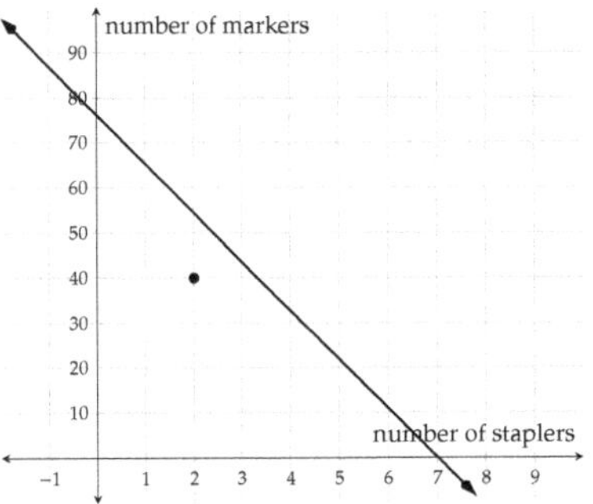
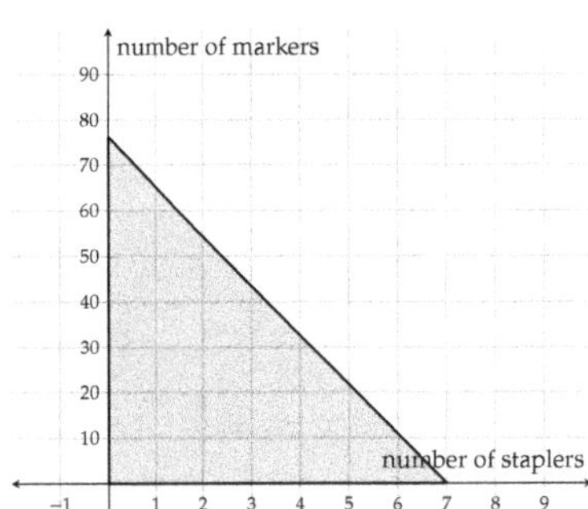

Figure 4.10.4: The line $19x + 1.75y = 133$ with a point identified that is within Isabel's budget.

Figure 4.10.5: Shading all points that solve the inequality.

Figure 4.10.6: Shading restricted to points that make physical sense in context.

Let's look at some more examples of graphing linear inequalities in two variables.

Example 4.10.7 Is the point $(1, 2)$ a solution of $y > 2x + 1$?

In the inequality $y > 2x + 1$, substitute x with 1 and y with 2, and we will see whether the inequality is true:

$$y > 2x + 1$$
$$2 \stackrel{?}{>} 2(2) + 1$$
$$2 \stackrel{no}{>} 5$$

Since $2 > 5$ is not true, $(1, 2)$ is not a solution of $y > 2x + 1$.

Example 4.10.8 Graph $y > 2x + 1$.

There are two steps to graphing this linear inequality in two variables.

1. Graph the line $y = 2x + 1$. Because the inequality symbol is $>$ (instead of $\geq$), the line should be dashed (instead of solid).

2. Next, we need to decide whether to shade the region above $y = 2x + 1$ or below it. We will choose a point to test whether $y > 2x + 1$ is true. As long as the line doesn't cross $(0, 0)$, we will use $(0, 0)$ to test because the number 0 is the easiest number for calculation.

$$y > 2x + 1$$
$$0 \stackrel{?}{>} 2(0) + 1$$

$$0 \overset{\text{no}}{>} 1$$

Because $0 > 1$ is not true, the point $(0, 0)$ is not a solution and should not be shaded. As a result, we shade the region *without* $(0, 0)$.

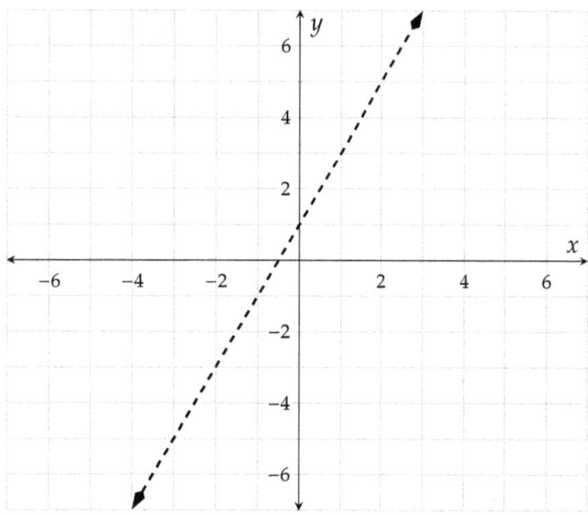

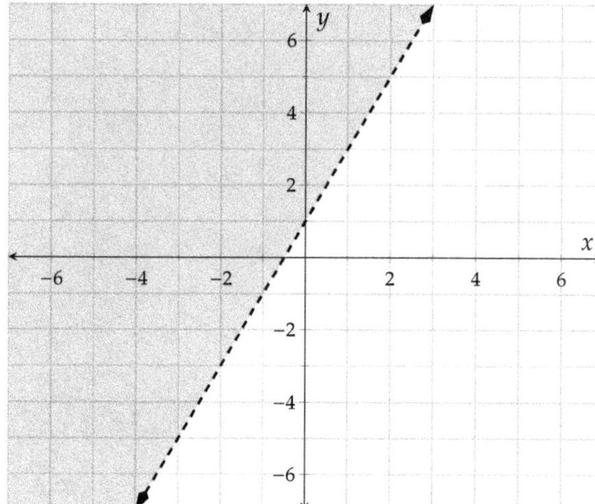

Figure 4.10.9: Step 1 of graphing $y > 2x + 1$ **Figure 4.10.10:** Complete graph of $y > 2x + 1$

Example 4.10.11 Graph $y \leq -\frac{5}{3}x + 2$.

There are two steps to graphing this linear inequality in two variables.

1. Graph the line $y = -\frac{5}{3}x + 2$. Because the inequality symbol is $\leq$ (instead of $<$), the line should be solid.

2. Next, we need to decide whether to shade the region above $y = -\frac{5}{3}x + 2$ or below it. We will choose a point to test whether $y \leq -\frac{5}{3}x + 2$ is true there. Using $(0, 0)$ as a test point:

$$y \leq -\frac{5}{3}x + 2$$
$$0 \overset{?}{\leq} -\frac{5}{3}(0) + 2$$
$$0 \overset{\checkmark}{\leq} 2$$

Because $0 \leq 2$ is true, the point $(0, 0)$ is a solution. As a result, we shade the region *with* $(0, 0)$.

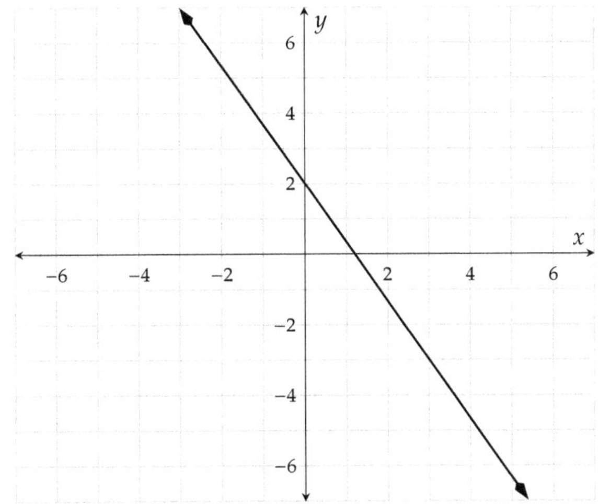

Figure 4.10.12: Step 1 of graphing $y \leq -\frac{5}{3}x + 2$

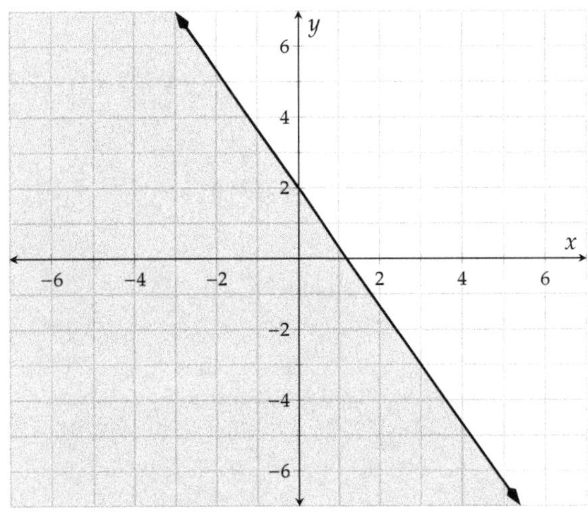

Figure 4.10.13: Complete graph of $y \leq -\frac{5}{3}x + 2$

Exercises

Review and Warmup   Find the line's slope and y-intercept.

1. A line has equation $y = 8x + 3$.

This line's slope is [   ].

This line's y-intercept is [   ].

2. A line has equation $y = 9x + 9$.

This line's slope is [   ].

This line's y-intercept is [   ].

3. A line has equation $y = -2x - 5$.

This line's slope is [   ].

This line's y-intercept is [   ].

4. A line has equation $y = -10x - 9$.

This line's slope is [   ].

This line's y-intercept is [   ].

5. A line has equation $y = -x - 6$.

This line's slope is [   ].

This line's y-intercept is [   ].

6. A line has equation $y = -x - 4$.

This line's slope is [   ].

This line's y-intercept is [   ].

7. A line has equation $y = -\dfrac{4x}{3} - 4$.

This line's slope is [   ].

This line's y-intercept is [   ].

8. A line has equation $y = -\dfrac{6x}{7} - 6$.

This line's slope is [   ].

This line's y-intercept is [   ].

Graphing Two-Variable Inequalities

9. Graph the linear inequality $y \geq -4x$.

10. Graph the linear inequality $y \leq -\frac{1}{2}x - 3$.

11. Graph the linear inequality $y < 3x + 5$.

12. Graph the linear inequality $y > \frac{4}{3}x + 1$.

13. Graph the linear inequality $2x + y \geq 3$.

14. Graph the linear inequality $3x + 2y < -6$.

15. Graph the linear inequality $y \geq 3$.

16. Graph the linear inequality $x < -1$.

Applications

17. You fed your grandpa's cat while he was on vacation. When he was back, he took out a huge bank of coins, including quarters and dimes. He said you can take as many coins as you want, but the total value must be less than \$30.00.

    (a) Write an inequality to model this situation, with q representing the number of quarters you will take, and d representing the number of dimes.

    (b) Graph this linear inequality.

18. A couple is planning their wedding. They want the cost of the reception and the ceremony to be no more than \$8,000.

    (a) Write an inequality to model this situation, with r as the cost of the reception (in dollars) and c as the cost of the ceremony (in dollars).

    (b) Graph this linear inequality.

4.11 Graphing Lines Chapter Review

4.11.1 Cartesian Coordinates

In Section 4.1 we covered the definition of the Cartesian Coordinate System and how to plot points using the x and y-axes.

Example 4.11.1 On paper, sketch a Cartesian coordinate system with units, and then plot the following points: $(3, 2), (-5, -1), (0, -3), (4, 0)$.

Explanation.

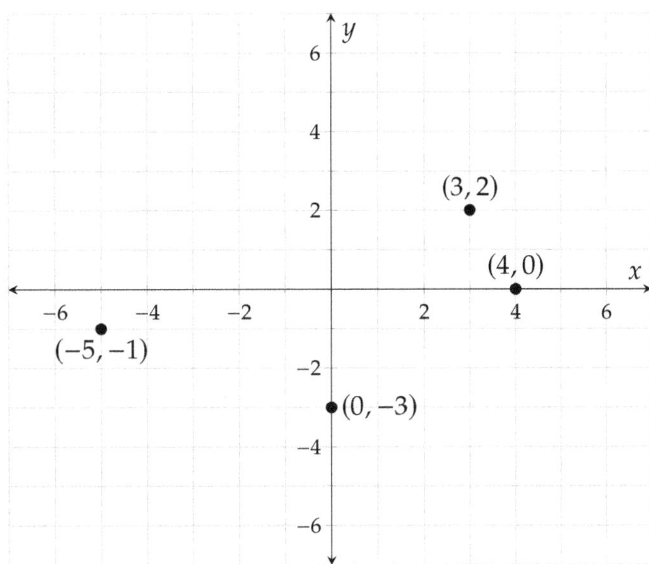

Figure 4.11.2: A Cartesian grid with the four points plotted.

4.11.2 Graphing Equations

In Section 4.2 we covered how to plot solutions to equations to produce a graph of the equation.

Example 4.11.3 Graph the equation $y = -2x + 5$.

Explanation.

x	$y = -2x + 5$	Point
-2		
-1		
0		
1		
2		

(a) Set up the table

x	$y = -2x + 5$	Point
-2	$-2(-2) + 5 = 9$	$(-2, 9)$
-1	$-2(-1) + 5 = 7$	$(-1, 7)$
0	$-2(0) + 5 = 5$	$(0, 5)$
1	$-2(1) + 5 = 3$	$(1, 3)$
2	$-2(2) + 5 = 1$	$(2, 1)$

(b) Complete the table

Figure 4.11.3: Making a table for $y = -2x + 5$

We use points from the table to graph the equation. First, plot each point carefully. Then, connect the points with a smooth curve. Here, the curve is a straight line. Lastly, we can communicate that the graph extends further by sketching arrows on both ends of the line.

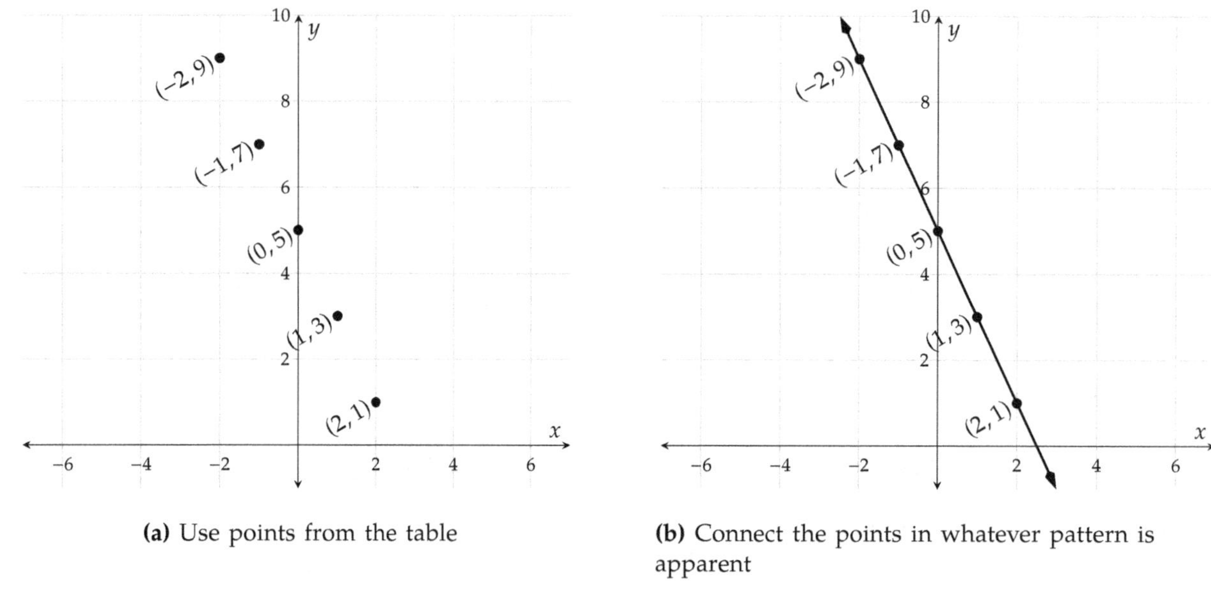

(a) Use points from the table

(b) Connect the points in whatever pattern is apparent

Figure 4.11.3: Graphing the Equation $y = -2x + 5$

4.11.3 Exploring Two-Variable Data and Rate of Change

In Section 4.3 we covered how to find patterns in tables of data and how to calculate the rate of change between points in data.

Exploring Two-Variable Data and Rate of Change   For a linear relationship, by its data in a table, we can see the rate of change (slope) and the line's y-intercept, thus writing the equation.

Write an equation in the form $y = \ldots$ suggested by the pattern in the table.

x	y
0	-4
1	-6
2	-8
3	-10

Table 4.11.5: A table of linear data.

Example 4.11.4 Explanation.

We consider how the values change from one row to the next. From row to row, the x-value increases by 1. Also, the y-value decreases by 2 from row to row.

	x	y	
	0	-4	
$+1 \rightarrow$	1	-6	$\leftarrow -2$
$+1 \rightarrow$	2	-8	$\leftarrow -2$
$+1 \rightarrow$	3	-10	$\leftarrow -2$

Since row-to-row change is always 1 for x and is always -2 for y, the rate of change from one row to another row is always the same: -2 units of y for every 1 unit of x.

We know that the output for $x = 0$ is $y = -4$. And our observation about the constant rate of change tells us that if we increase the input by x units from 0, the ouput should decrease by $\overbrace{(-2) + (-2) + \cdots + (-2)}^{x \text{ times}}$, which is $-2x$. So the output would be $-4 - 2x$.

So the equation is $y = -2x - 4$.

4.11.4 Slope

In Section 4.4 we covered the definition of slope 4.4.3 and how to use slope-triangles to calculate slope. There is also the slope formula (4.4.3) which helps find the slope through any two points.

Example 4.11.6 Find the slope of the line in the following graph.

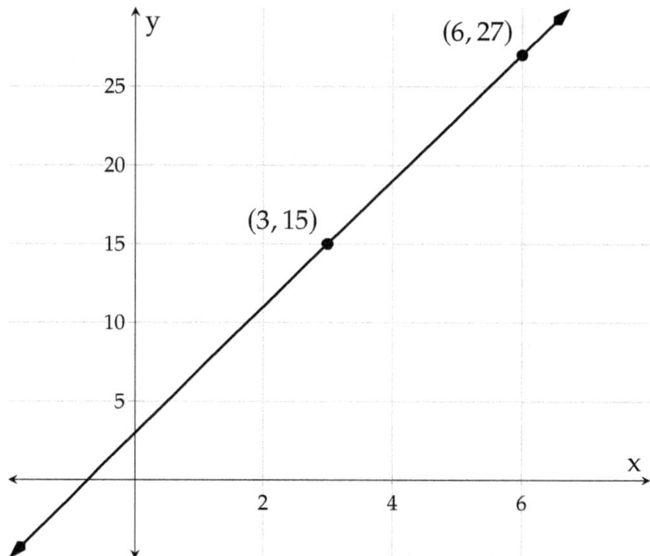

Figure 4.11.7: The line with two points indicated.

Explanation.

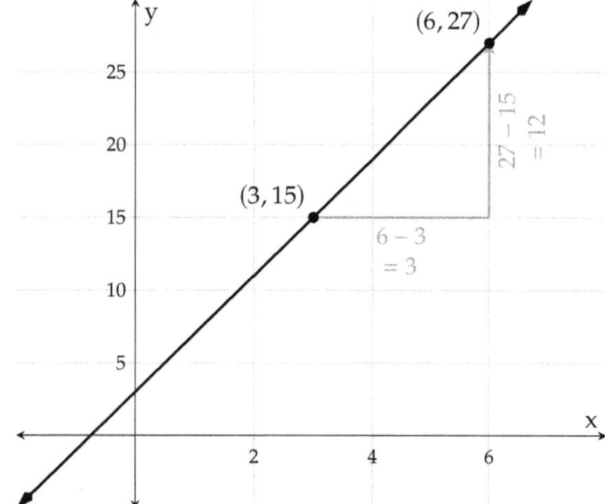

Figure 4.11.8: The line with a slope triangle drawn.

We picked two points on the line, and then drew a slope triangle. Next, we will do:

$$\text{slope} = \frac{12}{3} = 4$$

The line's slope is 4.

Example 4.11.9 Finding a Line's Slope by the Slope Formula. Use the slope formula (4.4.3) to find the slope of the line that passes through the points $(-5, 25)$ and $(4, -2)$.

Explanation.

$$\text{slope} = \frac{y_2 - y_1}{x_2 - x_1}$$

$$= \frac{-2 - (25)}{4 - (-5)}$$

$$= \frac{-27}{9}$$

$$= -3$$

The line's slope is -3.

4.11.5 Slope-Intercept Form

In Section 4.5 we covered the definition of slope intercept-form and both wrote equations in slope-intercept form and graphed lines given in slope-intercept form.

Example 4.11.10 Graph the line $y = -\frac{5}{2}x + 4$.

Explanation.

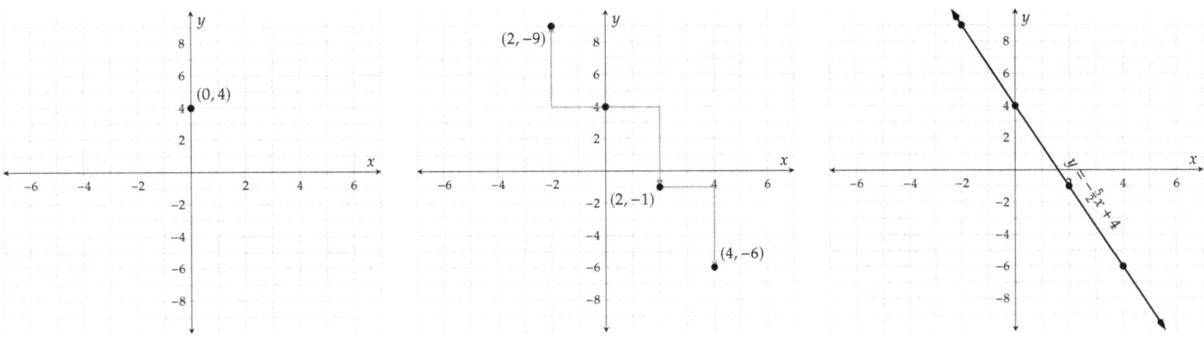

(a) First, plot the line's y-intercept, $(0, 4)$.

(b) The slope is $-\frac{5}{2} = \frac{-5}{2} = \frac{5}{-2}$. So we can try using a "run" of 2 and a "rise" of -5 or a "run" of -2 and a "rise" of 5.

(c) Arrowheads and labels are encouraged.

Figure 4.11.10: Graphing $y = -\frac{5}{2}x + 4$

Writing a Line's Equation in Slope-Intercept Form Based on Graph Given a line's graph, we can identify its y-intercept, and then find its slope by a slope triangle. With a line's slope and y-intercept, we can write its equation in the form of $y = mx + b$.

Find the equation of the line in the graph.

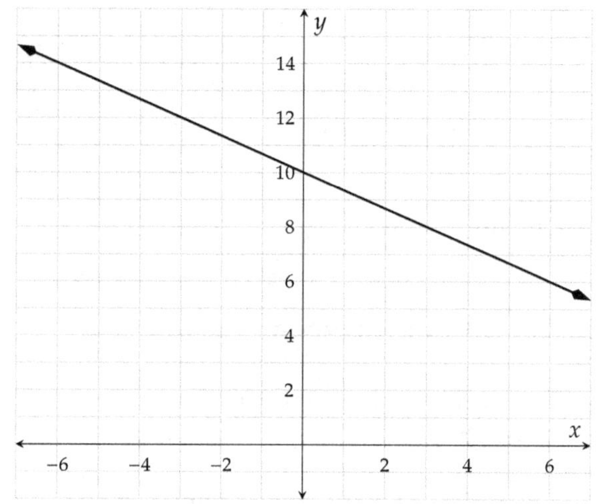

Figure 4.11.12: Graph of a line

Example 4.11.11 Explanation.

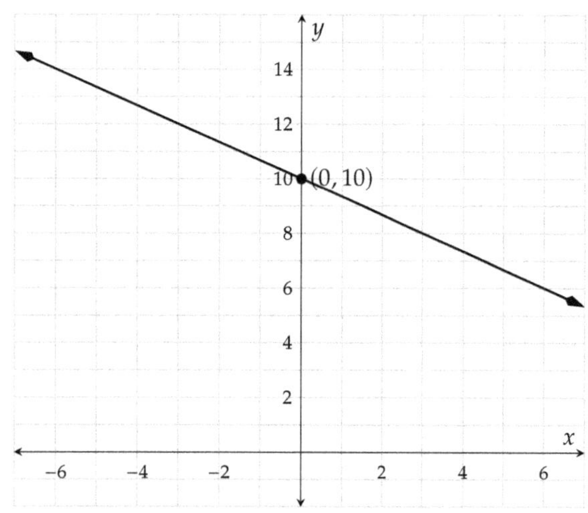

Figure 4.11.13: Identify the line's y-intercept, 10.

Figure 4.11.14: Identify the line's slope by a slope triangle. Note that we can pick any two points on the line to create a slope triangle. We would get the same slope: $-\frac{2}{3}$

With the line's slope $-\frac{2}{3}$ and y-intercept 10, we can write the line's equation in slope-intercept form: $y = -\frac{2}{3}x + 10$.

4.11.6 Point-Slope Form

In Section 4.6 we covered the definition of point-slope form and both wrote equations in point-slope form and graphed lines given in point-slope form.

Example 4.11.15 A line passes through $(-6, 0)$ and $(9, -10)$. Find this line's equation in point-slope.

Explanation. We will use the slope formula (4.4.3) to find the slope first. After labeling those two points as $(\overset{x_1}{-6}, \overset{y_1}{0})$ and $(\overset{x_2}{9}, \overset{y_2}{-10})$, we have:

$$
\begin{aligned}
\text{slope} &= \frac{y_2 - y_1}{x_2 - x_1} \\
&= \frac{-10 - 0}{9 - (-6)} \\
&= \frac{-10}{15} \\
&= -\frac{2}{3}
\end{aligned}
$$

Now the point-slope equation looks like $y = -\frac{2}{3}(x - x_0) + y_0$. Next, we will use $(9, -10)$ and substitute x_0 with 9 and y_0 with -10, and we have:

$$
\begin{aligned}
y &= -\frac{2}{3}(x - x_0) + y_0 \\
y &= -\frac{2}{3}(x - 9) + (-10) \\
y &= -\frac{2}{3}(x - 9) - 10
\end{aligned}
$$

4.11.7 Standard Form

In Section 4.7 we covered the definition of standard form of a linear equation. We converted equations from standard form to slope-intercept form and vice versa. We also graphed lines from standard form by finding the intercepts of the line.

Example 4.11.16

a. Convert $2x + 3y = 6$ into slope-intercept form.

b. Convert $y = -\frac{4}{7}x - 3$ into standard form.

Explanation.

a.

$$
\begin{aligned}
2x + 3y &= 6 \\
2x + 3y - 2x &= 6 - 2x \\
3y &= -2x + 6 \\
\frac{3y}{3} &= \frac{-2x + 6}{3}
\end{aligned}
$$

$$y = \frac{-2x}{3} + \frac{6}{3}$$

$$y = -\frac{2}{3}x + 2$$

The line's equation in slope-intercept form is $y = -\frac{2}{3}x + 2$.

b.

$$y = -\frac{4}{7}x - 3$$

$$7 \cdot y = 7 \cdot (-\frac{4}{7}x - 3)$$

$$7y = 7 \cdot (-\frac{4}{7}x) - 7 \cdot 3$$

$$7y = -4x - 21$$

$$7y + 4x = -4x - 21 + 4x$$

$$4x + 7y = -21$$

The line's equation in standard form is $4x + 7y = -21$.

To graph a line in standard form, we could first change it to slope-intercept form, and then graph the line by its y-intercept and slope triangles. A second method is to graph the line by its x-intercept and y-intercept.

Example 4.11.17 Graph $2x - 3y = -6$ using its intercepts. And then use the intercepts to calculate the line's slope.

Explanation.   We calculate the line's x-intercept by substituting $y = 0$ into the equation

$$2x - 3y = -6$$
$$2x - 3(0) = -6$$
$$2x = -6$$
$$x = -3$$

So the line's x-intercept is $(-3, 0)$.

Similarly, we substitute $x = 0$ into the equation to calculate the y-intercept:

$$2x - 3y = -6$$
$$2(0) - 3y = -6$$
$$-3y = -6$$
$$y = 2$$

So the line's y-intercept is $(0, 2)$.

With both intercepts' coordinates, we can graph the line:

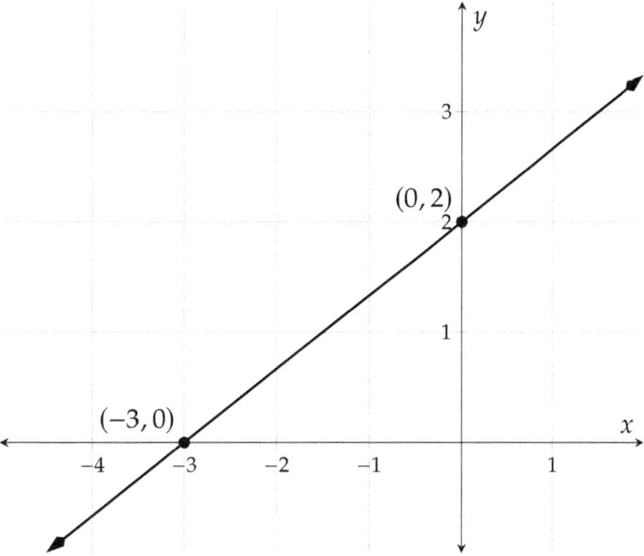

Figure 4.11.18: Graph of $2x - 3y = -6$

Now that we have graphed the line we can read the slope. The rise is 2 units and the run is 3 units so the slope is $\frac{2}{3}$.

4.11.8 Horizontal, Vertical, Parallel, and Perpendicular Lines

In Section 4.8 we covered what equations of horizontal and vertical lines. We also covered the relationships between the slopes of parallel and perpendicular lines.

Example 4.11.19 Line m's equation is $y = -2x + 20$. Line n is parallel to m, and line n also passes the point $(4, -3)$. Find an equation for line n in point-slope form.

Explanation. Since parallel lines have the same slope, line n's slope is also -2. Since line n also passes the point $(4, -3)$, we can write line n's equation in point-slope form:

$$y = m(x - x_1) + y_1$$
$$y = -2(x - 4) + (-3)$$
$$y = -2(x - 4) - 3$$

Two lines are perpendicular if and only if the product of their slopes is -1.

Example 4.11.20 Line m's equation is $y = -2x + 20$. Line n is perpendicular to m, and line q also passes the point $(4, -3)$. Find an equation for line q in slope-intercept form.

Explanation. Since line m and q are perpendicular, the product of their slopes is -1. Because line m's slope is given as -2, we can find line q's slope is $\frac{1}{2}$.

Since line q also passes the point $(4, -3)$, we can write line q's equation in point-slope form:

$$y = m(x - x_1) + y_1$$

$$y = \frac{1}{2}(x - 4) + (-3)$$

$$y = \frac{1}{2}(x - 4) - 3$$

We can now convert this equation to slope-intercept form:

$$y = \frac{1}{2}(x - 4) - 3$$

$$y = \frac{1}{2}x - 2 - 3$$

$$y = \frac{1}{2}x - 5$$

4.11.9 Linear Inequalities in Two Variables

In Section 4.10 we covered how to graph the solution set for an inequality with two variables as a region in the plane.

Example 4.11.21 Graph $y > 2x + 1$.

Explanation.   There are two steps to graph an inequality.

1. Graph the line $y = 2x + 1$. Because the inequality symbol is $>$, (instead of $\geq$) the line should be dashed (instead of solid).

2. Next, we need to decide whether to shade the region above $y = 2x + 1$ or below it. We will choose a point to test whether $y > 2x + 1$ is true. As long as the line doesn't cross $(0, 0)$, we will use $(0, 0)$ to test, because the number 0 is the easiest number for calculation.

$$y > 2x + 1$$

$$0 \overset{?}{>} 2(0) + 1$$

$$0 \overset{no}{>} 1$$

Because $0 > 1$ is not true, the point $(0, 0)$ is not a solution and should not be shaded. As a result, we shade the region without $(0, 0)$.

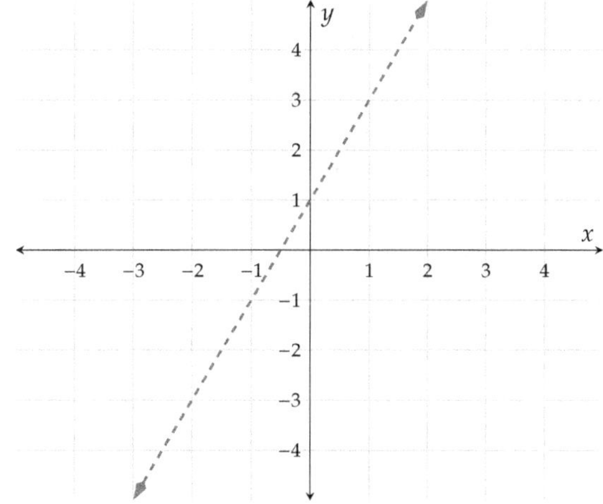

Figure 4.11.22: Step 1 of graphing $y > 2x + 1$

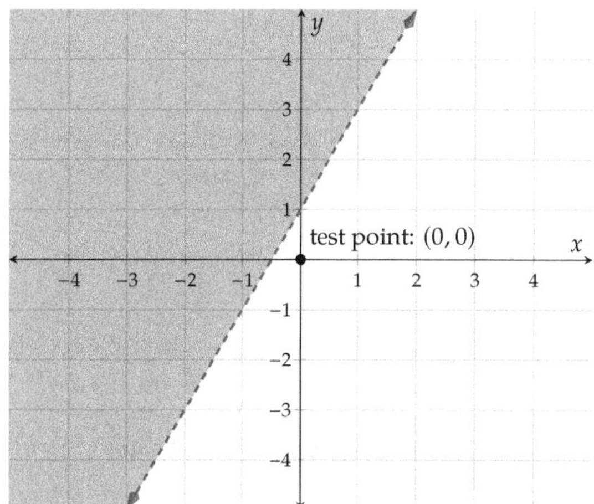

Figure 4.11.23: Step 2 of graphing $y > 2x + 1$

Exercises

1. Sketch the points $(8, 2)$, $(5, 5)$, $(-3, 0)$, $\left(0, -\frac{14}{3}\right)$, $(3, -2.5)$, and $(-5, 7)$ on a Cartesian plane.

2. Locate each point in the graph:

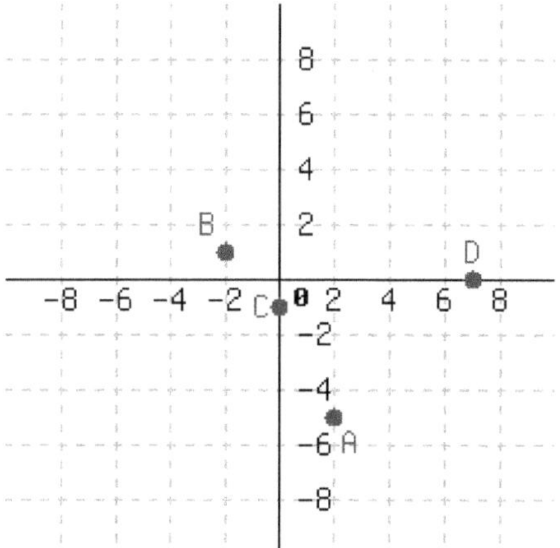

Write each point's position as an ordered pair, like $(1, 2)$.

$A = $ _______ $B = $ _______
$C = $ _______ $D = $ _______

3. Consider the equation

$$y = -\tfrac{3}{8}x - 3$$

Which of the following ordered pairs are solutions to the given equation? There may be more than one correct answer.

☐ $(-24, 9)$ ☐ $(32, -13)$ ☐ $(0, -3)$
☐ $(-40, 12)$

4. Consider the equation

$$y = -\tfrac{7}{8}x - 5$$

Which of the following ordered pairs are solutions to the given equation? There may be more than one correct answer.

☐ $(24, -25)$ ☐ $(0, -5)$ ☐ $(-16, 14)$
☐ $(-24, 16)$

5. Write an equation in the form $y = \ldots$ suggested by the pattern in the table.

x	y
0	−4
1	−2
2	0
3	2

6. Write an equation in the form $y = \ldots$ suggested by the pattern in the table.

x	y
0	3
1	−2
2	−7
3	−12

7. Below is a line's graph.

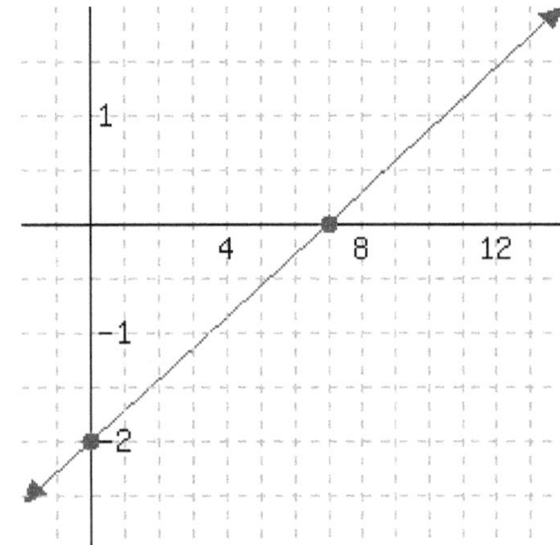

The slope of this line is ⬚.

8. Below is a line's graph.

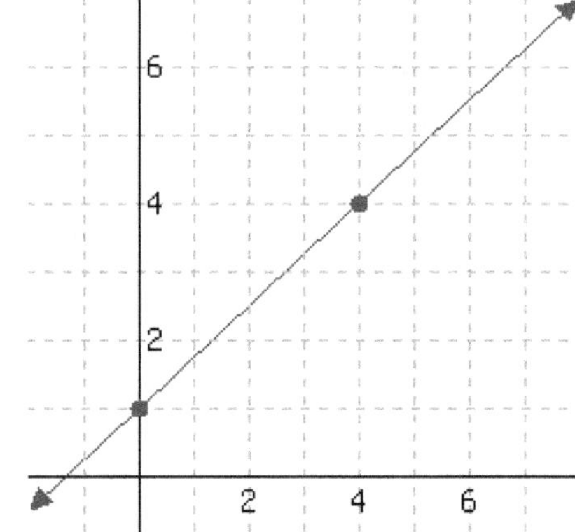

The slope of this line is ⬚.

9. Below is a line's graph.

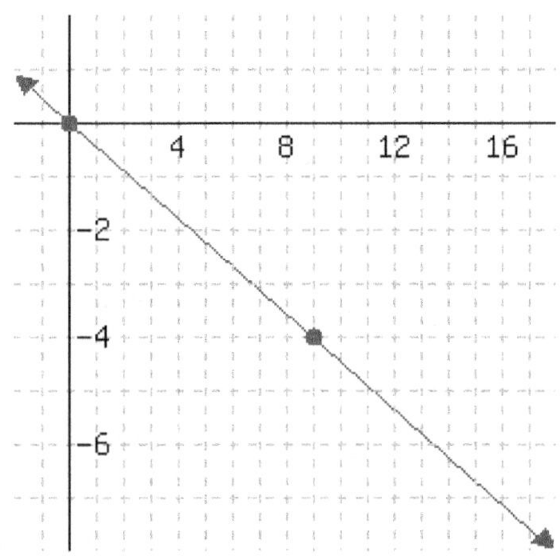

The slope of this line is ____________.

10. Below is a line's graph.

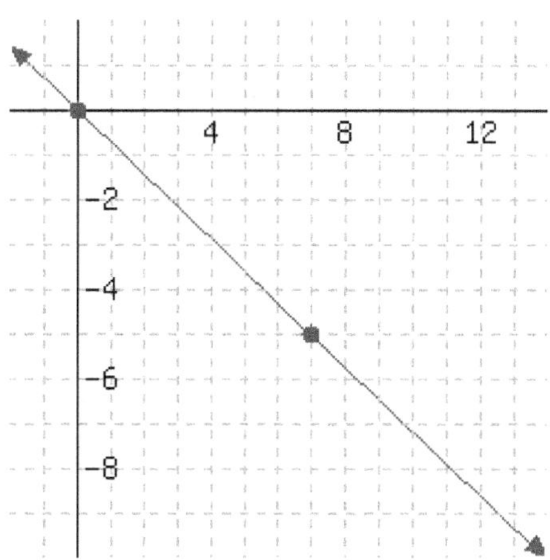

The slope of this line is ____________.

11. A line passes through the points $(-8, 23)$ and $(4, 2)$. Find this line's slope.

12. A line passes through the points $(-24, 23)$ and $(8, -13)$. Find this line's slope.

13. A line passes through the points $(4, 8)$ and $(-2, 8)$. Find this line's slope.

14. A line passes through the points $(2, 10)$ and $(-5, 10)$. Find this line's slope.

15. A line passes through the points $(-9, -1)$ and $(-9, 3)$. Find this line's slope.

16. A line passes through the points $(-6, -3)$ and $(-6, 5)$. Find this line's slope.

17. A line's graph is given.

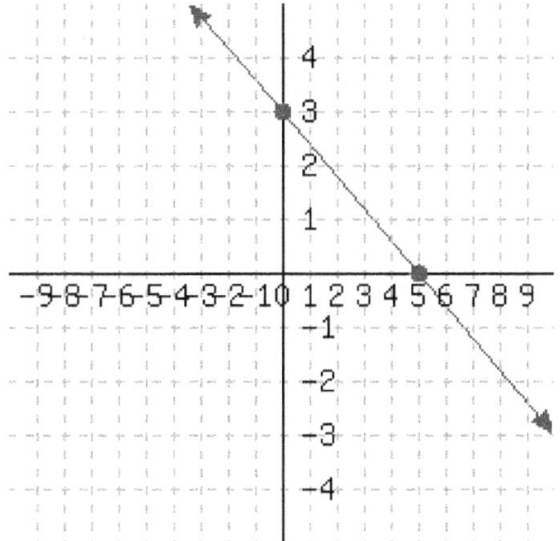

This line's slope-intercept equation is ____________

18. A line's graph is given.

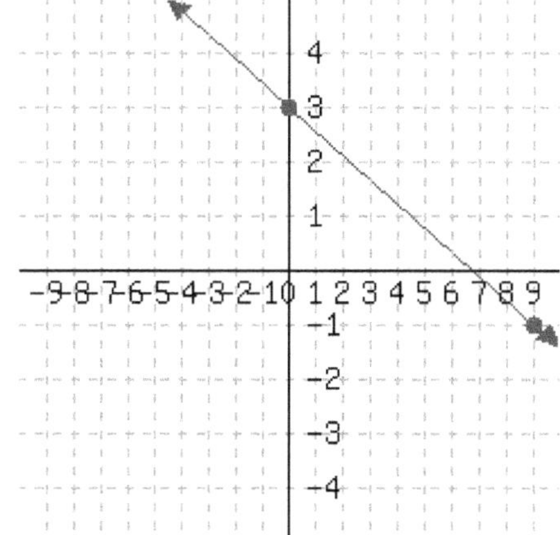

This line's slope-intercept equation is ____________

19. Find the line's slope and *y*-intercept.

A line has equation $3x - 5y = -15$.

This line's slope is ☐.

This line's *y*-intercept is ☐.

20. Find the line's slope and *y*-intercept.

A line has equation $5x - 6y = -30$.

This line's slope is ☐.

This line's *y*-intercept is ☐.

21. A line passes through the points $(8, -2)$ and $(24, 12)$. Find this line's equation in point-slope form.

Using the point $(8, -2)$, this line's point-slope form equation is ☐.

Using the point $(24, 12)$, this line's point-slope form equation is ☐.

22. A line passes through the points $(15, 22)$ and $(0, -2)$. Find this line's equation in point-slope form.

Using the point $(15, 22)$, this line's point-slope form equation is ☐.

Using the point $(0, -2)$, this line's point-slope form equation is ☐.

23. Scientists are conducting an experiment with a gas in a sealed container. The mass of the gas is measured, and the scientists realize that the gas is leaking over time in a linear way. Each minute, they lose 2.4 grams. Ten minutes since the experiment started, the remaining gas had a mass of 96 grams.

Let *x* be the number of minutes that have passed since the experiment started, and let *y* be the mass of the gas in grams at that moment. Use a linear equation to model the weight of the gas over time.

   a. This line's slope-intercept equation is ☐.

   b. 40 minutes after the experiment started, there would be ☐ grams of gas left.

   c. If a linear model continues to be accurate, ☐ minutes since the experiment started, all gas in the container will be gone.

24. Scientists are conducting an experiment with a gas in a sealed container. The mass of the gas is measured, and the scientists realize that the gas is leaking over time in a linear way. Each minute, they lose 8.2 grams. Six minutes since the experiment started, the remaining gas had a mass of 278.8 grams.

Let *x* be the number of minutes that have passed since the experiment started, and let *y* be the mass of the gas in grams at that moment. Use a linear equation to model the weight of the gas over time.

   a. This line's slope-intercept equation is ☐.

   b. 40 minutes after the experiment started, there would be ☐ grams of gas left.

   c. If a linear model continues to be accurate, ☐ minutes since the experiment started, all gas in the container will be gone.

25. Find the y-intercept and x-intercept of the line given by the equation. If a particular intercept does not exist, enter none into all the answer blanks for that row.

$$2x + 5y = -30$$

	x-value	y-value	Location
y-intercept	——	——	——
x-intercept	——	——	——

26. Find the y-intercept and x-intercept of the line given by the equation. If a particular intercept does not exist, enter none into all the answer blanks for that row.

$$4x + 3y = -12$$

	x-value	y-value	Location
y-intercept	——	——	——
x-intercept	——	——	——

27. Find the line's slope and y-intercept.

A line has equation $-5x + y = 5$.

This line's slope is ____.

This line's y-intercept is ____.

28. Find the line's slope and y-intercept.

A line has equation $-x - y = 1$.

This line's slope is ____.

This line's y-intercept is ____.

29. Find the line's slope and y-intercept.

A line has equation $8x + 10y = 1$.

This line's slope is ____.

This line's y-intercept is ____.

30. Find the line's slope and y-intercept.

A line has equation $10x + 12y = 5$.

This line's slope is ____.

This line's y-intercept is ____.

31. Fill out this table for the equation $x = -2$. The first row is an example.

x	y	Points
-2	-3	$(-2, -3)$
______	-2	____________
______	-1	____________
______	0	____________
______	1	____________
______	2	____________

32. Fill out this table for the equation $x = -1$. The first row is an example.

x	y	Points
-1	-3	$(-1, -3)$
______	-2	____________
______	-1	____________
______	0	____________
______	1	____________
______	2	____________

33. A line's graph is given.

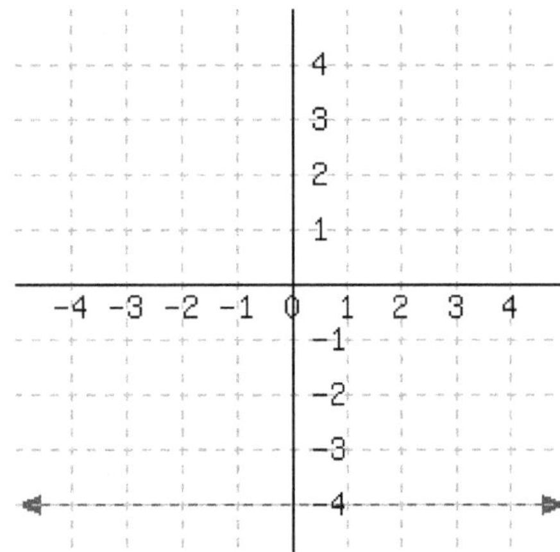

This line's equation is [   ]

34. A line's graph is given.

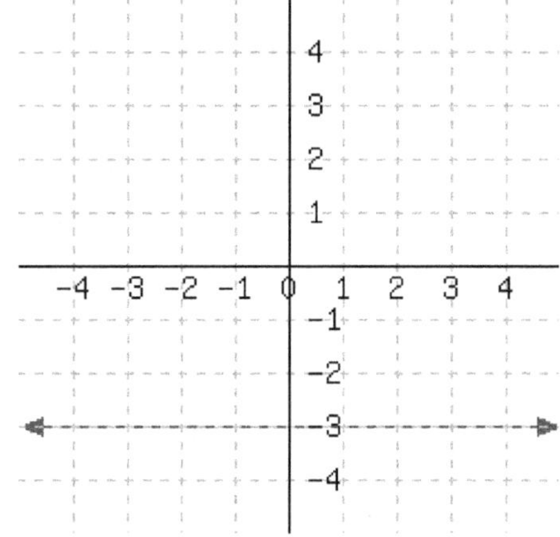

This line's equation is [   ]

35. Line m passes points $(-4, 10)$ and $(-4, 9)$.

Line n passes points $(7, 6)$ and $(7, -10)$.

Determine how the two lines are related.

These two lines are

⊙ parallel

⊙ perpendicular

⊙ neither parallel nor perpendicular

36. Line m passes points $(-2, 3)$ and $(-2, -4)$.

Line n passes points $(1, 6)$ and $(1, 0)$.

Determine how the two lines are related.

These two lines are

⊙ parallel

⊙ perpendicular

⊙ neither parallel nor perpendicular

37. Line k's equation is $y = -\frac{6}{7}x + 3$.

Line ℓ is perpendicular to line k and passes through the point $(-6, -2)$.

Find an equation for line ℓ in both slope-intercept form and point-slope forms.

An equation for ℓ in slope-intercept form is:

An equation for ℓ in point-slope form is: .

38. Line k's equation is $y = -\frac{7}{9}x + 5$.

Line ℓ is perpendicular to line k and passes through the point $(7, 12)$.

Find an equation for line ℓ in both slope-intercept form and point-slope forms.

An equation for ℓ in slope-intercept form is:

An equation for ℓ in point-slope form is: .

39. Graph the linear inequality $y > \frac{4}{3}x + 1$.

40. Graph the linear inequality $y \leq -\frac{1}{2}x - 3$.

41. Graph the linear inequality $y \geq 3$.

42. Graph the linear inequality $3x + 2y < -6$.

Index

CPSIA information can be obtained
at www.ICGtesting.com
Printed in the USA
LVHW062301170319
610989LV00006B/56/P